D1274146

CHROMATIN
and
CHROMOSOME STRUCTURE

Edited by

HSUEH JEI LI

Division of Cell and Molecular Biology
State University of New York at Buffalo
Buffalo, New York

RONALD A. ECKHARDT

Department of Biology
Brooklyn College of The City University of New York
Brooklyn, New York

Academic Press New York San Francisco London 1977

A Subsidiary of Harcourt Brace Jovanovich, Publishers

ACADEMIC PRESS, INC.
111 Fifth Avenue, New York, New York 10003

United Kingdom Edition published by
ACADEMIC PRESS, INC. (LONDON) LTD.
24/28 Oval Road, London NW1

Library of Congress Cataloging in Publication Data

Main entry under title:

Chromatin and chromosome structure.

  Papers presented at a Ph. D. seminar course
given at City University of New York, 1975.
  Bibliography: p.
  Includes index.
  1. Chromosomes—Congresses. 2. Chromatin—
Congresses. 3. Histones—Congresses. I. Li,
Hsueh Jei. II. Eckhardt, Ronald A. III. New
York (City). City University of New York.
QH600.C48   574.8'732   76-54757
ISBN 0–12–450550–3

CONTENTS

LIST OF CONTRIBUTORS

Numbers in parentheses indicate the pages on which the authors' contributions begin.

Vincent G. Allfrey (167), The Rockefeller University, New York, New York 10021

Jen-Fu Chiu (193), Department of Biochemistry, Vanderbilt University School of Medicine, Nashville, Tennessee 37232

Gerald D. Fasman (71), Graduate Department of Biochemistry, Brandeis University, Waltham, Massachusetts 02154

A. Gayler Harford (315), Division of Cell and Molecular Biology, State University of New York at Buffalo, Buffalo, New York 14214

Lubomir S. Hnilica (193), Department of Biochemistry, Vanderbilt University School of Medicine, Nashville, Tennessee 37232

Ru Chih C. Huang (299), Department of Biology, Johns Hopkins University, Baltimore, Maryland 21218

Hsueh Jei Li (1, 37, 143), Division of Cell and Molecular Biology, State University of New York at Buffalo, Buffalo, New York 14214

Herbert C. Macgregor (339), Department of Zoology, School of Biological Sciences, University of Leicester, Leicester LE1 7RH, England

B. W. O'Malley (255), Department of Cell Biology, Baylor College of Medicine, Houston, Texas 77030

Ming-Jer Tsai (255), Department of Cell Biology, Baylor College of Medicine, Houston, Texas 77030

PREFACE

During the past decade, great progress has been made in our knowledge of the chemistry and interactions of chromosomal components as well as in the physical structure and biological functions of chromatin and chromosomes. Chromatin and chromosomes in eukaryotic cells have been found to control growth, mitosis, differentiation, hormone action, aging, cancer, and many other phenomena in a higher organism. Such control is accomplished through interactions among chromosomal macromolecules, DNA, histones, nonhistone proteins, and RNA, and through interactions between chromatin and nonchromosomal molecules.

For the benefit of both faculty and students in the Biology Ph.D. Program of the City University of New York (CUNY), a seminar course was offered in the spring of 1975. The speakers emphasized the importance of the various subjects to research investigators and students in biology, biochemistry, biophysics, and biomedical sciences in general. This series of seminars was presented at Brooklyn College and was televised throughout the various campuses of CUNY and other institutions in the metropolitan area.

The coordinators and contributors to this seminar course prepared their seminars as chapters for this book. Most of the chapters were prepared in the spring and summer of 1976 and included many new observations reported after the seminars. In order to include most of the subjects originally presented in the series, Dr. A. G. Harford contributed a chapter on polytene chromosome structure, originally given by Dr. C. Laird, and Dr. H. J. Li contributed two chapters on conformational studies of histones and chromatin subunits, originally covered by Drs. E. M. Bradbury and G. Felsenfeld, respectively.

We are grateful to our former colleagues, Professors L. G. Moriber, D. D. Hurst, and M. Gabriel who helped in making possible the seminar series. We thank our former students Drs. C. Chang, I. M. Leffak, M. F. Pinkston, R. M. Santella, S. S. Yu, and Mr. J. C. Hwan and N. Rubin who have provided assistance in the preparation of lecture transcripts. Our thanks to Mrs. R. Bellamy and Mrs. D. Galeno, who spent many hours in typing the manuscripts and lecture transcripts.

NOMENCLATURE FOR HISTONE FRACTIONS

Lysine-rich histones	H1	I	F1
	H5	V	F2c
Slightly lysine-rich histones	H2A	IIb1	F2a2
	H2B	IIb2	F2b
Arginine-rich histones	H3	III	F3
	H4	IV	F2a1

CHROMATIN
and
CHROMOSOME
STRUCTURE

Chapter 1

CONFORMATIONAL STUDIES OF HISTONES

HSUEH JEI LI

Division of Cell and Molecular Biology
State University of New York at Buffalo
Buffalo, New York 14214

I. Introduction

The regular supercoiled structure with a pitch of 120Å and a diameter of 100Å, as originally proposed by Pardon *et al.* (1), is not compatible with either the observations of a string of beads in chromatin (2-5) or the separation of chromatin into nuclease-susceptible and nuclease-resistant fragments (6-19). A new concept of chromatin structure, namely a chromatin with distinct subunits, has been developed. Recently, quite a few models have been proposed to describe these subunits (20-24). The subject of chromatin structure has been dealt with extensively in a recent review (25).

In the past six years, studies of histone conformation and histone-histone interactions in solution have greatly increased the understanding of the structures and interactions among histones; this information is a necessary prerequisite for research on detailed structures of histones and on histone assembly into chromatin subunits. Such studies include kinetic and equilibrium interactions and conformations using circular dichroism (CD), fluorescence and nuclear magnetic resonance (NMR).

II. Circular Dichroism (CD) and Fluorescence Studies of Histone H4

Initially, formation of α-helix in histones upon the addition of salt or DNA were investigated by optical rotatory dispersion (ORD) (26-28). Jirgenson and Hnilica (26) reported phosphate to be more efficient than chloride in inducing an ordered structure in histones. Subsequently, more extensive studies on the conformation and interactions of histone H4, using CD and fluorescence polarization, were combined

1

with kinetics and equilibrium methods, as described below (29).

The fluorescence anisotropy of histone H4 tyrosine residues in water, r_w, and at time t after the addition of phosphate, $r_p(t)$, is depicted in Fig. 1, which shows a dependence of this anisotropy on both time and phosphate concentration.

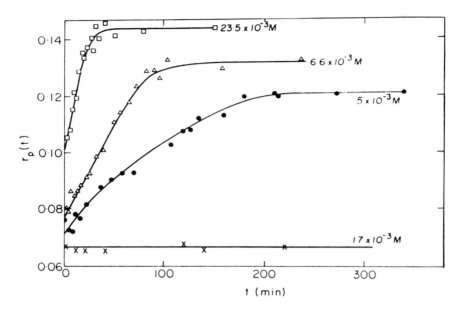

Figure 1. *Fluorescence anisotropy of histone H4 as a function of time and phosphate concentrations. $H_O = 0.8 \times 10^{-5}$ moles per liter, pH 7.4. Phosphate concentrations are shown in the figure. The anisotropy of histone H4 in water, r_W is 0.050 ± 0.003. Li, H. J., Wickett, R., Craig, A. H., and Isenberg, I. (1972)* Biopolymers 11, 375. *Reprinted with the permission of John Wiley and Sons, Inc.*

At low phosphate concentration, 1.7×10^{-3}M, for example, after a rapid increase from $r_w = 0.050$ to $r_p = 0.067$, the fluorescence anisotropy remains unchanged for more than 8 hours. At higher phosphate concentrations, in addition to the rapid increase in r, there is a slow and time-dependent increase which finally reaches a plateau, $r_p(\infty)$. The slow step can be described, approximately, as a single exponential function of time.

$$\frac{r_p(\infty) - r_p(t)}{r_p(\infty) - r_p(0)} = e^{-t/\tau_F} \tag{1}$$

where τ_F is the time constant for this step of the reaction as measured by fluorescence. The plot of eq. (1), shown in Fig. 2a, yields both the time constant τ_F and $r_p(0)$, the extrapolated anisotropy at t = 0 after the addition of phosphate. With these factors determined, it is then possible to calculate $r_p(0) - r_w$ of the fast step and $r_p(\infty) - r_p(0)$ of the slow step.

As with the anisotropy, the CD was also observed to be a function of time after the addition of phosphate. The CD results can be described by an equation similar to eq. (1). Define $\Delta\varepsilon_w(\lambda)$ and $\Delta\varepsilon_p(t,\lambda)$, respectively, as the CD of amide groups of histone H4 in water and at time t after the addition of phosphate, both measured at wavelength λ. The slow step of the conformational change measured by CD can be described by the following equation:

$$\frac{\Delta\varepsilon_p(\infty,220) - \Delta\varepsilon_p(t,220)}{\Delta\varepsilon_p(\infty,220) - \Delta\varepsilon_p(0,220)} = e^{-t/\tau_{CD}} \tag{2}$$

where τ_{CD} is the time constant of the slow step measured by CD (29). Fig. 2b shows the results from a plot of eq. (2).

Thus, conformational changes in histone H4 can be separated into a fast and a slow step by measuring either the fluorescence of tyrosine residues or the CD of amide groups. Based upon difference CD spectra, it was further demonstrated that, following the addition of phosphate, the fast step involves α-helix formation and the slow step formation of β-sheet in histone H4 (Fig. 3) (29).

The conformational changes in both fast and slow steps depend not only upon phosphate concentration (Fig. 4) but also upon histone concentration (Fig. 5).

The fast-step conformational changes in Fig. 4 can be shown to result from binding of phosphate to histone H4 by use of the following equations:

$$B + X \overset{K}{\underset{\leftarrow}{\rightarrow}} BX \tag{3}$$

$$K = \frac{[BX]}{[B]\,[X]} \tag{4}$$

where [B] is the concentration of binding sites, [X] the concentration of ligand (phosphate in Fig. 4), and [BX] the concentration of bound sites. Table I summarizes the binding

3

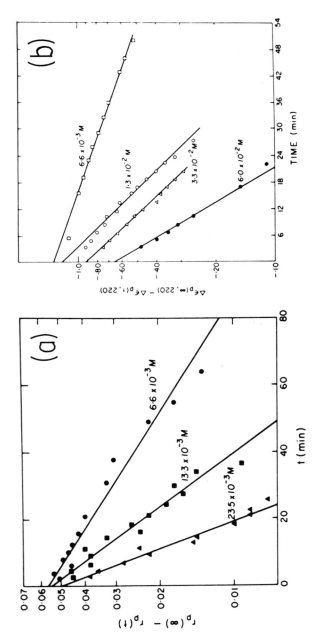

Figure 2a. Semilogarithmic plot of $r_p(\infty) - r_p(t)$ for the slow step as a function of time.

2b. Semilogarithmic plot of $\Delta\varepsilon_p(\infty,220) - \Delta\varepsilon_p(t,220)$ as a function of time. $H_O = 0.8 \times 10^{-5}M$, pH 7.4; phosphate concentrations are shown. CD results are given as $\Delta\varepsilon = \varepsilon(left) - \varepsilon(right)$ in cm^{-1} liter per mole of residue. Li, H. J., Wickett, R., Craig, A. M., and Isenberg, I. (1972)Biopolymers 11, 375. Reprinted with the permission of John Wiley and Sons, Inc.

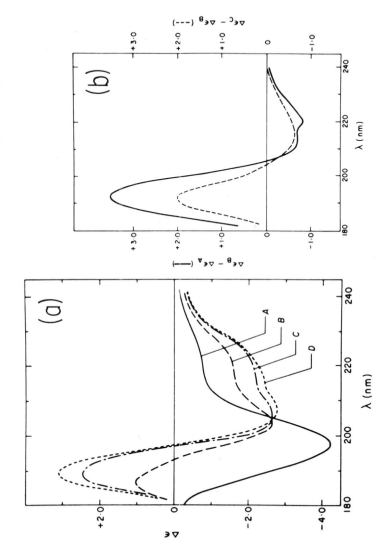

Figure 3a. CD spectra of histone H4 in water (Curve A), in $3.3 \times 10^{-3}M$ phosphate, pH 7.4, recorded from t = 25 to 55 min (Curve B), from t = 290 to 320 min (Curve C), and from t = 960 to 990 min (Curve D). $H_O = 0.8 \times 10^{-5}M$, optical path length was 1 mm.

3b. *Difference CD spectra computed from Figure 3a. Li, H. J., Wickett, R., Craig, A. M., and Isenberg, I. (1972) Biopolymers 11, 375. Reprinted with the permission of John Wiley and Sons, Inc.*

5

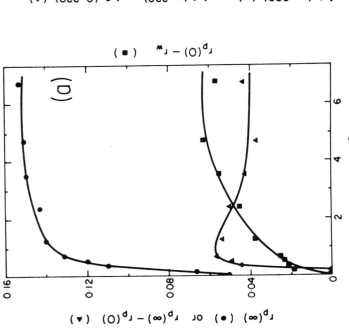

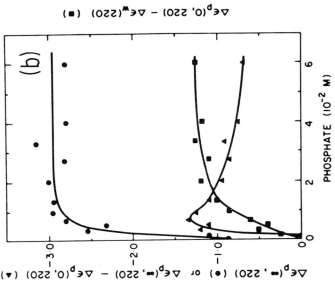

Figure 4a. Phosphate concentration dependence of the final anisotropy and anisotropy changes of the fast and slow processes. $r_w = 0.05 \pm 0.003$.

4b. Phosphate concentration dependence of the final CD and CD changes of the fast and the slow steps. $H_0 = 0.8 \times 10^{-5}M$, pH 7.4, measured at 220 nm. $\Delta\varepsilon_W(220) = 0.85 \pm 0.1$. Li, H. J., Wickett, R., Craig, A. M., and Isenberg, I. (1972) Biopolymers 11, 375. Reprinted with the permission of John Wiley and Sons, Inc.

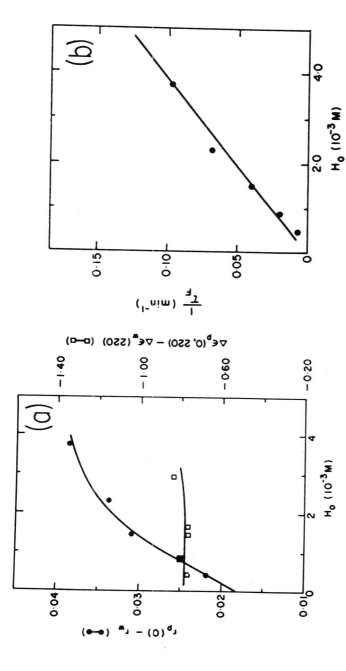

Figure 5. Histone concentration dependence of the anisotropy change and the CD change of the fast step (a) and the rate constant of slow step (b). Phosphate concentration is 6.6 x 10⁻³M, pH 7.4. The unit of histone concentration in this figure should be divided by 102, the total number of residues in histone H4. Li, H. J., Wickett, R., Craig, A. M., and Isenberg, I. (1972) Biopolymers 11, 375. Reprinted with the permission of John Wiley and Sons, Inc.

7

constants for histones with various salts as ligands. In
general, the binding affinity for histone is greater with

Table I. Binding constants between histones and various
salts.

Histones	Salt	K^*_{CD} (M^{-1})	K^*_F (M^{-1})	Reference
H4	Sodium phosphate (pH 7.4)	80	85	29
	NaH_2PO_4 (PO_4^-)	10	12	29,30
	Na_2HPO_4 ($PO_4^=$)	120	130	29
	Na_2SO_4	105	95	30
	$NaClO_4$	32	16	30
	NaCl	2.2	2.6	30
	$MgCl_2$	2.5	3.3	30
	NaF	2.0	2.2	30
H2B	Sodium phosphate (pH 7.4)	150	220	31
	NaCl	5.4	4.3	31
H3	Sodium phosphate	660	710	32

$^*K_D$ is measured by CD and K_F measured by fluorescence anisotropy.

divalent anions, $SO_4^=$ or $PO_4^=$,,than with monovalent, PO_4^- or
ClO_4^-,,and these monovalent anions bind with greater affinity
than do Cl^- and F^-. The physical meaning of these differences
will be discussed later, but it is noted here that the
greater efficiency of phosphates over chloride in inducing
conformational changes in histones confirms the earlier observation based upon optical rotatory dispersion studies (26).
     In the case of histone H4, the dependence of anisotropy
on histone concentration (Fig. 5a) was attributed to dimer
formation, while the dependence of $1/\tau_F$ on histone H4 concentration (Fig. 5b) was explained only as intermolecular inter-

action in the slow step (29).

III. Mechanism of Conformational Changes in Histone H4

Despite an extensive study on histone H4 conformation by Li et al. (29) and Wickett et al. (30) and on other histones by Isenberg and co-workers (31-37), the experimental results are still fragmentary and not well integrated. For instance, the exact physical meaning of the plot of eqs. (1) and (2) (Fig. 2a and 2b) and of the linear relationship between $1/\tau_F$ and histone concentration (Fig. 5b) cannot be fully understood in terms of molecular interaction. However, the following approach, made in 1973, may still be useful for this discussion and for those interested in research into more detailed mechanisms of conformational changes in histones in the future.

Define the following terms for the equations:

H: a histone H4 molecule.

X: a salt molecule, presumably an anion.

HX_i: a histone molecule bound by i molecules of X at a certain salt concentration.

$(HX_i)_2$: a dimer of HX_i with α-helical structure but no β-sheet.

$(HX_i)_{2,\beta}$: a dimer of HX_i after β-sheet formation.

$k_{12,i}$ and $k_{21,i}$: forward and backward rate constants of the first step of reaction (dimerization).

$k_{23,i}$: forward rate constant of the second step of reaction.

$K_i = \dfrac{k_{12,i}}{k_{23,i}}$: equilibrium constant of the first step of reaction (dimerization).

It will be shown below that the following equation can describe the observations reported by Li et al. (29):

$$2HX_i \; \underset{k_{21,i}}{\overset{k_{12,i}}{\rightleftarrows}} \; (HX_i)_2 \; \xrightarrow{k_{23,i}} \; (HX_i)_{2,\beta} \qquad (5)$$

Assume that the first step (dimerization) is rapid compared with the second step (β-sheet formation) and that $K_i (HX_i) \ll 1$. The first assumption has been shown to be true (29); the second will be tested below. Based upon these two assumptions, eqs. (6) and (7) can be derived (see Appendix):

9

$$(HX_i)_{2,\beta} = \frac{\frac{1}{4}k_{app}(H_o)t}{1 + \frac{1}{2}k_{app}t} \tag{6}$$

where

$$k_{app} = 4k_{23,i} \, K_i(H_o) \tag{7}$$

Eq. (8) defines f, the fraction of histone H4 molecules in β-sheet structure, in terms of $(HX_i)_{2,\beta}$ and (H_o), which are, respectively, the concentration of H4 dimer with β-sheet and the total histone concentration in monomer:

$$f = \frac{2(HX_i)_{2,\beta}}{(H_o)} \tag{8}$$

f can be expressed in terms of changes in either CD or fluorescence anisotropy as a result of the slow step (29). As shown in the Appendix, for CD and for fluorescence anisotropy, the following two equations can be obtained:

$$\frac{1}{\Delta\varepsilon_p(\infty,220) - \Delta\varepsilon_p(t,220)} =$$

$$\frac{1}{\Delta\varepsilon_p(\infty,220) - \Delta\varepsilon_p(0,220)} \, [1 + \frac{1}{2}(k_{app})_{CD}t] \tag{9}$$

$$\frac{1}{r_p(\infty) - r_p(t)} = \frac{1}{r_p(\infty) - r_p(0)} \, [1 + \frac{1}{2}(k_{app})_F t] \tag{10}$$

As t is small, both eqs. (9) and (10) can be rewritten as:

$$\frac{\Delta\varepsilon_p(\infty,220) - \Delta\varepsilon_p(t,220)}{\Delta\varepsilon_p(\infty,220) - \Delta\varepsilon_p(0,220)} = \exp[-\frac{1}{2}(k_{app})_{CD}t] \tag{11}$$

and

$$\frac{r_p(\infty) - r_p(t)}{r_p(\infty) - r_p(0)} = \exp[-\frac{1}{2}(k_{app})_F t] \tag{12}$$

Eqs. (11) and (12) would be identical to eqs. (2) and (1), respectively, if

$$\frac{1}{\tau_{CD}} = \frac{1}{2}(k_{app})_{CD} \tag{13}$$

10

$$\frac{1}{\tau_F} = \frac{1}{2}(k_{app})_F \tag{14}$$

or

$$\frac{1}{\tau} = \frac{1}{2}k_{app} = 2k_{23,i} K_i (H_o) \tag{15}$$

The data from the original report (29) were shown to fit both eqs. (9) and (10) (Fig. 6) as well as the approximate equations (11) and (12) (29,30).

Eq. (15) predicts a linear relationship between $\frac{1}{\tau}$ and (H_o), which has been confirmed in phosphate (Fig. 5b). From Fig. 5b, it is calculated that $k_{23,i} K_i = 1.3 \times 10^{-3} M^{-1}$ min^{-1} at $6.6 \times 10^{-3} M$ phosphate, pH 7.4. Under these conditions, $k_{23,i} K_i = 1.2 \times 10^{-3} M^{-1}$ min^{-1} based upon τ measured from Fig. 4 in Li *et al.* (29). These two values are in agreement with each other. Under these same conditions, the dimerization constant K_i was determined to be $1.5 \times 10^4 M^{-1}$ (29). Thus, $K_i (HX_i) < K_i (H_o) = 0.12 \ll 1.0$, which satisfies one of the two assumptions used before for the derivation of eq. (5).

Though the rate constants of $k_{12,i}$ and $k_{21,i}$ of the dimerization step cannot be determined from the present data, $k_{23,i}$ of β-sheet formation can be determined from eq. (15) if K_i is known. For instance, at $6.6 \times 10^{-3} M$ phosphate, pH 7.4, K_i was determined to be $1.5 \times 10^4 M^{-1}$ (29), from which $k_{23,i}$ is calculated as 8×10^{-2} min^{-1}, which is twice as large as the rate determined by the approximation method of eq. (5) (38). In either case, the small rate constant implies that β-sheet formation is indeed a slow process.

Although the mechanism utilized in eq. (5) has been shown to be satisfactory for describing certain phenomenological observations (29), the following comments on the above derivations must be considered:

(a) No serious attempt has been made to search for other mechanisms which would also fit the data; in other words, the mechanism of eq. (5) should be considered only as a possible, even likely, mechanism for conformational changes in histone H4; (b) It is assumed that the changes in CD and fluorescence anisotropy in the slow step are directly proportional to the concentration of histone H4 molecules in β-sheet within the dimer; such assumption is only an approximation, because the fluorescence anisotropy of each of the four tyrosine residues and the CD of the amide groups along the whole molecule may not follow exactly the same rate of reaction, since some regions of histone H4 may form β-sheet more easily or faster than the other regions; (c) It should be noted that the

11

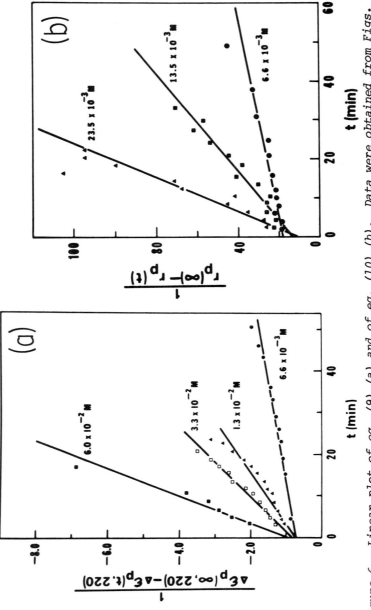

Figure 6. Linear plot of eq. (9) (a) and of eq. (10) (b). Data were obtained from Figs. 2a and 2b.

linear plot shown in Figs. 2a and 2b is obeyed for experiments done near room temperature, but that deviation occurs at higher temperature (32°, for example) (27); perhaps the above approximation does not hold at higher temperatures, or perhaps further reactions, such as aggregation also occur in addition to dimerization and β-sheet formation within the dimer as described in eq. (5).

The success in explaining the many observations originally reported (29,30) by the simple mechanism of eq. (5) is remarkable. Despite the above comments, the mechanism of eq. (5) could provide a guideline for the qualitative description of interaction among histone H4 molecules (Fig. 7).

Immediately after the addition of salt to histone H4, ionic interaction is expected to neutralize the charge in a histone molecule and encourage the rapid formation of more ordered and compact structures, such as α-helix. The same ionic interaction can also reduce charge repulsion between histone molecules so that two histone H4 molecules could rapidly form a dimer through non-specific contact in their hydrophobic regions (carboxyl terminus). When these two molecules are brought into close proximity as a dimer, specific hydrogen bonds between the amide groups of those amino acid residues favoring the β-sheet would form in parallel. The reason for the suggestion of a parallel β-sheet (Fig. 7), as opposed to an anti-parallel arrangement, divalent anions in the solution can link two monovalent cations on both chains of the dimer more effectively than can the monovalent anions (29,30).

Although the fast step (α-helix formation and dimerization) can occur at any salt concentration, the occurrence of the slow step (β-sheet formation) requires a critical salt concentration, a threshold concentration below which only a fast process is evident (Fig. 4). Since the slow process is associated with the β-sheet formation rather than the α-helix formation or dimerization, such a threshold value would probably imply that the cooperative β-sheet formation within the dimer (a very close packing of two molecules) requires a minimum charge neutralization on each molecule to reduce charge repulsion and/or form sufficient salt bridges in the dimer with which to stabilize the β-sheet.

IV. Conformational Changes in Other Histones and Histone Self-Interactions

Phosphate-induced conformational changes in arginine-rich histone H3 are very similar to those reported for arginine-rich histone H4 (29) in that (a) there are both fast

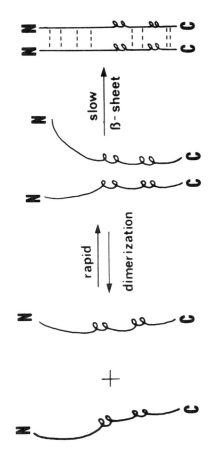

Figure 7. Schematic drawing of the fast step with dimerization and the slow step with β-sheet formation in histone H4.

14

and slow types of conformational changes, formation of α-helix in the fast step and β-sheet in the slow step; and (b) the rate of the slow step depends upon both histone concentration and temperature (32). Although no plot similar to that of Fig. 4 for histone H4 was reported for histone H3, the results in Fig. 1 of D'Anna and Isenberg (32) indicate that the threshold phosphate concentration for the occurrence of the slow process is between 3×10^{-3}M and 6×10^{-3}M, a value close to that for histone H4 (Fig. 4). Despite these similarities, the equilibrium constant in the fast process of histone H3-phosphate interaction is about one order of magnitude greater than in that of H4 (Table I).

For slightly lysine-rich histone H2B, D'Anna and Isenberg (31) reported only the fast process of α-helix formation when sodium phosphate or sodium chloride was added to the histone. No slow process and β-sheet structure were observed. The equilibrium constant of the fast process in eq. (1) for histone H2B was reported to be about two to three times greater than for histone H4 (Table I).

After the addition of phosphate to slightly lysine-rich histone H2A, D'Anna and Isenberg (34) also reported only the fast α-helix formation, but this histone was reported to be more sensitive to salt and pH changes than histone H2B or H4. Furthermore, eq. (3), which was used successfully to describe the fast process of interaction between salt and histone H4 (29,30), was reported to be non-applicable in the use of histone H2A, since no equilibrium constant could be obtained for this histone (34). Presumably, histone-salt interactions or salt-induced conformational changes in histone H2A are different from those in the other three histones.

The studies of salt-induced conformational changes in these four histones suggest the following conclusions: (a) arginine-rich histones H3 and H4 form a rapid α-helix and a slow β-sheet; probably the mechanism in eq. (5) and the schematic description of parallel dimer with β-sheet in Fig. 7 proposed for histone H4 could be applied to histone H3; (b) slightly lysine-rich histones H2A and H2B primarily demonstrate a fast process with α-helix formation but lack the slow process with β-sheet formation; (c) there is a large variation in binding constants of eq. (4) for interaction between salt and various species of histone (Table I); histone H2A, in particular, does not follow the same mechanism of salt interaction or salt-induced conformational changes as do histones H4, H3 and H2B.

V.   Histone Cross-Complexing

The analytical methods and equations developed for histone H4 (29) have been applied to other histones individually as described above. However, they cannot be applied directly to a mixture of two different species of histones. D'Anna and Isenberg (33) then developed methods and derived equations to study these problems. Fig. 8a shows changes in tyrosine anisotropy and Fig. 8b, the changes in CD of histone H4 alone, H2B alone and a mixture of H2B and H4 after phosphate was added. Strikingly, the presence of H2B in the mixture prohibits the slow process which would have occurred if histone H4 were present alone. Presumably, some interaction occurred between histone H2B and histone H4 which prevented H4 from forming the stable β-sheet structure.

If A and B represent two species of histones, such as H2B and H4, their interaction can be described by:

$$A + nB \underset{\leftarrow}{\overset{K_{AB}}{\rightarrow}} AB_n \qquad (16)$$

where K_{AB} is the association constant. These authors assumed no self-interaction (e.g., H2B-H2B or H4-H4) under their conditions, otherwise interactions among A or among B molecules have to be considered in conjunction with eq. (16). If it is assumed that there exist intensive properties, $\Delta\epsilon_A'$, $\Delta\epsilon_B'$ and $\Delta\epsilon_{AB_n}'$, these authors derived the following equation:

$$\Delta\epsilon' - \Delta\epsilon_I' = \frac{\Delta\epsilon_{ABn}' - \Delta\epsilon_A' - n\Delta\epsilon_B'}{C_o} [AB_n] \qquad (17)$$

where C_o is the sum of starting concentrations of A and B, $\Delta\epsilon'$ the measured CD on the basis of C_o and $\Delta\epsilon_I'$ the calculated CD assuming non-interacting species A and B in mixed solutions. The mole fraction of component A, X_A, at which a maximum occurs in $\Delta\epsilon' - \Delta\epsilon_I'$ is related to n by:

$$n = \frac{1 - X_A}{X_A} \qquad (18)$$

Fig. 9 shows the plot of calculated $\Delta\epsilon'$ and $\Delta\epsilon_I'$ as a function of molar ratio, X_{H2B}, in a mixture of H2B and H4. Since the peak of $\Delta\epsilon' - \Delta\epsilon_I'$ occurred at $X_{H2B} = 0.5$, D'Anna and Isenberg (33) concluded that n = 1.0, i.e., the stoichiometry of the reaction between H2B and H4 was 1 : 1. The equilibrium constant K_{AB} was then calculated under various conditions for H2B-H4 (33), H2A-H2B and H2A-H4 (35) and H3-H4 and other histone pairs (36). Their results are summarized in Table II. These authors concluded a cross-

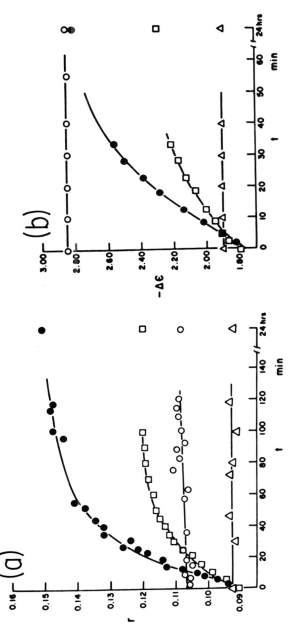

Figure 8. Fluorescence anisotropy (a) and CD (220 nm) of histone solutions as a function of time at 0.016M phosphate, pH 7.0. 0.90 x 10⁻⁵M histone H4 (●); 0.90 x 10⁻⁵M histone H2B (△); 0.90 x 10⁻⁵M histone H4 plus 0.90 x 10⁻⁵M histone H2B, measured (○) and calculated (□) for a mixture of noninteracting histones. D'Anna, J. A., Jr., and Isenberg, I. (1973) Biochemistry 12, 1035. Reprinted with permission of the American Chemical Society.

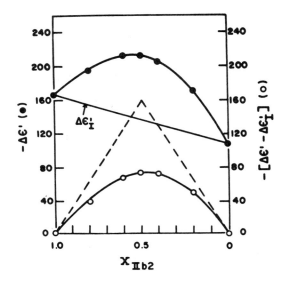

Figure 9. CD (220 nm) continuous variation curves, $\Delta\varepsilon'$ - $\Delta\varepsilon_I'$ vs. X_{IIb2} (o) at 0.0031M phosphate, pH 7.0, $(H_O) = 0.5$ x $10^{-5}M$. Dotted lines are extrapolations of the slopes of the $\Delta\varepsilon'$ - $\Delta\varepsilon_I'$ curve at $X_{IIb2} = 0$ and $X_{IIb2} = 1.0$. Histone IIb2 is the same as histone H2B. X_{IIb2} is the mole fraction of histone H2B in the mixed solution of histones H2B and H4. D'Anna, J. A., Jr. and Isenberg, I. (1973) Biochemistry 12, 1035. Reprinted with permission of the American Chemical Society.

Table II. Association Constants Between Histones[a]

Complex	Molar Ratio	K_{AB}	Time-dependent β-sheet formation
H3-H4[b]	1 : 1	$0.7 \times 10^{21} M^{-3}$	No
H2A-H2B	1 : 1	$10^6 M^{-1}$	No
H2B-H4	1 : 1	$10^6 M^{-1}$	No
H3-H2A	1 : 1	$0.1-1 \times 10^6 M^{-1}$	Yes
H2A-H4	1 : 1	$0.04 \times 10^6 M^{-1}$	Partially
H3-H2B	?	?	Yes

[a] Data obtained from D'Anna and Isenberg (36).

[b] The association constant for H3-H4 is that for monomer-tetramer equilibrium.

complexing pattern among histones with strong interactions in H2A-H2B, H2B-H4, H3-H4, and H2A-H3, and with weak interactions in H2A-H4 and H2B-H3.

Although the experimental results of D'Anna and Isenberg (33,35,36) showed some kind of interaction among various species of histones, the quantitative analysis for obtaining the stoichiometry and equilibrium constants of these reactions may need to be re-evaluated. Their results shown in Table II should, perhaps, be questioned for the following reasons:

(1) As shown in Fig. 8a and 8b, the mixture contained 0.90×10^{-5}M of histone H2B and H4 with a total concentration of 1.8×10^{-5}M in histone molecules, twice as much as when either H2B or H4 was used alone. The phosphate to histone ratio in the mixture was only one half that present when individual histone alone was used. Thus, the average of the two values when individual histone alone was used (Fig. 8a and 8b) cannot be the theoretical value of a non-interacting mixture in the mixed solution, since conformational changes in histones are extremely sensitive to both salt and histone concentrations (Figs. 1-5 and Table I). Therefore, the plot of eq. (17), which includes the theoretical value of non-interacting mixture $\Delta\varepsilon_T'$, becomes questionable. The same comments can be applied to their analysis of fluorescence anisotropy data.

(2) Theoretically, one has to consider the equilibrium equations, simultaneously, between salt and each histone species, eq. (3), and between two histones, eq. (16). Since the association constant between salt and histone varies greatly among histone species (Table I), the method used by these authors for obtaining the CD or fluorescence anisotropy of a non-interacting mixture would still be incorrect even if a constant ratio of salt to histone were maintained for a series of mixtures with varied composition.

(3) When eq. (16) was used, these authors assumed no self-interaction, for example, H2B-H2B or H4-H4 interactions. This assumption should be questioned because the lack of a slow process in the mixture does not necessarily exclude the formation of a dimer from the same species, H2B-H2B, H4-H4, etc. For instance, dimerization in histone H4 (Fig. 5 and eq. 5) is a fast process. It might not be impossible for self-interaction to occur rapidly without the formation of β-sheet in a mixed solution. Therefore, the authors have to apply more critical tests in order to justify the assumption of no self-interaction in the mixture.

(4) According to Table IV in D'Anna and Isenberg (33) $\Delta\varepsilon(\infty)$

is $-2.65M^{-1}$ cm^{-1} for a measured mixture of histones H2B and H4 and $-2.23M^{-1}$ cm^{-1} for a culculated non-interacting mixture. The difference is $0.42M^{-1}$ cm^{-1} which is only about 5% of the CD of an α-helical protein (39-42). Such a small difference is detectable above the noise level of the intrument. However, it might be difficult to measure variation in CD within this order of magnitude by changing histone composition and to obtain from it an equilibrium constant of a reaction.

Because of the complexity involved in histone-salt interactions, (for example, rapid α-helix formation and dimerization and slow β-sheet formation), and histone-histone interactions (self-interaction and cross-complexing), as well as the small difference in measurable signals between an interacting and non-interacting mixture of histones, thermodynamic analysis of histone-histone interactions may need a more intensive and more sophisticated plan of research than has been required in the past (29-38).

VI. Conformational Studies of Histones Using Nuclear Magnetic Resonance (NMR)

High resolution nuclear magnetic resonance (NMR) has also been used as an analytical tool to investigate conformational changes in histones and histone-histone interactions (43-51).

In a high resolution NMR spectrum, protons of amino acid residues with close chemical shifts appear as a separate resonance peak. The intensity (or area) of each peak is proportional to the concentration of those residues which have sufficient freedom of motion. By measuring the loss of "visible" intensity of a peak, it is possible to probe the participation of one or more particular amino acid residues in the formation of a more rigid local structure (β-sheet, α-helix, or others) and aggregation.

Fig. 10a and 10b show, respectively, high field and low field 100 MHz NMR spectra of histone H4 in D_2O, in the absence and presence of NaCl (43). The addition of NaCl to histone H4 leads to a greater loss in "visible" intensity from the peak at 0.93 ppm (parts per million), due to the $-CH_3$ of Val, Leu and Ile, than from other peaks in the high field (Fig. 10a) and greater diminution in the peak at 6.81 and 7.14 ppm due to tyrosine protons than in the histidine peak in the low field (Fig. 10b). These results were used as the basis for a model proposed by these authors suggesting that the carboxyl half of histone H4, most likely in the 55 to 72 region of the chain, forms α-helices which participate

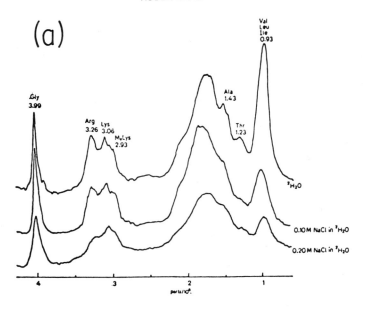

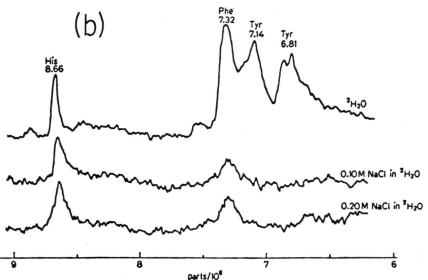

Figure 10. 100 MHz proton magnetic resonance spectra of histone H4 in D_2O and $D_2O/NaCl$ in high field (a) and low field (b). Boublik, M., Bradbury, E. M., and Crane-Robinson, C. (1970) Eur. J. Biochem. 14, 486. Reprinted with permission of the Federation of European Biochemical Societies.

in histone-histone interaction while the amino half of the histone remains in a non-interacting, unordered coil (43).

Fig. 11 shows the effect of histone concentration on the 220-MHz NMR spectrum of histone H4 in the high field region (48). Peak I corresponds to the $-CH_3$ of Val, Leu and Ile, and Peaks VI, VII and VIII, primarily to $\gamma-CH_2$ of Glu, $\varepsilon-CH_2$ of Lys and $\delta-CH_2$ of Arg, respectively. Fig. 11 shows that, at higher histone concentrations, peaks I, VI and VIII lose their intensity faster than do the others. Presumably, histone-histone interaction involves not only hydrophobic residues, Val, Leu and Ile of Peak I, but also ionic residues, Glu of Peak VI and Arg of Peak VIII. Thus, in this model, both hydrophobic contact and charge-charge interaction play

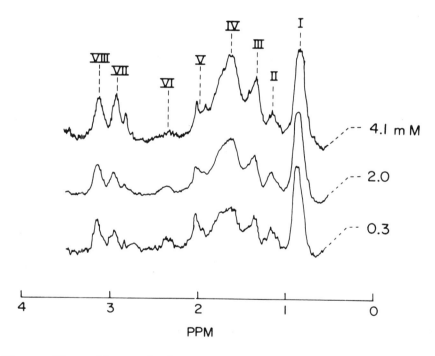

Figure 11. Effect of histone concentration on the 220-MHz proton magnetic resonance spectrum of histone H4 in the high field region. Histone concentration in mM, pD 3.7. Pekary, A. E., Li, H. J., Chan, S. I., Hsu, C. J. and Wagner, T. E. (1975) Biochemistry 14, 1177. Reprinted with permission of The American Chemical Society.

23

important roles in histone-histone interaction, in contrast to the model of Boublik *et al.* (43) which emphasizes only hydrophobic contact.

Table III summarizes quantitative results of the NaCl-induced effect on NMR spectrum of histone H4. At high NaCl concentration, a faster reduction in residual intensity occurs not only for Peak I of hydrophobic residues, but also for Peak VIII of Arg and Peak VII of Lys (48). At the same time, conformational changes in histone H4 induced by NaCl occur not only in the hydrophobic C-terminal half but also in the basic N-terminal half of this molecule. The model shown in Fig. 7 seems to be more compatible with these NMR results than does the one originally proposed by Boublik *et al.* (43) and later revised by Lewis *et al.* (47) (Fig. 13).

Table III[a]. Effect of NaCl Concentration on the Percentage of "Visible" Histone H4 Protons, $([H_{obsd}]/[H_o]) \times 100\%$, within Each High-Field Region PMR Band[b]

PMR Band	NaCl (M)				
	0	0.1	0.2	0.4	0.8
I	62	43	23	20	15
II	80	56			
III	82	70	61	61	61
IV	67	55	37	36	31
V	68	48	38	34	
VI	75				
VII	57	46	31	30	
VIII	77	59	32	21	

[a]From Pekary, A. E., Li, H. J., Chan, S. I., Hsu, C. J. and Wagner, T. E. (1975), *Biochemistry 14*, 1177. Reprinted with permission of the American Chemical Society.

[b]$[H_o] = 4.2 \times 10^{-4}M$ histone H4.

Fig. 12 shows 220 MHz NMR spectra of histone H2B in the high field region in both the absence and presence of NaCl (44). A greater loss of intensity is seen for the peak of $-CH_3$ of Val, Leu and Ile and of $\delta-CH_2$ of Arg than for the other peaks at higher NaCl concentration. This phenomenon is similar to that observed in histone H4 (Figs. 10 and 11)

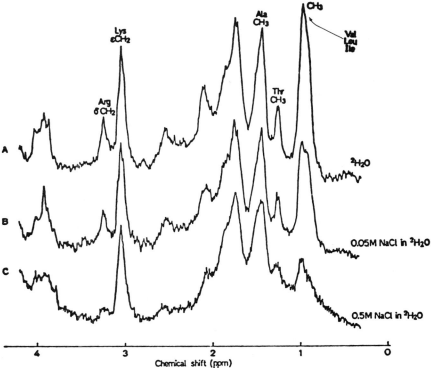

Figure 12. High field 220 MHz proton magnetic resonance spectra of histone H2B at various ionic strengths in aqueous solution. Boublik, M., Bradbury, E. M. and Crane-Robinson, C. (1970) Eur. J. Biochem. 17, 151. Reprinted with permission of the Federation of European Biochemical Societies.

and implies similar structural roles for the hydrophobic and arginine residues of both histones.

Histone self-interaction has also been studied in histone H1 (43) using proton (^{1}H) magnetic resonance, and in histone H2B, H2A and H3 using ^{13}C and ^{1}H nuclear magnetic resonance (50,51). The general conclusions from these studies stress the role played by non-polar residues, although some reports mention the structural roles played by other residues, particularly of Glu and Arg (48).

Using proton magnetic resonance and sedimentation, Pekary *et al.* (49) studied two histone H4 fragments cleaved at Met-84 by cyanogen bromide. The N-peptide was found to be in an extended random coil which undergoes transition to a more rigid structure with increase in the concentration of either peptide or NaCl in a manner similar to that observed

25

for intact histone H4 (49). Presumably the N-peptide plays an active role both in the formation of rigid secondary structure and in histone-histone interaction. On the other hand, the C-peptide is in a rigid conformation at low peptide concentration without NaCl, and is insensitive to increases in the concentration of either peptide or salt.

Lewis *et al.* (47) cleaved histone H4 into fragments and investigated these fragments using proton magnetic resonance, CD and infrared spectroscopy. They reported that intact histone H4 exhibited a fast structural change, involving the formation of α-helix and aggregation, and a slow change involving β-sheet formation, which confirmed the report by Li *et al.* (29) on experiments using CD and fluorescence anisotropy. However, among the fragments (1-23, 25-67, 69-84 69-102 and 86-102) studied, only fragment 25-67 showed structural changes and interaction after the addition of salt. The others were reported to be non-interacting. The fragment 25-67 showed a fast change with α-helix formation and aggregation but no time-dependent structural changes. These authors then proposed a model (Fig. 13) (a) that addition of salt to histone H4 causes fast formation of α-helix and suggesting parallel polymers through interaction among these α-helical regions and (b) that the slow step involves β-sheet formation in the C-terminal regions (73-100), while the N-regions (1-33) are non-interacting and, presumably, in extended coiled conformation.

Although, Lewis *et al.* (47) concluded that β-structure must occur in the region between 74 and 102 (Fig. 13), their experimental data were unable to support it. In fact, no β-sheet structure was observed for fragments 1-23, 69-84, 86-102 and 69-102; the last three fragments included the region of 73-102 shown with β-sheet structure in Fig. 13 (47). On the other hand, the fragment 25-67 possessed not only α-helix (about 12 residues) but also β-sheet (about 9 residues). Fig. 13 suggested that α-helix occurs only in the region of 33-73 which then makes it difficult to explain the presence of only about 12 α-helical residues in the fragment of 25-67 while there are approximately 25 α-helical residues in an intact histone H4 (47).

The mechanism of eq. (5) and the model suggested in Fig. 7 stress that dimers with β-sheet structure are the main structural subunits, and that these dimers may form larger aggregates. However, the kinetics demonstrated in Sections II and III strongly suggest that β-sheet structures within the dimers are not distorted to a detectable level if larger aggregates are formed from the dimers. The mechanism of eq. (5) and the model in Fig. 7, therefore, are quite

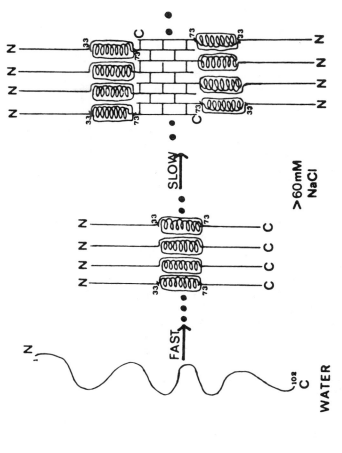

Figure 13. Schematic representation of structural changes induced by salt (NaCl > 60mM) in a 1 mM solution of histone H4. No residual structure is shown in water due to uncertainty as to its nature. Both the fast and slow step products are shown for clarity in a very extended form; they may, however, have a more compact tertiary structure. Lewis, P. N., Bradbury, E. M., and Crane-Robinson, C. (1975) Biochemistry 14, 3391. Reprinted with permission of the American Chemical Society.

different from the model shown in Fig. 13 which suggests the presence of β-sheet in multiple chains of histone H4.

Histone H4 is the only histone molecule having been studied so extensively. Although some understanding of conformational changes and interactions in this histone have been developed, a detailed description of its structure is still difficult. The studies of fragments of histone H4 (47,49) are interesting and useful. Nevertheless, a direct extrapolation from the structure of fragments to that of whole molecules should be carefully undertaken because the behavior of one portion of an intact molecule is likely to be affected by the presence of another portion and vice versa.

Very recently, Bradbury and co-workers (52,53) have extended their NMR studies of pure histones to histone mixtures, such as histone H3/H4 (52) and H2A/H2B complexes (53).

Histone H3/H4 complexes were dissociated by salt extraction of chromatin. A tetrameric subunit was implicated based upon its molecular weight (54-56). Fig. 14 shows NMR spectra of H3/H4 complexes in the presence or absence of 0.1M NaCl. The spectra of native complexes (both in the upfield and downfield regions) show sharp resonance peaks which differ from those of globular proteins. The latter are characterized by line broadening due to the presence of a range of chemical shifts of each given type of amino acid residue in the protein caused by various microenvironments through rigid secondary and tertiary structures. Presumably, the conformation of H3/H4 complexes is not compact and is different from that of globular proteins. CD spectra indicate the presence of $29 \pm 2\%$ α-helices but no β-sheets in H3/H4 complexes, in agreement with that of an equimolar mixture of H3/H4 in 0.01M phosphate (57) which was shown to form a tetramer (56). Specific β-sheet formation in the dimer of H4 (eq. 5 and Fig. 7) (29) or H3 (32) is prohibited by the other species of histone, a phenomenon first demonstrated by D'Anna and Isenberg (36). Such prohibition of formation of β-sheet structure in H3/H4 complexes, however, does not necessarily exclude the formation of dimers $(H3)_2$ and $(H4)_2$ in the tetramer, since, according to eq. (5), a dimer can be formed in the rapid step no matter whether there is a slow step of β-sheet formation or not.

Quantitative analysis of NMR spectra of H3/H4 complexes suggests a loss of about 30% of "visible" intensity in their native state (52) which could be related to the presence of 20-30% of molecules sedimented as aggregates with sedimentation coefficients greater than 20S. In other words, histone H3/H4 complexes could be in equilibrium between tetramers and higher aggregates. Whether the higher aggregates are polymers

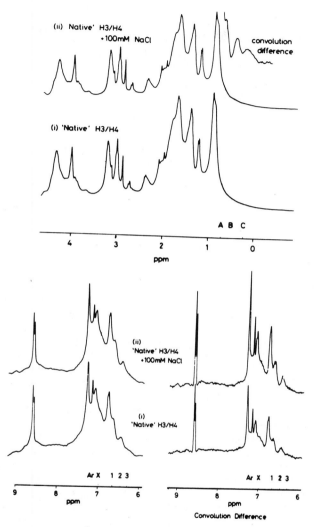

Figure 14. 270 MHz ¹H NMR spectra of the "native" H3/H4 histone at 0.5 mM H3/H4, (a) upfield (top) and (b) downfield
(bottom); (i) 50 mM sodium acetate - d₃, 50 mM sodium bisul-
fide - d, pH 4.8; (ii) + 100 mM NaCl, pH 4.8, added as concentrated NaCl in the above buffer. Moss, T., Cary, P. D.,
Crane-Robinson, C., and Bradbury, E. M. (1976) Biochemistry
15, 2261. Reprinted with permission of the American Chemical
Society.

of H3, H4, or mixtures of H3 and H4 is still unclear.

Interaction between histone H3 and H4 in H3/H4 complexes further reduces salt-induced aggregation of H4 (29,30,43) and H3 (32). For instance, severe NMR line broadening and area loss occur in histone H4 at 0.1M NaCl (Fig. 10) (43,48) but not in H3/H4 complexes under the same ionic conditions (Fig. 14). Such reduction in self-aggregation of H4 or H3 is attributed by the authors (52) to the existence of some kind of specific tertiary interaction in H3/H4 complexes. This interaction is also implicated by the presence of perturbed resonances (see convolution difference spectra in Fig. 14) caused by close contacts between different chemical groups of the proteins.

As discussed earlier in Figs. 10-12 and Table III, histone-histone interactions involve not only hydrophobic residues, Val, Leu and Ile, but also ionic residues, Glu and Arg, because the "visible" intensity of these residues were reduced to a much greater extent than that of other residues when histones were placed in high salt, 0.9-1.0M NaCl, for example. Presumably, both non-polar and ionic interactions play key roles in interactions in histone H4 (Fig. 10, 11 and Table III), H2B (Fig. 12), and also in H3/H4 (52). In fact, most of the negatively charged acidic residues (Glu and Asp) in histones H2B, H3 and H4 are found in the C-terminal halves of the molecules where a majority of hydrophobic residues are also located (58-63). Even in histone H2A, half of its acidic residues are located in the C-terminal half of this molecule (58,64,65). Perhaps hydrophobic interaction among non-polar residues is closely coupled with ionic interaction among acidic and basic residues in histone-histone interactions.

In studies similar to those of H3/H4 complexes, Moss *et al.* (53) have observed perturbed NMR resonance peaks both from aromatic and non-polar residues of H2A/H2B complexes. These perturbed resonances were lost when H2A/H2B complexes were titrated to pH 3.0 or when H2A or H2B alone was examined. These results imply tertiary interactions between H2A and H2B in the complexes, presumably in the central regions of these two molecules where the majority of aromatic and hydrophobic residues are located (53).

As illustrated in this chapter, conformation and interactions in histones are complicated. Nevertheless, when combined with thermodynamics, physical techniques, such as CD, fluorescence and NMR have indeed provided valuable structural information about histones.

ACKNOWLEGEMENT: Our research was supported in part by National Science Foundation grant PCM76-032 and National Institutes of Health grants GM23079 and GM23080.

APPENDIX

$$2HX_i \underset{k_{21,i}}{\overset{k_{12,i}}{\rightleftarrows}} (HX_i)_2 \xrightarrow{k_{23,i}} (HX_i)_{2,\beta} \tag{I}$$

$$K_i = \frac{k_{12,i}}{k_{21,i}} \tag{II}$$

$$\frac{d(HX_i)_{2,\beta}}{dt} = k_{23,i} \, K_i (HX_i)_2 \tag{III}$$

If dimerization step is much faster than β-sheet formation, monomer-dimer equilibrium is always reached in the time scale of β-sheet formation. Therefore

$$(HX_i)_2 = K_i (HX_i)^2 \tag{IV}$$

$$\frac{d(HX_i)_{2,\beta}}{dt} = k_{23,i} \, K_i (HX_i)^2 \tag{V}$$

Now

$$(H_o) = (HX_i) + 2[(HX_i)_2 + (HX_i)_{2,\beta}] \tag{VI}$$

From equation (IV)

$$(H_o) = (HX_i) [1 + 2K_i (HX_i)] + 2(HX_i)_{2,\beta}$$

or

$$(H_o) \approx (HX_i) + 2(HX_i)_{2,\beta} \qquad\qquad (VII)$$

if

$$2K_i (HX_i) << 1$$

This is approximately true in histone H4 as discussed in the text. Combining both equations (V) and (VII), we obtain

$$\frac{d(HX_i)_{2,\beta}}{dt} = k_{23,i} K_i [(H_o) - 2(HX_i)_{2,\beta}]^2 \qquad (VIII)$$

or

$$\frac{d(HX_i)_{2,\beta}}{dt} = \frac{k_{app}}{(H_o)} [(HX_i)_{2,\beta} - \tfrac{1}{2}(H_o)]^2 \qquad (IX)$$

where

$$k_{app} = 4k_{23,i} K_i (H_o) \qquad\qquad (X)$$

Using the initial condition that $(HX_i)_{2,\beta} = 0$ at $t = 0$, the solution of equation (IX) is

$$(HX_i)_{2,\beta} = \frac{\tfrac{1}{4}k_{app} (H_o) t}{1 + \tfrac{1}{2}k_{app}t} \qquad\qquad (XI)$$

Define $f = \dfrac{2(HX_i)_{2,\beta}}{(H_o)}$, where f is the fraction of histone H4 in β-sheet structure. Equation (XI) becomes

$$\frac{1}{1-f} = 1 + \tfrac{1}{2}k_{app}t \qquad\qquad (XII)$$

As previously described in Li et al. (29), f can be measured by circular dichroism (f_{CD}) or fluorescence anisotropy (f_F), namely in CD

$$f_{CD} = \frac{\Delta\varepsilon_p (t, 220) - \Delta\varepsilon_p (o, 220)}{\Delta\varepsilon_p (\infty, 220) - \Delta\varepsilon_p (o, 220)} \qquad (XIII)$$

and in fluorescence anisotropy

$$f_F = \frac{r_p(t) - r_p(0)}{r_p(\infty) - r_p(0)} \tag{XIV}$$

By combining equations (XII) and (XIII) or (XII) and (XIV), we obtain

$$\frac{1}{\Delta\varepsilon_p(\infty,220) - \Delta\varepsilon_p(t,220)} =$$

$$\frac{1}{\Delta\varepsilon_p(\infty,220) - \Delta\varepsilon_p(0,220)} [1 + \tfrac{1}{2}(k_{app})_{CD}t] \tag{XV}$$

$$\frac{1}{r_p(\infty) - r_p(t)} = \frac{1}{r_p(\infty) - r_p(0)} [1 + \tfrac{1}{2}(k_{app})_F t] \tag{XVI}$$

As t is small, both equations (XV) and (XVI) approximately become

$$\ln \frac{\Delta\varepsilon_p(\infty,220) - \Delta\varepsilon_p(t,220)}{\Delta\varepsilon_p(\infty,220) - \Delta\varepsilon_p(0,220)} = -\tfrac{1}{2}(k_{app})_{CD}t \tag{XVII}$$

and

$$\ln \frac{r_p(\infty) - r_p(t)}{r_p(\infty) - r_p(0)} = \tfrac{1}{2}(k_{app})_F t \tag{XVIII}$$

Comparing equations (XVII) and (XVIII) here with equations (2) and (1) in the text, it is concluded that

$$\frac{1}{\tau_F} = \tfrac{1}{2}(k_{app})_F \tag{XIX}$$

and

$$\frac{1}{\tau_{CD}} = \tfrac{1}{2}(k_{app})_{CD} \tag{XX}$$

where τ_F and τ_{CD} are, respectively, the time constants of the slow process determined by fluorescence anisotropy (F) and circular dichroism (CD).

REFERENCES

1. Pardon, J. F., Wilkins, M. H. F., and Richards, B. M., *Nature 215,* 508 (1967).
2. Olins, A. L., and Olins, D. E., *Science 183,* 330 (1974).
3. Oudet, P., Cross-Bellard, M., and Chambon, P., *Cell 4,* 281 (1975).
4. Griffith, J. D., *Science 187,* 1202 (1975).
5. Langmore, J. P., and Wooley, J. C., *Proc. Nat. Acad. Sci. U.S.A. 72,* 2691 (1975).
6. Clark, R. J., and Felsenfeld, G., *Nature New Biol. 229,* 101 (1971).
7. Hewish, D. R., and Burgoyne, L. A., *Biochem. Biophys. Res. Commun. 52,* 504 (1973).
8. Rill, R., and Van Holde, K. E., *J. Biol. Chem. 248,* 1080 (1973).
9. Sahasrabuddhe, C. G., and Van Holde, K. E., *J. Biol. Chem. 249,* 152 (1974).
10. Weintraub, H., *Proc. Nat. Acad. Sci. U.S.A. 72,* 1212 (1975).
11. Sollner-Webb, B., and Felsenfeld, G., *Biochemistry 14,* 2915 (1975).
12. Axel, R., *Biochemistry 14,* 2921 (1975).
13. Rill, R. L., Osterhof, D. K., Hozier, J. C., and Nelson, D. A., *Nucl. Acids Res. 2,* 1525 (1975).
14. Finch, J. T., Noll, M., and Kornberg, R. D., *Proc. Nat. Acad. Sci. U.S.A. 72,* 3320 (1975).
15. Gorovsky, M. A., and Keevert, J. B., *Proc. Nat. Acad. Sci. U.S.A. 72,* 3536 (1975).
16. Pardon, J. F., Worcester, D. L., Wooley, J. C., Tatchell, K., Van Holde, K. E., and Richards, B. M., *Nucl. Acids Res. 2,* 2163 (1975).
17. Simpson, R. T., and Whitlock, J. P., Jr., *Nucl. Acids Res. 3,* 117 (1976).
18. Varshavsky, A. J., Bakayev, V. V., and Georgiev, G. P., *Nucl. Acids Res. 3,* 477 (1976).
19. Shaw, B. R., Herman, T. M., Kovacic, R. T., Beaudreau, G. S., and Van Holde, K. E., *Proc. Nat. Acad. Sci. U.S.A. 73,* 505 (1976).
20. Kornberg, R. D., *Science 184,* 868 (1974).
21. Van Holde, K. E., Sahasrabuddhe, C. G., Shaw, B. R., Van Bruggen, E. F. J., and Annberg, A. C., *Biochem. Biophys. Res. Commun. 60,* 1365 (1974).
22. Baldwin, J. P., Boseley, P. G., Bradbury, E. M., and Ibel, K., *Nature 253,* 245 (1975).
23. Hyde, J. E., and Walker, I. O., *FEBS Lett. 50,* 150 (1975).

24. Li, H. J., *Nucl. Acids Res. 2,* 1275 (1975).

25. Li, H. J., *Chromatin Structure* in *Hormone Receptors, I. Steroid Hormones.* Eds. B. W. O'Malley and L. Birn aumer, Academic Press (in press) (1977).

26. Jirgensons, B., and Hnilica, L. S., *Biochim. Biophys. Acta 109,* 241 (1965).

27. Bradbury, E. M., Crane-Robinson, C., Phillips, D. M. P., Johns, E. W., and Murray, K., *Nature 205,* 1315 (1965).

28. Tuan, D. Y. H., and Bonner, J., *J. Mol. Biol. 45,* 59 (1969).

29. Li, H. J., Wickett, R., Craig, A. M., and Isenberg, I., *Biopolymers 11,* 375 (1972).

30. Wickett, R., Li, H. J., and Isenberg, I., *Biochemistry 11,* 2952 (1972).

31. D'Anna, J. A., Jr., and Isenberg, I., *Biochemistry 11,* 4017 (1972).

32. D'Anna, J. A., Jr., and Isenberg, I., *Biochemistry 13,* 4987 (1974).

33. D'Anna, J. A., Jr., and Isenberg, I., *Biochemistry 12,* 1035 (1973).

34. D'Anna, J. A., Jr., and Isenberg, I., *Biochemistry 13,* 2093 (1974).

35. D'Anna, J. A., Jr., and Isenberg, I., *Biochemistry 13,* 2098 (1974).

36. D'Anna, J. A., Jr., and Isenberg, I., *Biochemistry 13,* 4992 (1974).

37. Smerdon, M. J., and Isenberg, I., *Biochemistry 13,* 4046 (1974).

38. Li, H. J., *Fed. Proc. Abs. No. 2098* (1973).

39. Holzwarth, G., and Doty, P., *J. Amer. Chem. Soc. 87,* 218 (1965).

40. Greenfield, N., and Fasman, G. D., *Biochemistry 8,* 4108 (1969).

41. Saxena, V. P., and Wetlaufer, D. B., *Proc. Nat. Acad. Sci. U.S.A. 68,* 969 (1971).

42. Chen, Y. H., Yang, J. T., and Martinez, H. M., *Biochemistry 11,* 4120 (1972).

43. Boublik, M., Bradbury, E. M., and Crane-Robinson, C., *Eur. J. Biochem. 14,* 486 (1970).

44. Boublik, M., Bradbury, E. M., Crane-Robinson, C., and Johns, E. W., *Eur. J. Biochem. 17,* 151 (1970).

45. Boublik, M., Bradbury, E. M., Crane-Robinson, C., and Rattle, H. W. E., *Nature New Biol. 229,* 149 (1971).

46. Bradbury, E. M., and Rattle, H. W. E., *Eur. J. Biochem. 27,* 270 (1972).

47. Lewis, P. N., Bradbury, E. M., and Crane-Robinson, C., *Biochemistry 14,* 3391 (1975).

48. Pekary, A. E., Li, H. J., Chan, S. I., Hsu, C. J., and Wagner, T. E., *Biochemistry 14,* 1177 (1975).

49. Pekary, A. E., Chan, S. I., Hsu, C. J., and Wagner, T. E., *Biochemistry 14,* 1184 (1975).

50. Clark, V. M., Lilley, D. M. J., Howarth, O. W., Richards, B. M., and Pardon, J. F., *Nucl. Acids Res. 1,* 865 (1974).

51. Lilley, D. M. J., Howarth, O. W., Clark, V. M., Pardon, J. F., and Richards, B. M., *Biochemistry 14,* 4590 (1974).

52. Moss, T., Cary, P. D., Crane-Robinson, C., and Bradbury, E. M., *Biochemistry 15,* 2261 (1976).

53. Moss, T., Cary, P. D., Abercrombie, B. D., Crane-Robinson, C., and Bradbury, E. M., *Eur. J. Biochem.* (1976) (in press).

54. Kornberg, R. D., and Thomas, J., *Science 184,* 865 (1974).

55. Roark, D. E., Geoghegan, T. E., and Keller, G. H., *Biochem. Biophys. Res. Commun. 59,* 542 (1974).

56. D'Anna, J. A., Jr., and Isenberg, I., *Biochem. Biophys. Res. Commun. 61,* 343 (1974).

57. Yu, S. S., Li, H. J., and Shih, T. Y., *Biochemistry 15,* 2034 (1976).

58. Hnilica, L. S., *The Structure and Biological Functions of Histones,* The Chemical Rubber Co. Press, Cleveland, Ohio (1972).

59. Iwai, K., Ishikawa, K., and Hayashi, H., *Nature 226,* 1056 (1970).

60. Olson, M. O. J., Jordan, J., and Busch, H., *Biochem. Biophys. Res. Commun. 46,* 50 (1972).

61. DeLange, R. J., Hopper, J. A., and Smith, E. L., *J. Biol. Chem. 248,* 3261 (1973).

62. DeLange, R. J., Fambrough, D. M., Smith, E. L., and Bonner, J., *J. Biol. Chem. 244,* 319 (1969).

63. Ogawa, Y., Quagliaroti, G., Jordan, J., Taylor, C. W., Starbuck, W. C., and Busch, H., *J. Biol. Chem. 244,* 4387 (1969).

64. Yeoman, L. C., Olson, M. O. J., Sugano, H., Jordan, J. J., Taylor, C. W., Starbuck, W. C., and Busch, H., *J. Biol. Chem. 247,* 6018 (1972).

65. Sautiere, P., Tyrou, D., Laine, B., Mizon, J., Ruffin, P., and Biserte, G., *Eur. J. Biochem. 41,* 563 (1974).

Chapter 2

HISTONE-DNA INTERACTIONS:

THERMAL DENATURATION STUDIES

HSUEH JEI LI

*Division of Cell and Molecular Biology
State University of New York at Buffalo
Buffalo, New York 14214*

I.   Introduction

     Both histones and DNA are macromolecules. Their inter-
actions are complex and should be examined extensively from
all possible angles, using various techniques. DNA is the
central molecule in chromatin and all protein·DNA complexes.
Any physical chemical method which can reveal information
about DNA will have its potential application in the studies
of protein·DNA interactions, such as those of histone·DNA in
chromatin or of other histone·DNA complexes. Physical
methods used in studies of nucleic acids include thermal denaturation, circular dichroism (CD), nuclear magnetic resonance (NMR), x-ray diffraction, viscosity, sedimentation,
electron microscopy and neutron diffraction, all of which
have been applied to chromatin and histone·DNA complexes.
In this chapter, the use of thermal denaturation in the investigation of histone-DNA interactions will be discussed.

II.   Thermal Denaturation in DNA

     Denaturation of a double-helical DNA is accompanied by
an increase in absorbance. This phenomenon is called hyper-
chromism and is attributed to the destruction of stacking
interaction among nucleotides in DNA (1-4). Denaturation of
DNA can occur by adding acid or base to the solution, or by
raising its temperature. In the latter case, it is called
thermal denaturation. It was observed that denaturation of
DNA occurs in a small temperature range (5). The temperature at which 50% of native DNA is denatured is called the
melting temperature (T_m) of the DNA. The T_m is a linear
function of the G + C (guanine + cytosine) content of the

DNA (5) and is greatly increased in a solution of higher ionic strength (5-11). Statistical thermodynamics has been used successfully in treating helix-coil transition in DNA (12-16).

Based upon both theory and experiment, helix-coil transition in DNA consists of the following key elements:

a) A nucleation step, which is the denaturation of the first base pair in a helical segment. In this step, both hydrogen bonds in the base pair and two stacking interactions between the base pairs on either side of the denatured base pair are destroyed.

b) A propagation step, which is the breaking of the hydrogen bonds in a base pair next to a denatured (coiled) segment. Only one stacking interaction with the adjacent base pair in the helical segment has to be disrupted in this step; because of this difference in stacking interactions, there is a tendency for a coiled segment to grow instead of generating more short coiled segments which is the basis for cooperative transition when a DNA is denatured.

c) A strand separation step, which involves a physical separation of both complementary strands of DNA.

d) Since a G·C pair is thermodynamically more stable than an A·T pair, more free energy is needed to denature a G·C pair than an A·T pair (5,13,14,17).

e) Electrostatic repulsion between phosphates on the negatively charged phosphate lattice of DNA tends to destabilize the helical structure; charge neutralization on phosphates by cations stabilizes greatly the double helical DNA (5-11).

Such helix-coil transition in DNA has been discussed in many articles and treated very extensively in the books of Poland and Scheraga (18) and of Bloomfield, Crothers and Tinoco (19).

III. Thermal Denaturation in Nucleoproteins - Early Development

Thermal denaturation of protein·DNA complexes (nucleoproteins) was first studied more than a decade ago, with experiments on histone-DNA complexes (20) and nucleohistones (21). The reported melting curves were broad and not definitively explained at that time.

In other simpler systems, such as polylysine·DNA and polyarginine·DNA complexes, well-defined biphasic melting curves were observed. One phase of melting occurs at a lower temperature, corresponding to the T_m of free DNA; another phase at a higher temperature, corresponding to the T_m' of polypeptide-bound DNA (22-27). Experiments showed that the binding of these polypeptides to DNA is cooperative and that the complexes could be separated into two fractions: free DNA and that bound by polylysine. Because of cooperative melting in pure DNA, and what was apparently cooperative binding of polylysine to DNA, it was considered that each separate phase of melting in a nucleoprotein must result from cooperative binding of protein to DNA. In other words, it seemed likely that, in order to induce a separate phase of melting in a nucleoprotein, the protein must bind DNA in a cooperative manner, so that every protein-covered segment which melts independently from the neighboring free DNA must be longer than several hundred base pairs (21,24).

The above concept of the conditions for biphasic melting was found to be incompatible with the thermal denaturation results found both in pea bud chromatin and in partially dehistonized chromatin. In 1969 (28), it was observed that, at low ionic strength (2.5×10^{-4}M EDTA, pH 8.0), both native chromatin and partially dehistonized chromatin from pea bud show two chracteristic melting bands at 66 and 81°C. The amplitude of these two melting bands was proportionately decreased when histones were removed by H_2SO_4, NaCl or $MgCl_2$. The two characteristic phases of melting in histone-bound DNA in chromatin could not be satisfactorily interpreted using the concept of cooperative binding of histones to DNA in native and in salt or acid-treated chromatin. Because of this difficulty, further thermal denaturation experiments were performed on pea bud chromatin and reconstituted nucleo-histones using histone H2A + H2B fractions from calf thymus or the two half-molecules of histone H2B cleaved by cyanogen bromide. All these results were in accord with an interpre-tation which attributed the melting bands at 66 and 81°C to DNA segments bound respectively by the less-basic and the more basic regions of individual histone molecules (29). This interpretation contradicted the concept which attributed a biphasic melting profile to cooperative binding of proteins to DNA. Initially such interpretation was received less favorably because one half of a histone molecule was consid-ered to be too short to cover DNA with an independent phase of melting. However, the length of DNA covered by proteins, in many cases, may not be an important factor in the phen-omenon of biphasic melting, as will be discussed more exten-sively below.

IV.   Thermal Denaturation in Nucleoprotein - A Theoretical
      Model

The basic assumption, which, in turn, restricts the
applications, is that proteins bind DNA very tightly, in
fact so tightly that, within one to two hours of thermal
denaturation experiments, no significant dissociation of
proteins from DNA has occurred. In other words, the proteins
can be considered to be irreversibly bound to DNA, whether at
room temperature or at elevated temperatures before the
protein-bound DNA regions are denatured (30). The above
assumption also covers those cases in which protein structure
is changed before the protein-bound DNA is thermally denatured.

If a protein·DNA complex satisfies the above condition,
the DNA base pairs in the complex can be divided into two
groups, protein-free and protein-bound DNA segment. Each
segment can be considered to be denatured independently from
its neighboring segments, so that distinct phases of melting,
with definitive melting temperatures, can be observed for
free and protein-bound DNA segments.

The percentage increase of hyperchromicity at temperature T is defined as $h(T)$, h_{max}, the maximum hyperchromicity
obtained. Although the hyperchromicity of protein-free and
-bound DNA segments do vary in some systems, such as directly
mixed polylysine·DNA (31), protamine·DNA (32) and polyargi-
nine·DNA complexes (33), no substantial variation in hyperchromicity has been observed in chromatin (21,29,31).
Furthermore, the correction in hyperchromicity for polyly-
sine-free and -bound DNA does not seem to change the important parameters of nucleoprotein, such as the average number
of amino acid residues per nucleotide in polylysine-bound
regions (34). Consequently, the same hyperchromicity, h, is
used in calculations for free and all protein-bound DNA segments.

A plot of $h(T)$ yields an ordinary melting profile of a
nucleoprotein. The derivative melting profile, $dh(T)/dT$, can
be obtained by the following equation (29):

$$\frac{dh(T)}{dT} = \frac{h(T + 1) - h(T - 1)}{2} \qquad (1)$$

As shown in many reports (29-37), such a derivative
melting profile, $dh(T)/dT$, is more informative than the
ordinary melting profile of $h(T)$. For instance, Fig. 1
shows a biphasic melting profile $h(T)$, and its derivative
profile, $dh(T)/dT$, for a nucleoprotein. In the derivative
plot, each phase of transition is represented by a melting

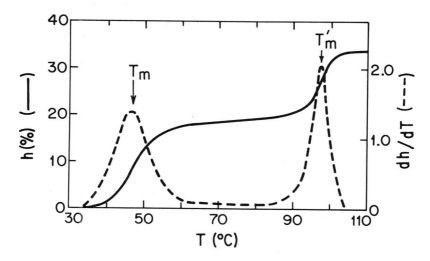

Figure 1. A typical biphasic melting profile and its deriva-tive profile of a nucleoprotein. h is hyperchromicity (per-cent increase in absorbance). T_m and T'_m are, respectively, the melting temperatures of free and protein-bound DNA base pairs.

temperature of that transition, T_m for the low melting band, and T'_m for the high melting band.

The biphasic melting curve can be schematically descri-bed as in Fig. 2a (38). The free DNA segments in the complex are denatured in the first phase of transition of T_m. The protein-bound DNA segments remain intact until the second phase of transition at T'_m.

Since the hyperchromicity changes in both transitions represented by the area under the melting band at T_m (A_{T_m}) and T'_m ($A_{T'_m}$), are proportional to the faction of base pairs free or bound by the protein, the following equation can be used to calculate the fraction (F) of base pairs in the com-plex bound by the protein (30):

$$F = \frac{A_{T'_m}}{A_{T_m} + A_{T'_m}} = \frac{A_{T'_m}}{A_T} \qquad (2)$$

where A_T is the total melting area which is equal to h_{max}.

If r is the input ratio of protein to DNA (amino acid/nucleotide) in a complex, the following equation describes the relation between r and the melting area (30):

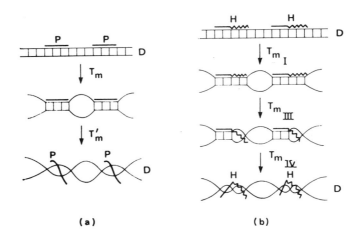

(a)   (b)

Figure 2a. Model of thermal denaturation of basic polypeptide·DNA complex. P is the polypeptide and D is the DNA. Free base pairs are first melted at T_m while polypeptide-bound base pairs remain intact at T_m and are disrupted at T_m'.

Figure 2b. Model of thermal denaturation of chromatin. H represents histones and D, the DNA. —— and ᨆ are, respectively, the more basic and the less basic half-molecules of histones. T_{mI}, T_{mIII} and T_{mIV} are, respectively, the melting temperatures of free segments, DNA segments bound by the less basic and the more basic half-molecules of histones. Both polypeptide P and histone H are shown as covering certain regions of DNA. They do not mean the binding of one strand of DNA or the number of base pairs covered. Li, H. J. (1972) Biopolymers 11, 835. Reprinted with permission of John Wiley & Sons, Inc.

$$r = \beta \; \frac{A_{T_m'}}{A_T} \qquad\qquad (3)$$

where β is the average number of amino acid residues per nucleotide in protein-bound regions in the complex. A linear plot of r against $A_{T_m'}/A_T$ measured from its melting curve yields a slope which is equal to β.

Both equations 2 and 3 demonstrate the use of thermal denaturation as a means for measuring two important parameters, F and β, in a nucleoprotein complex. The same method can be extended to a complex showing a multiphasic melting curve such as that seen in chromatin. A more general treatment of this problem has been dealt with elsewhere (30,31).

The best way to judge the above method for analyzing thermal denaturation curves of nucleoprotein lies in testing the dependence of melting curves on the size of DNA segments bound by proteins.

Polylysine added to DNA at low ionic strength binds DNA irreversibly and non-cooperatively (22,31,39). It is considered likely that polylysine-bound segments are scattered independently along the complex molecule rather than clustered together in a cooperative manner. Complexes prepared by this method showed biphasic melting curves when the chain length of polylysine is greater than 14-18 residues (25,26). Using the same solution medium for thermal denaturation of different complexes, the same melting temperature (T_m') of polylysine-bound regions has been obtained for complexes made with polylysine of various chain lengths ranging from 14 to 1000 residues (25,26,40-43). Furthermore, both equations 2 and 3 have been successfully used for all DNA complexes, whether complexed with histones (30,31), polylysine (31), polyarginine (33), protamine or basic copolypeptides containing lysine and non-basic amino acid residues (32,34, 44-46). These facts seem to indicate that the above method of analysis of thermal denaturation of nucleoprotein is correct for those basic proteins or model proteins which tend to bind DNA irreversibly.

V.   Thermal Denaturation of Chromatin

In 2.5×10^{-4}M EDTA, pH 8.0, both chromatin and partially dehistonized chromatin isolated from pea bud manifest multiphasic melting bands (Fig. 3) (29). Each melting curve can be resolved into four melting bands, T_{m_I} at 42°C, $T_{m_{II}}$ at 52°C, $T_{m_{III}}$ at 66°C and $T_{m_{IV}}$ at 81°C. Both the 66 and 81°C melting bands correspond to base pairs bound by histones, because their amplitudes are reduced proportionately

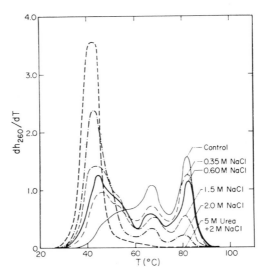

Figure 3. Derivative melting profiles of chromatin and NaCl-treated partially dehistonized chromatin from pea bud. Buffer used for melting was 2.5 x 10⁻⁴M EDTA, pH 8.0. Li, H. J., and Bonner, J. (1971) Biochemistry 10, 1461. Reprinted with permission of the American Chemical Society.

as histones are removed by NaCl, $MgCl_2$ or H_2SO_4. T_{mI} at 42°C was designated as the melting band of free base pairs, with T_{mII} at 52°C as that of small free DNA gaps between two histone-bound DNA segments or that of nonhistone protein-bound base pairs (29). Supported by thermal denaturation results of reconstituted nucleohistone using H2A + H2B, H2B-N, or H2B-C-half-molecules as well as by differential effects of ionic strength on melting bands T_{mIII} and T_{mIV}, and by the amino acid sequences of histone H1 (47), H2B (48) and H4 (49, 50), a model was proposed ·to describe melting bands III and IV as shown in Fig. 2b (29,38); the T_{mIII} and T_{mIV} bands correspond, respectively, to those base pairs bound by the less basic and the more basic halves of histones. This model implies that every histone molecule can be separated into a more basic and a less basic half, which has been shown to be true by the amino acid sequence in histones H1 (51), H2A (52,53), H2B (48), H3 (54) and H4 (49,50).

Melting curves similar to those shown in Fig. 3 were also reported for chromatin isolated from rat thymus but melted in 3.6M urea (37). However, these authors interpreted the two highest melting bands (corresponding to T_{mIII} and T_{mIV} in Fig. 3) as being due to two or more different types of histone·DNA complexes in chromatin instead of the binding of the two halves of histones as proposed by Li and Bonner (29). Recently, melting curves for chromatin have been reported from calf thymus (31,35), sea cucumber and sea urchin (56), chick erythrocytes (57) and chick embryo brain (58), which are qualitatively similar to those reported earlier (29,37).

Fig. 4 shows derivative melting profiles of rat thymus chromatin treated by trypsin and melted in 3.6M urea, as first reported by Ansevin et al. in 1971 (37). The addition of trypsin to chromatin preferentially reduces the highest melting band. This is also true for calf thymus chromatin treated by trypsin but melted in 2.5 x 10^{-4}M EDTA, pH 8.0 (43). Such results were first interpreted by Li et al. in 1973 (31) as being due to preferential digestion of the more

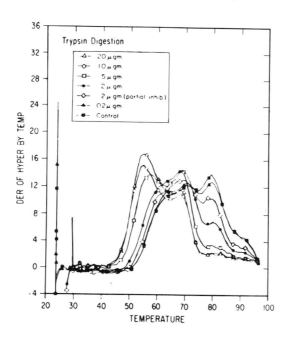

Figure 4. Derivative profiles of trypsin-treated rat thymus chromatin. Amounts of trypsin added per mg of chromatin are indicated on the figure insert. Melting medium is 3.6M urea, 0.005M cacodylate buffer. Ansevin, A. T., Hnilica, L. S., Spelsberg, T. C., and Kehn, S. L. (1971) Biochemistry 10, 4793. Reprinted with permission of the American Chemical Society.

basic regions of histones by trypsin; the interpretation was later confirmed by Weintraub and Van Lente (59) using peptide mapping and electrophoresis. Since trypsin preferentially digests the more basic regions of histones in chromatin, it can be used to study the structures of both the more basic and the less basic regions of histones as well as their effects on DNA structure. For instance, circular dichroism spectra of chromatin treated by trypsin indicate that the less basic regions of histones in chromatin have more α-helical structures than do the more basic regions. Their binding to DNA in chromatin also induces a greater structural alteration in DNA than does that of the more basic regions (43).

Equation 2 can be used for calculating the fraction of DNA base pairs bound by histones in chromatin or in partially dehistonized chromatin. For this purpose, $A_{T_m'}$ is replaced by $A_{T_{m_{III}}} + A_{T_{m_{IV}}}$, since, in chromatin, both melting bands III and IV (see Fig. 3) are induced by histone binding. It was determined that, in pea bud chromatin 75 ± 8% of DNA is covered by histones (30); in calf thymus, the coverage is 79 ± 3% (31). The difference between these two chromatin may not be significant.

The above value of approximately 80% DNA in chromatin bound by histones is much higher than the previous value of 50% which was estimated from results of polylysine binding to chromatin and nuclease digestion of chromatin (39). As shown in Fig. 5, polylysine binding to chromatin primarily depresses melting bands I, II and III with a concomitant appearance and increase of a new band V at 95-98°C which corresponds to the melting of polylysine-bound base pairs. The results suggest that polylysine can still bind the base pairs already bound by the less basic regions of histones. In other words, part of the 50% of DNA in chromatin titratable by polylysine (31,39) is also covered by the less basic regions of histones. This 50% represents an over-estimation of histone-free regions in chromatin.

The 50% of DNA in chromatin digestable by staphylococcal nuclease (39) is close to the 50-60% of DNA bound by histones in 0.6M NaCl-treated chromatin (31). At 0.6M NaCl, histone H1 is selectively removed from chromatin (21). Recently, it was suggested that perhaps histone H1, which has much less α-helical structure than do the other histones in chromatin, does not protect DNA against nuclease digestion as well as do the other histones (43). In other words, histone H1-bound regions do not belong to the 50% of nuclease-resistant fractions (39) which represent those DNA regions bound only by histones H2A, H2B, H3 and H4. This value of 50% will then

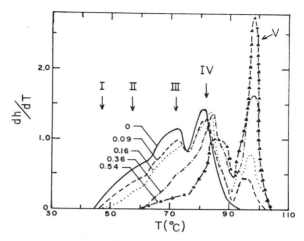

Figure 5. Derivative melting profiles of polylysine·calf thymus chromatin complexes. The input ratios of polylysine to chromatin (lysine/nucleotides) are given in the figure. Melting regions I, II, III and IV from chromatin are described in the text. Melting region V corresponds to the melting of base pairs directly bound by polylysine. Li, H. J., Chang, C., and Weiskopf, M. (1973) Biochemistry 12, 1763. Reprinted with permission of the American Chemical Society.

agree with 50-60% bound by these latter histones as determined by thermal denaturation (31). In fact, the above suggestion is supported by the recent reports (60a,60b) that histone H1 does not belong to the basic subunit in nuclease-resistant particles as to be discussed in Section VIII.

Based upon the above evidence and discussion, the figure of 80% for DNA bound by histones in chromatin as determined by thermal denaturation (at least in pea bud and calf thymus) (30,31) is a more reliable value than that of the 50% determined by polylysine binding and nuclease digestion (39).

Another important factor with regard to chromatin structure, which is obtainable by thermal denaturation, is the length of DNA bound by each histone molecule. According to equation 3, the slope of the linear plot of r (the input ratio of histone/DNA or amino acid/nucleotide) against $(^AT_{mIII} + ^AT_{mIV})/^AT$ is β, which is the number of amino acid residues per nucleotide in histone-bound regions in chromatin. Fig. 6 shows such a plot for chromatin and partially dehistonized chromatin isolated from pea bud (30). The β value is 3.2 amino acid residues/nucleotide. A similar plot for

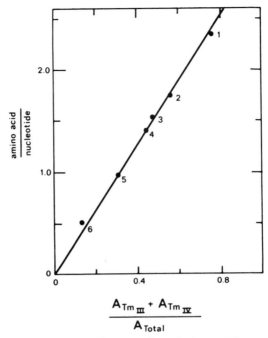

Figure 6. *Linear plot of equation 3 in native and partially dehistonized chromatin from pea bud. In equation 3, $A_{T'_m}$ is replaced by $A_{T_{mIII}} + A_{T_{mIV}}$. Chromatin was, respectively, pretreated with NaCl of 0.0M (1); 0.35M (2); 0.60M (3); 1.0M (4); 1.5M (5) and 2.0M (6). Li, H. J. (1973) Biopolymers 12, 287. Reprinted with permission of John Wiley & Sons, Inc.*

calf thymus chromatin yields a β value of 3.7 amino acid residues/nucleotide (31). In the average, there are 3.5 amino acid residues per nucleotide or 7 amino acid residues per base pair. Since the total number of amino acid residues of each histone molecule is known, it is possible to calculate the length of DNA covered by each histone molecule, since it is equal to the total number of amino acid residues of a histone divided by seven. If each histone molecule exists as a monomer in chromatin, the length of DNA bound by a histone molecule will be 31 base pairs for H1, 18 base pairs for H2A, H2B and H3, and 15 base pairs for H4 (31). If a dimer is the basic subunit, the length of DNA bound by each dimer will be double the monomer length. It is interesting to note that the total length of DNA bound by a subunit of octamer formed from $(H2A)_2$, $(H2B)_2$, $(H3)_2$ and $(H4)_2$ would be 140 base pairs (61), which is close to the length of DNA within

a nuclease-resistant particle (60,62-66) or a ν-body seen
under electron microscopy (67). In addition, if a histone Hl
monomer binds to DNA between two adjacent subunits, as suggested recently (43,60,61), the length of this spacer will be
31 base pairs, or about 105Å, which again closely approaches
the 110Å obtained from neutron diffraction (68). Such close
agreement between the results obtained from thermal denaturation and those from other methods (electron microscopy,
hydrodynamic measurement of nuclease-resistant particles and
neutron diffraction) further demonstrates the usefulness of
thermal denaturation for studies of histone-DNA interaction
and chromatin structure.

Since the thermal denaturation properties of native
chromatin have been well characterized, they can be used as
criteria for determining whether a particular histone·DNA
complex indeed has the same type of histone-DNA interaction
as found in native chromatin. This application to reconstituted nucleohistones will be discussed more extensively later,
but one example of such application to chromatin will be
given here. It is known that the α-helical structure of histones in chromatin can be disrupted by urea, as manifested
by a reduction of the negative CD at 220 nm (69-71). Nevertheless, no simultaneous increase in the free DNA melting
band has been observed under these conditions, which suggests
that the histones are not dissociated from DNA by urea (71),
a conclusion in agreement with results obtained from electrophoresis (69,70). Furthermore, the subsequent removal of
urea from the chromatin solution restores both thermal
denaturation and CD properties of the chromatin, indicating
the reversibility both of structure (of both histones and
DNA) and of binding characteristics as histones complex to
DNA in chromatin following urea denaturation of the histone
structure (71). This reversibility implies that both the
initial binding of histones to DNA in chromatin and the
original chromatin conformation are thermodynamically the
most stable.

VI. Other Thermal Denaturation Studies of Chromatin

For a given DNA, with a fixed G + C content, the melting
temperature is greatly influenced by ionic and other environmental conditions in the vicinity of the DNA molecule.
Although the stabilization or destabilization of the DNA
helix by interactions between proteins and DNA due to factors
other than ionic bonding has not been studied extensively,
the effect of ionic bonding on the melting temperature of DNA
is well understood.

With pure DNA, the addition of NaCl to a DNA solution raises the melting temperature (5-11). Conceptually, this can be explained as a result of stabilization of the DNA helix through charge neutralization of the phosphate lattice by Na^+. In other words, a very close ionic bonding between the cationic sodium and the anionic phosphate provides the principal stabilization of the helix. The chloride anions which surround the DNA-Na^+ have a much smaller effect on the thermal stability of DNA, because electrostatic interaction is reduced as a function of the reciprocal of the distance between two ions; this gives a much weaker interaction between chloride and phosphate than that between sodium and phosphate. Such a concept can also be applied to a protein· DNA complex.

Every histone molecule contains both cationic and anionic residues. If a histone molecule is fully flexible (in random coil conformation, for example), it would be expected that all cationic residues would be bound to phosphates, while the anionic residues would not be directly bound to DNA, and that there would be only one melting temperature for histone-bound regions. This T_m would either be close to that of polylysine·DNA or polyarginine·DNA or somewhere between them, depending upon the ratio of lysine to arginine. In reality, histones in chromatin are not bound in unordered fashion. Instead, because they possess substantial amounts of α-helical structure (72-74) which are not distributed evenly along the histone molecule (43), and because any ordered structure would impose some rigidity on a histone molecule, not all the available cationic residues could interact directly with phosphates on DNA. This results in incomplete charge neutralization of phosphates and a melting temperature in histone-bound regions of chromatin (66-73°C for T_{mIII} and 82°C for T_{mIV}) lower than in either polylysine·DNA (98-100°C) (27,31) or in polyarginine·DNA (92-100°C) (27,33). Furthermore, the less basic regions of histones contain a smaller percentage of cationic residues, which allows for more α-helical structure than is found in the more basic regions. Both factors would reduce charge neutralization of phosphates in the segments bound by these less basic regions of histones and result in the lower melting temperature (66-73°C for T_{mIII}) observed for these segments, compared to that of segments bound by the more basic regions of histones (83°C for T_{mIV}). Considering this effect, in reverse, since there are more free phosphates in those segments bound by the less basic regions of histones, a greater response in thermal stabilization of these segments would be expected when Na^+ in the chromatin solution is

50

increased. Experimentally, the increase in T_{mIII} induced by salt is indeed greater than that in T_{mIV}, when NaCl is added to the solution of pea bud chromatin (29). A similar result has also been reported for sea urchin and sea cucumber chromatin (56).

At intermediate ionic strength, 0.1 - 0.2M NaCl, a DNA can be renatured to a large extent if effected before strand separation (75). The kinetics of renaturation of DNA, both before and after strand separation, have been reported before (76-84). At low ionic strength, renaturation of DNA is not quite reversible (80). Although the kinetics of renaturation of DNA at low ionic strength still are not well understood, conceptually it is reasonable to consider that if the two complementary strands of a denatured DNA segment were not too far apart, they could be renatured more easily. To be more specific, in a protein·DNA complex, a denatured loop from a shorter free DNA segment could be renatured more easily than could one from a longer segment. Even in protein-bound segments, if the protein could still hold both complementary strands in close proximity after denaturation, a greater renaturation in these segments might be expected. Therefore, renaturation studies of nucleoproteins also provide important information about the complex.

Observations on the renaturation of chromatin have been reported from several laboratories (29,42,56,85,86). Renaturation of free and protein-bound regions has been extensively examined in chromatin, polylysine·DNA and polylysine·chromatin complexes (42).

In a polylysine·DNA complex, the amount of renaturation of free DNA regions is proportional to the input ratio, r, of the complex, and therefore inversely proportional to the size of free DNA loops in the complex. Agreement between expectation and observation in polylysine·DNA complexes indicates that renaturation of free DNA regions can be used as a means of comparing the sizes of free DNA loops in partially dehistonized chromatin. For instance, 75% renaturation of free DNA regions has been observed for 0.6M NaCl-treated chromatin, which treatment removes only histone H1 (21). Such a result implies that the removal of histone H1 yields short free DNA segments in chromatin, which, in turn implies that histone H1 does not form big clusters in chromatin. Such a conclusion is in agreement with the recent reports concerning the location of histone H1 in chromatin structure, namely, one H1 molecule between two subunits, each containing the other four histones (43,60,61). In 1.6M NaCl-treated chromatin, from which about 60% of the histones have been removed (31), including all of histone H1 and portions of other histones, 60% of the free DNA regions can be renatured (42). Again, this result implies a non-cooperative removal of other histones by 1.6M NaCl. Although a quantitative estimation of the size of free DNA segments by renaturation is still not possible, this method can be used for a quick, qualitative comparison of the size of free DNA regions in protein·DNA complexes.

If a directly-mixed polylysine·DNA complex is fully denatured after complexing, about 60-70% of polylysine-bound DNA can be renatured (42). This implies that, despite denaturation, both complementary strands are still held in close proximity by polylysine. However, after the chromatin is fully denatured, only 10-15% of histone-bound DNA can be renatured (29,42). Apparently histones themselves, in contrast to polylysine, do not hold together the two complementary strands of DNA in histone-bound regions after denaturation. It is possible that histones are partially dissociated from DNA after both strands are opened and denatured. It is also possible that, after denaturation, some histone molecules are associated with one strand of DNA and the others with another strand. The above two possibilities may bear some relationship to the opening of histone-bound DNA regions in chromatin during replication.

VII. Thermal Denaturation of Histone·DNA Complexes

Thermal denaturation has also been used by many workers as a method for studying the interaction between DNA and purified histones.

Fig. 7 shows the melting profiles of reconstituted nucleohistone H1 (87) formed by NaCl-gradient dialysis without urea, a method first developed by Huang et al. (20). Biphasic melting curves are developed in the complexes. In addition to the phase of melting at 50-55°C, histone H1 binding to DNA induces another phase of melting at 76°C. Similar melting results have also been obtained when the reconstituted nucleohistone H1 was formed by NaCl-gradient dialysis with urea (35). In either case, the calculated β value in protein-bound regions is 2.9-3.3 amino acid residues/nucleotide (30). However, when histone H1 and DNA were complexed directly in 3.6M urea in the absence of NaCl and melted in the same solvent, different melting curves were obtained, with a major melting band at 70°C and a minor one (shoulder) at 80°C (36). The use of equation 3 yields a much lower β value of 2.1 amino acid residues/nucleotide for these complexes. Apparently, the complexes formed by direct mixing in urea are different from those formed by NaCl-gradient

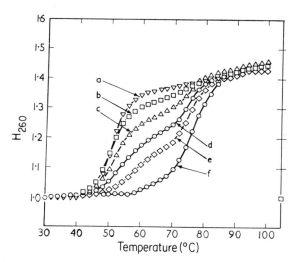

Figure 7. Thermal denaturation profiles of histone H1·DNA complexes. The solvent is 0.001M sodium cacodylate, pH 7.0. The input ratio of histone to DNA (+/-) is 0 for curve a, 0.17 for b, 0.33 for c, 0.50 for d, 0.67 for e and 1.0 for f. Olins, D. E. (1969) J. Mol. Biol. 43, 439. Reprinted with permission of Academic Press.

dialysis with or without urea.

For reconstituted nucleohistone H2B, formed by step-wise NaCl-gradient dialysis in urea, a major melting band at 80-83°C and a minor one at 60-65°C are induced by histone H2B binding (35). Continuous NaCl-gradient dialysis in urea, on the other hand, yields nucleohistone H2B with two melting bands at 70 and 90°C (88), representing melting curves which, despite differing T_m's, are qualitatively similar to those reported by Shih and Bonner (35). In both cases, the complexes showed strong light scattering. However, if a mixture of histones H2A and H2B was used for reconstitution with DNA by continuous NaCl-gradient dialysis in urea, well-defined melting curves, similar to those in chromatin, were obtained; namely, one band at 66°C and the other at 82°C (29). In addition, light scattering in these complexes is low. These results imply that histones H2A and H2B possibly interact with each other to form a more natural subunit, which then binds DNA in a more natural way (61).

If the two halves of histone H2B are prepared by cyanogen bromide cleavage, the N-half molecule (more basic portion) of H2B stabilizes DNA to 70°C, while the C-half molecule

(less basic portion) stabilizes only to 57°C. These results
were used as one type of evidence in support of the model
which suggests that melting band III (66-72°C) in chromatin
corresponds to the melting of DNA bound by the less basic
halves of histones, while melting band IV (82°C) corresponds
to that of DNA bound by the more basic halves (29).

If histone·DNA complexes are formed in 3.6M urea in the
absence of NaCl and melted in the same medium, histone H2B
binding to DNA induces only one low melting band at 68°C;
histone H2A binding, on the other hand, induces two melting
bands at 65 and 80°C. In this respect, the results for
nucleohistone H2B prepared by this method differ from those
prepared by NaCl-gradient dialysis with urea (35,88) and
from chromatin (29,37), while those of nucleohistone H2A are
similar to chromatin results (29,37).

In the case of reconstituted nucleohistone H4, prepared
by step-wise NaCl-gradient dialysis with urea, in addition
to the free DNA melting band at 45°C, there is only one
melting band at a higher temperature of 83°C induced by the
binding of histone H4 (35). Similar results have been obtained when complexes were prepared by continuous NaCl-
gradient dialysis, whether with or without urea, except that
the complexes formed from NaCl-gradient dialysis without
urea show stronger light scattering and lower efficiency in
histone binding to DNA, indicating that aggregation of H4
occurs in NaCl without urea (89). Contrary to the thermal
denaturation properties observed with nucleohistone H4 prepared by NaCl-gradient dialysis with or without urea (35,89),
directly mixed histone H4·DNA complexes in 3.6M urea without
NaCl show two histone H4-induced melting bands at 64 and
84°C (36) and have a calculated β value of 5.0 amino acid
residues/nucleotide. Although the two induced melting bands
resemble those observed with chromatin, the β value is much
higher than the 3.5 found in chromatin (30,31). If histone
H4 is added directly to DNA in 2.5 x 10^{-4}M EDTA, pH 8.0,
without urea, only one melting band at 87°C is induced,
yielding a β value of 5.2 amino acid residues/nucleotide
(89) which figure is still larger than the 3.5 of native
chromatin (30,31). Although histone H4 alone bound to DNA
can induce melting properties qualitatively resembling those
of chromatin, there is still some quantitative difference
between reconstituted nucleohistone H4 and native chromatin.

When histone H3 is added to DNA directly in 3.6M urea
in the absence of NaCl and melted in the same medium, a major
melting band at 78°C and a minor one (shoulder) at 62°C are
induced (36). A linear plot of equation 3, using the data
in the report of Ansevin and Brown (36) yields a β value of

5.2 amino acid residues/nucleotide, which is again much greater than the 3.5 obtained from chromatin (30,31). If the complexes between histone H3 and DNA are formed instead by direct mixing in 2.5 x 10^{-4}M EDTA, pH 8.0, without urea, only one melting band, at 88-90°C, is induced; and the value varies from 5.6 to 6.5 amino acid residues/nucleotide depending upon whether complexing occurred with a histone H3 monomer, dimer or oligomers from calf thymus or with an H3 monomer or dimer from duck erythrocytes (89). Again, although qualitatively partial agreement in the melting properties of nucleohistone H3 and chromatin has been obtained, quantitatively, the results still are not satisfactory.

A tetramer of histones H3 and H4 was first demonstrated by electrophoresis (90). It can be formed in 0.01M phosphate pH 7.0 shown by sedimentation (91). If this tetramer is formed and subsequently complexed with DNA in phosphate buffer, and finally dialyzed into 2.5 x 10^{-4}M EDTA buffer, pH 8.0, two melting bands at 65-70°C and 85-87°C are induced and a β value of 3.0 amino acid residues/nucleotide is obtained (89). As far as the melting properties are concerned, the (H3 + H4) tetramer forms complexes which yield melting curves quantitatively similar to those derived from native chromatin. These results imply that the (H3 + H4) tetramer is a basic subunit in native chromatin.

Studies of the binding of chicken erythrocyte histone H5 to DNA have also made extensive use of thermal denaturation (92). For these, when NaCl-gradient dialysis with urea was used for complex formation, reconstituted nucleohistone H5 yielded two melting bands at 75-79°C and 90-93°C; the β value determined was 1.5 amino acid residues/nucleotide. Characteristic melting properties such as these have been obtained with nucleohistone H5 using DNA purified from chicken erythrocytes, from calf thymus and from *M. luteus*. These results imply that the specific type of binding between histone H5 and DNA is determined primarily by the histone itself and not by the A + T content or the base sequence of the particular DNA molecule (92). A similar conclusion was also obtained both with histone H2B, based upon thermal denaturation results using DNA from calf thymus or *M. luteus* (93), and with reconstituted chromatin, based upon polyacrylamide gel electrophoresis pattern, using the distribution of products of nuclease digestion as a criterion (94).

The selective binding of polylysine to (A + T)-rich DNA in 1.0M NaCl was first reported by Leng and Felsenfeld (23).

Since the melting bands of two DNAs with varied G + C content can be distinguished by differences in melting temperature, selective binding of a protein to one DNA will

result in a greater proportionate decrease of its free DNA
melting band when compared to that of the other. Using this
method, it was observed that polylysine selectively binds
(A + T)-rich DNA during NaCl-gradient dialysis (95), which
concurs with an earlier report using direct mixing in 1.0M
NaCl (23).

Figure 8 shows application of the above method to studies
of selective binding of histone H5 to DNAs of varied G + C

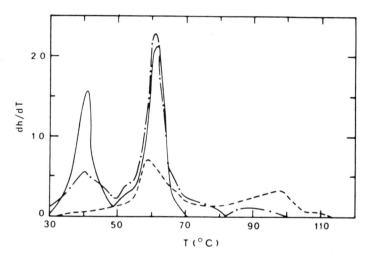

*Figure 8. Selective binding of histone H5 to (A + T)-rich
DNA. An equimolar mixture of C. perfringens DNA and M. luteus DNA was complexed with histone H5 from chick erythrocytes
by the method of reconstitution. The input ratio of histone
to DNA, r, is 0 (———); 0.5 (-·-·-) and 1.0 (----). C. perfringens DNA with 69% A +T melts at 40.5°C while M. luteus
DNA with 30% A + T melts at 61.5°C. Hwan, J. C., Leffak,
I. M., Li, H. J., Huang, P. C., and Mura, C. (1975) Biochemistry 14, 1390. Reprinted with permission of the American
Chemical Society.*

content (92). It is clear that histone H5 binds C. perfringens DNA, with 69% A + T (40°C band), more readily than
M. luteus DNA, with 30% A + T (60°C band). The same method
has also been used to study the competition between chicken
DNA and C. perfringens DNA for histone H5 binding (Figs. 9
and 10). Chicken DNA has a lower A + T content (56%) than
C. perfringens DNA (69%) but contains natural base sequences
for histone H5 binding which the latter does not have. The

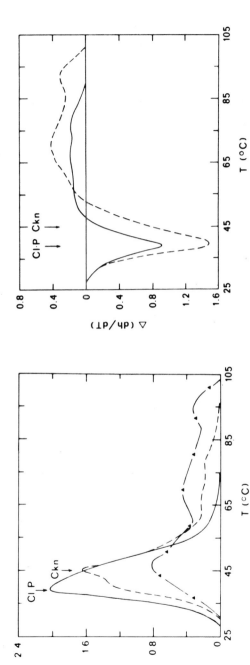

Figure 9. Competition between chicken and C. perfringens DNA for histone H5 binding. The arrows indicate the melting temperatures for chicken and C. perfringens DNA. An equimolar mixture of chicken and C. perfringens DNA was complexed with histone H5 by reconstitution. r = 0 (——), 0.5 (---) and 1.0 (-▲-). Hwan, J. C., Leffak, I. M., Li, H. J., Huang, P. C., and Mura, C. (1975) Biochemistry 14, 1390. Reprinted with permission of the American Chemical Society.

Figure 10. Difference derivative melting profiles of Fig. 9. $\Delta(dh/dT) = (dh/dT)_{complex} - (dh/dT)_{DNA}$. For the complex r = 0.5 (——) and 1.0 (---). Hwan, J. C., Leffak, I. M., Li, H. J., Huang, P. C., and Mura, C. (1975) Biochemistry 14, 1390. Reprinted with permission of the American Chemical Society.

results shown in Figs. 9 and 10 clearly indicate preferential
binding of histone H5 to *C. perfringens* DNA. The natural
base sequence in chicken DNA is therefore less important than
the A + T content in the competition for histone H5 binding.

At least two different types of binding have to be considered when a ligand is complexed with a macromolecule:
isolated binding, when the ligand binds to a site at which
the neighboring sites are still unoccupied; and binding of
the ligand to a site where an adjacent site is already occupied. The binding can be cooperative, anti-cooperative or
non-cooperative. Selectivity of binding to a particular DNA
by a protein could, therefore, result from two sources: a
stronger binding affinity between the protein and the favored
DNA, or a greater cooperativity in binding between the protein and the favored DNA, or a combination of both (19). In
those cases where selective binding had been reported [i.e.,
polylysine for (A + T)-rich DNA both in 1.0M NaCl (23) and
in NaCl-gradient dialysis (95)], the binding of protein to
DNA is highly cooperative (23,24,95). In the DNA complexes
made with histone H2B (88) or H5 (92), where strong light
scattering had been observed, the reconstituted complexes
could be fractionated into a pellet and a supernatant by
centrifugation. The pellet showed more histone binding than
did the supernatant, an indication of some cooperative
binding of histone to DNA in the pellet fraction. Although
the reason for selectivity of DNA for protein binding is not
well understood, the data available suggest a close relation
between selectivity and cooperativity.

VIII.   Chromatin Structure

The supercoil model of chromatin structure (96) had
been considered favorably by scientists in the field for
many years until the report from Olins and Olins (67) of a
string of beads in chromatin as seen by electron microscopy.
Since then, many models of chromatin structure have been
published (61,64,97-99). For example, Kornberg (99) proposed a string of contact beads as the basic structure for
chromatin, each bead consisting of 200 base pairs of DNA and
an octamer of histones. Van Holde *et al.* (64) proposed a
particulate model of chromatin composed of a histone core
(an octamer) wound by DNA from the outside.

Although each model is supported by certain experimental data on histones, histone·DNA complexes and chromatin,
they do not reveal some details of chromatin structure which
are suggested by thermal denaturation. Utilizing results
obtained from thermal denaturation of chromatin and histone·

DNA complexes, as well as from other techniques, a model of chromatin structure was recently proposed which is shown in Fig. 11 (61). The key points for this model of chromatin

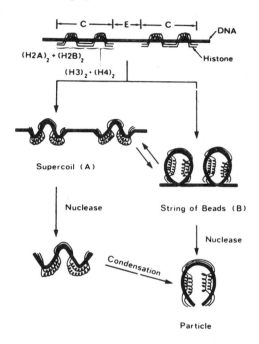

Figure 11. A model for chromatin structure. The more basic region (———) and the less basic or the more hydrophobic region of a histone molecule (⌇⌇⌇⌇⌇). The parallel dimer of each histone species is represented by ═⌇⌇⌇⌇, while the tetramer formed from the two dimers by ═⌇⌇⌇⌇═ . The drawing simply represents the regions of histones and does not represent any secondary or tertiary structures of histones. The histones on both sides of the DNA represent the binding of the more basic and the less basic regions of histones in the opposite grooves of DNA. Each C region represents a chromatin subunit composed of a DNA segment bound by an octamer of histones. Each E region represents a DNA segment bound by histone H1 or nonhistone proteins or free of protein. C regions are expected to be more condensed than E regions. The shapes of coiled or beaded regions in (A) or (B) are schematic. Foldings of DNA in three-dimensional space are likely and not presented here due to the lack of information. Li, H. J. (1975) Nucleic Acid Res. 2, 1275. Reprinted with permission of Information Retrieval Limited.

structure, specifically in histone-bound regions, are as follows: (a) formation of an octamer composed of four parallel dimers, $(H2A)_2$, $(H2B)_2$, $(H3)_2$ and $(H4)_2$; (b) subsequent formation of a histone subunit from one octamer and one H1; (c) cooperative binding of these subunits to DNA, an octamer bound on a segment of 130-150 base pairs adjacent to another segment of 30-40 base pairs bound by one H1; (d) the more basic regions of histones primarily bind the minor groove of DNA, the less basic regions of histones, the major groove; (e) histone-bound regions can be either in an open and extended coil or in a closed and compact bead, depending primarily upon the external conditions; (f) histone H1 provides less protection of DNA against nuclease digestion than does the octamer that the end product of nuclease-resistant fragment contains 130-150 base pairs of DNA with an octamer of histone.

As discussed earlier, the above model is compatible with the thermal denaturation results of chromatin, in β value of histone-bound regions, in the length of DNA bound by histone molecules in chromatin, and in the role of the more-basic and the less-basic regions of histones. The suggestion of only one H1 per subunit, and its binding to base pairs not directly bound by other histones, is in agreement with thermal denaturation and renaturation results of H1-depleted chromatin (29-31,47,61) and the recent reports of Varshavsky *et al.* (60a) and Shaw *et al.* (60b). These authors showed two types of nuclease-resistant fragments: one type contains all five histones and a DNA segment of 180-200 base pairs; and the other type, a DNA segment of 140-170 base pairs with all histones minus H1 or minus H1 + H5.

IX.   Thermal Denaturation of Other Protein·DNA Complexes

Recently, copolypeptides of basic and non-basic amino acid residues have been used as model proteins for investigation of protein-DNA interactions (45,46,100-103). Thermal denaturation is one of the major techniques used to obtain information about these interactions. For instance, it was found that under the same experimental conditions, poly (Lys^{50}, Tyr^{50}) stabilizes DNA to the same temperature as does polylysine, indicating that the tyrosine residues in the model protein neither stabilize nor destabilize the DNA helix (45). However, destabilization of the DNA helix has been observed with poly (Lys^{58}, Phe^{42}), and this has been explained as resulting from partial or full intercalation of the phenylalanine chromophore between the DNA bases (102a). In the case of poly (Lys^{40}, Ala^{60}), although there is an apparent decrease in the T'_m of the complex from that observed with polylysine·DNA, this reduction in T'_m could be explained as a result of incomplete charge neutralization on phosphate lattice of DNA due to rigid α-helical structures in poly (Lys[40], Ala[60]). The presence of α-helical structures somehow prohibits full interaction between the lysine residues and the phosphates, since increasing the Na^+ concentration in solution raises the T'_m of the complex toward that of the polylysine-bound DNA (46). Similar results have been observed for other poly (Lys[m],Ala[n]), where m + n = 100% (102b). The hydrophobic side chains of alanine in the protein apparently do not in themselves destabilize DNA helix to any detectable degree.

Nonhistone protein—DNA interactions have also been studied using thermal denaturation and circular dichroism (104). The two nonhistone proteins isolated from calf thymus chromatin by 0.35M NaCl, HMG1 and HMG2, have 30 mole percent basic amino acid residues and 25 mole percent acidic residues (105,106). Nevertheless, the binding of these two proteins to DNA stabilizes the DNA from 47°C to 60-65°C (104). This result implies that the role of ionic interaction between the basic residues of these two nonhistone proteins and phosphates of DNA is an important one. Perhaps these nonhistone proteins have folded in such a way that the basic residues are distributed mainly on the side of the protein which binds directly to DNA while the acidic residues are distributed primarily on the outside of the complex. Whether or not such ionic bonding is a general phenomenon for nonhistone proteins is still unclear. The binding results at least indicate the possibility that an acidic protein might fold in such a way as to have maximum ionic bonding between phosphates on the DNA and basic residues of the protein. It also indicates that acidic proteins do not necessarily destabilize the DNA helix as might be expected from their acidic nature.

Interactions between DNA and other proteins, such as the lactose repressor (107-111), ribonuclease (15,112,113), and others, have also been studied, and the subjects have been reviewed extensively by von Hippel and McGhee (114). These interactions have also been examined using thermal denaturation (111,112). For instance, non-specific binding of lac repressor to poly (dA-dT) stabilizes the latter from 30°C to two temperatures, one major transition at 45°C and a minor one at 60°C. These two additional phases of melting are independent of the extent of binding and can be assigned to the melting of repressor-bound regions in the complex. Using a method of analysis similar to that employed with equation 3, it was determined that one tetrameric repressor binds about 20 base pairs of DNA (111). In the case of RNase, the binding of this protein to DNA results in destabilization (lowering of the melting temperature of free DNA)(112); this unusual phenomenon has been explained as resulting from a stronger binding affinity of RNase for denatured than for native DNA (15). Under such circumstances, a different theoretical model (15,16) is required, because the protein must be considered as a ligand which binds DNA reversibly, rather than irreversibly.

Interactions between DNA and individual amino acids or oligopeptides have also been studied using thermal denaturation (115-117). Monophasic melting curves, similar to those obtained with RNase·DNA complexes, showed an upward shift in the melting temperature. Again, the theoretical model (15, 16), applicable to reversible binding between ligands and DNA, should be used for describing these melting properties. The model described earlier in this chapter does not apply to these systems.

X. Other Applications of Thermal Denaturation

A separate phase of melting is induced only by irreversible binding of protein to DNA. Digestion of the protein is expected to reduce the area under this second melting band in a manner similar to that observed in trypsin-treated chromatin (37,43) wherein the band attributable to free DNA grows at the expense of bound DNA. Thermal denaturation, therefore, can be used for studying the amount of protection against enzymatic digestion of a protein when it is bound. For instance, a polylysine molecule bound to DNA is found to be digested readily by trypsin, because the melting band of polylysine-bound DNA (T_m' at 98°C) disappears entirely after the digestion. However, a polyarginine molecule is well protected against trypsin digestion by its binding on DNA, since a substantial amount of the melting band due to polyarginine-bound DNA (T_m' at 92°C) still remains after trypsin digestion. This implies better protection of the peptide bonds in polyarginine against trypsin digestion than of those in polylysine when bound to DNA. In other words, it is possible that each polypeptide binds DNA differently, with polylysine in the minor groove and polyarginine in the major groove, for example (118).

Since thermal denaturation can provide a means of calculating the fraction of DNA bound by a protein in each complex, some physical properties, such as circular dichroism (CD) spectrum, of the protein-bound region in a complex can also be calculated, assuming that these properties are linear combinations of those properties from protein-free and protein-bound regions. This method has been used for determining the circular dichroism spectra of DNA bound by various proteins, polylysine (119), polyarginine (120), protamine (32), histone H3, H4 or H3 + H4 (121) and copolypeptides (46,102a, 102b). Based upon this same method, it was calculated that, in chromatin, the CD of DNA segments bound by whole histones minus H1 was close to the C-type spectrum (43). In other wors, the average conformation of DNA within the chromatin subunit (DNA bound by the octamer)(61,64,90,99) is close to the C conformation (19,122,123). This conclusion was used as a basis for the hypothesis that a distorted helix (124) rather than a kinky helix (125) is the fundamental conformation of DNA within the chromatin subunit.

The use of thermal denaturation in combination with circular dichroism adds another dimension to its applications with respect to protein·DNA complexes. Thus far, thermal denaturation has been measured at 260 nm, focussing on changes in the absorbance of DNA. If only the DNA hyperchromicity is measured, no information about the protein in the complex can be revealed by thermal denaturation. However, if the CD spectrum of a protein·DNA complex is measured at various temperatures, it is possible to follow the structural transition of both DNA and protein during the thermal denaturation. For instance, if a protein·DNA complex exhibits a large negative CD at 220 nm, CD changes at this wavelength during thermal denaturation can be considered as an approximate measurement of transition in protein structure because, at this wavelength, the CD contribution from the DNA is small by comparison to the protein contribution. On the other hand, the protein makes a negligible contribution to the CD near 275 nm, compared to that made by the DNA. CD changes measured at this wavelength during thermal denaturation, therefore, can be used to monitor the structural transition in DNA. CD melting curves have been studied in pure DNA (126-128) and in protein·DNA complexes, such as chromatin (57), copolypeptide·DNA (45,101,102,103), repressor·DNA (111) and RNase·DNA complexes (112).

ACKNOWLEDGEMENT: The author wishes to express appreciation to his former co-workers: Drs. C. Chang, R. M. Santella and S. S. Yu and B. Brand, P. Epstein, J. C. Hwan, I. M. Leffak, M. F. Pinkston, A. Ritter and M. Weiskopf for their contribution to the materials used in this chapter. Our research was supported in part by National Science Foundation Grant PCM76-03268 and National Institutes of Health Grant GM23079 and 23080.

REFERENCES

1. Tinoco, I., Jr., *J. Amer. Chem. Soc. 82*, 4785 (1960).
2. Tinoco, I., Jr., *J. Amer. Chem. Soc. 83*, 5047 (1961).
3. DeVoe, H., *J. Chem. Phys. 43*, 3199 (1965).
4. Felsenfeld, G., and Hirschman, S. Z., *J. Mol. Biol. 13*, 407 (1965).
5. Marmur, J., and Doty, P., *J. Mol. Biol. 5*, 109 (1962).
6. Dove, W. F., and Davidson, N., *J. Mol. Biol. 5*, 467 (1962).
7. Schildkraut, C., and Lifson, S., *Biopolymers 3*, 195 (1965).
8. Record, M. T., Jr., *Biopolymers 5*, 975 (1967).
9. Gruenwedel, D. W., and Hsu, C. H., *Biopolymers 7*, 557 (1969).
10. Gruenwedel, D. W., Hsu, C. H., and Lu, D. S., *Biopolymers 10*, 47 (1971).
11. Manning, G. S., *Biopolymers 11*, 937 (1972).
12. Lifson, S., and Zimm, B. H., *Biopolymers 1*, 15 (1963).
13. Crothers, D. M., Kallenbach, N. R., and Zimm, B. H., *J. Mol. Biol. 11*, 802 (1965).
14. Crothers, D. M., *Biopolymers 6*, 1391 (1968).
15. Lazurkin, Y. S., Frank-Kamentskii, M. D., and Trifonov, E. N., *Biopolymers 9*, 1253 (1970).
16. Crothers, D. M., *Biopolymers 10*, 2147 (1971).
17. Tinoco, I., Jr., Borer, P. N., Dengler, B., Levine, M. C., Uhlenbeck, O. C., Crothers, D. M., and Gralla, J., *Nature New Biol. 246*, 40 (1973).
18. Poland, D., and Scheraga, H. A., *Theory of Helix-Coil Transitions in Biopolymers*, Academic Press, N.Y. (1970).
19. Bloomfield, V. A., Crothers, D. M., and Tinoco, I., Jr., *Physical Chemistry of Nucleic Acids*, Harper and Row Publishers, Inc., N.Y. (1974).
20. Huang, R. C. C., Bonner, J., and Murray, K., *J. Mol. Biol. 8*, 54 (1964).
21. Ohlenbusch, H., Olivera, B. M., Tuan, D., and Davidson, N., *J. Mol. Biol. 25*, 299 (1967).
22. Tsuboi, M., Matsuo, K., and Ts'o, P. O. P., *J. Mol. Biol. 15*, 256 (1966).
23. Leng, M., and Felsenfeld, G., *Proc. Nat. Acad. Sci. U.S.A. 56*, 1325 (1966).
24. Olins, D. E., Olins, A. L., and von Hippel, P. H., *J. Mol. Biol. 24*, 157 (1967).
25. Olins, D. E., Olins, A. L., and von Hippel, P. H., *J. Mol. Biol. 33*, 265 (1968).
26. Kawashima, S., Inoue, S., and Ando, T., *Biochim. Biophys. Acta 186*, 145 (1969).

27. Shih, T. Y., Ph.D. Thesis, California Institute of Technology (1969).
28. Li, H. J., *Biology Annual Report, California Institute of Technology* (1969).
29. Li, H. J., and Bonner, J.,,*Biochemistry 10,* 1460 (1971).
30. Li, H. J., *Biopolymers 12,* 287 (1973).
31. Li, H. J., Chang, C., and Weiskopf, M., *Biochemistry 12,* 1763 (1973).
32. Yu, S. S., and Li, H. J., *Biopolymers 12,* 2777 (1973).
33. Epstein, P., Yu, S. S., and Li, H. J., *Biochemistry 13,* 3706 (1974).
34. Li, H. J., Herlands, L., Santella, R. M., and Epstein, P., *Biopolymers 14,* 2401 (1975).
35. Shih, T. Y., and Bonner, J., *J. Mol. Biol. 48,* 469 (1970).
36. Ansevin, A. T., and Brown, B. W., *Biochemistry 10,* 1133 (1971).
37. Ansevin, A. T., Hnilica, L. S., Spelsberg, T. C., and Kehn, S. L., *Biochemistry 10,* 4793 (1971).
38. Li, H. J., *Biopolymers 11,* 835 (1972).
39. Clark, R. J., and Felsenfeld, G., *Nature New Biol. 229,* 101 (1971).
40. Inoue, S., and Ando, T., *Biochemistry 9,* 388 (1970).
41. Inoue, S., and Ando, T., *Biochemistry 9,* 395 (1970).
42. Li, H. J., Chang, C., Weiskopf, M., Brand, B., and Rotter, A., *Biopolymers 13,*
43. Li, H. J., Chang, C., Evagelinou, Z., and Weiskopf, M., *Biopolymers 14,* 211 (1975).
44. Burckhardt, G., Zimmer, C., and Luck, G., *Nucleic Acids Res. 3,* 537 (1976).
45. Santella, R. M., and Li, H. J., *Biopolymers 13,* 1909 (1974).
46. Pinkston, M. F., and Li, H. J., *Biochemistry 13,* 5227 (1974).
47. Bustin, M., Rall, S. C., Stellwagen, R. H., and Cole, R. D., *Science 163,* 391 (1969).
48. Iwai, K., Ishikawa, K., and Hayashi, H., *Nature 226,* 1056 (1970).
49. Delange, R. J., Fambrough, D. M., Smith, E. L., and Bonner, J., *J. Biol. Chem. 244,* 319 (1969).
50. Ogawa, Y., Qualiarotti, G., Jordan, J., Taylor, C. W., Starbuck, W. C., and Busch, H., *J. Biol. Chem. 244,* 4387 (1969).
51. Rall, S. C., and Cole, R. D., *J. Biol. Chem. 246,* 7175 (1971).
52. Sautiere, M. P., Tyrou, D., Laine, B., Mizone, J., Lambelin-Breynaert, M. D., Ruggin, P., and Biserte, G.,

C. R. Acad. Sci. 274, 1422 (1972).

53. Yeoman, L. C., Olson, M. O. J., Sugano, N., Jordan, J. J., Taylor, C. W., Starbuck, W. C., and Busch, H., *J. Biol. Chem. 247,* 6018 (1972).

54. DeLange, R. J., Hopper, J. A., and Smith, E. L., *J. Biol. Chem. 248,* 3261 (1973).

55. Hanlon, S., Johnson, R. S., Wolf, B., and Chan, A., *Proc. Nat. Acad. Sci.U.S.A. 69,* 3263 (1972).

56. Subirana, J. A., *J. Mol. Biol. 74,* 363 (1973).

57. Wilhelm, F. X., de Murcia, G. M., Champagne, M. H., and Daune, M. P., *J. Biochem. 45,* 431 (1974).

58. Hjelm, R. P., Jr., and Huang, R. C. C., *Biochemistry 13,* 5275 (1974).

59. Weintraub, H., and Van Lente, F., *Proc. Nat. Acad. Sci. U.S.A. 71,* 4249 (1974).

60a. Varshavsky, A. J., Bakayev, V. V., and Georgiev, G. P., *Nucleic Acids Res. 3,* 477 (1976).

60b. Shaw, B. R., Herman, T. M., Kovacic, R. T., Beaudreau, G. S., and Van Holde, K. E., *Proc. Nat. Acad. Sci. U.S.A. 73,* 505 (1976).

61. Li, H. J., *Nucleic Acids Res. 2,* 1275 (1975).

62. Noll, M., *Nature 251,* 249 (1974).

63. Sahasrabuddhe, C. G., and Van Holde, K. E., *J. Biol. Chem. 249,* 152 (1974).

64. Van Holde, K. E., Sahasrabuddhe, C. G., and Shaw, B. R., *Nucleic Acids Res. 1,* 1579 (1974).

65. Sollner-Webb, B., and Felsenfeld, G., *Biochemistry 14,* 2915 (1975).

66. Axel, R., *Biochemistry 14,* 2921 (1975).

67. Olins, A. L., and Olins, D. E., *Science 183,* 330 (1974).

68. Baldwin, J. P., Boseley, P. G., Bradbury, E. M., and Ibel, K., *Nature, 253,* 245 (1975).

69. Bartley, J., and Chalkley, R., *Biochemistry 12,* 468 (1973).

70. Shih, T. Y., and Lake, R. S., *Biochemistry 11,* 4811 (1972).

71. Chang, C., and Li, H. J., *Nucleic Acid Res. 1,* 945 (1974).

72. Shih, T. Y., and Fasman, G. D., *J. Mol. Biol. 52,* 125 (1970).

73. Simpson, R. B., and Sober, H. A., *Biochemistry 9,* 3103 (1970).

74. Permogorov, U. I., Debavov, U. G., Sladkova, I. A., and Rebentish, B. A., *Biochim. Biophys. Acta 199,* 556 (1970).

75. Marmur, J., and Doty, P., *J. Mol. Biol. 3,* 585 (1961).

76. Crothers, D. M., *J. Mol. Biol. 9,* 712 (1964).

77. Spatz, H.-ch., and Crothers, D. M., *J. Mol. Biol. 42*, 191 (1969).
78. Massie, H. R., and Zimm, B. H., *Biopolymers 7*, 475 (1969).
79. Hickey, T. M., and Hamori, E., *J. Mol. Biol. 57*, 359 (1971).
80. Hoff, A. J., and Roos, A. L. M., *Biopolymers 11*, 1289, (1972).
81. Cohen, R. J., and Crothers, D. M., *Biochemistry 9*, 2533 (1970).
82. Britten, B. J., and Kohne, D. E., *Yearbook Carnegie Inst.* 78 (1966).
83. Wetmur, J. G., and Davidson, N., *J. Mol. Biol. 31*, 349 (1968).
84. Wetmur, J. G., *Biopolymers 10*, 601 (1971).
85. Wilhelm, F. X., Champagne, M. H., and Daune, M. P., *Eur. J. Biochem. 15*, 321 (1970).
86. Henson, P., and Walker, I. O., *Eur. J. Biochem. 16*, 524 (1970).
87. Olins, D. E., *J. Mol. Biol. 43*, 439 (1969).
88. Leffak, I. M., Hwan, J. C., Li, H. J., and Shih, T. Y., *Biochemistry 13*, 1116 (1974).
89. Yu, S. S., Li, H. J., and Shih, T. Y., *Biochemistry* (1976).
90. Kornberg, R. D., and Thomas, J. D., *Science 184*, 865 (1974).
91. D'Anna, J. A., Jr., and Isenberg, I., *Biochem. Biophys. Res. Commun. 61*, 343 (1974).
92. Hwan, J. C., Leffak, I. M., Li, H. J., Huang, P. C., and Mura C., *Biochemistry 14*, 1390 (1975).
93. Leffak, I. M., and Li, H. J., *(unpublished results)*.
94. Axel, R., Melchoir, W., Jr., Sollner-Webb, B., and Felsenfeld, G., *Proc. Nat. Acad. Sci. U.S.A. 71*, 4101 (1974).
95. Li, H. J., Brand, B., and Ritter, A., *Nucleic Acids Res. 1*, 257 (1974).
96. Pardon, J. F., Wilkins, M. H. F., and Richards, B. M., *Nature 215*, 508 (1967).
97. Kornberg, R. D., *Science 184*, 868 (1974).
98. Hyde, J. E., and Walker, I. O., *Nucleic Acids Res. 2*, 405 (1975).
99. Pardon, J. F., Worcester, D. L., Wooley, J. C., Tatchell, K., Van Holde, K. E., and Richards, B. M., *Nucleic Acids Res. 2*, 2163 (1975).
100. Sponar, J., Stokrova, S., Koruna, I., and Blaha, K., *Collect. Czech. Chem. Commun. 39*, 1625 (1974).
101. Mandel, R., and Fasman, G. D., *Biochem. Biophys. Res.*

Commun. 59, 672 (1974).

102a. Santella, R. M., and Li, H. J., *Biochemistry 14,* 3604 (1975).

102b. Pinkston, M. F., Ritter, A., and Li, H. J., *Biochemistry 15,* 1676 (1976).

103a. Ong, E. C., Snell, C., and Fasman, G. D., *Biochemistry 15,* 468 (1976).

103b. Ong, E. C., and Fasman, G. D., *Biochemistry 15,* 477 (1976).

104. Yu, S. S., Li, H. J., Goodwin, G. H., and Johns, E. W. *(in preparation).*

105. Goodwin, G. H., and Johns, E. W., *Eur. J. Biochem. 40,* 215 (1973).

106. Shooter, K. V., Goodwin, G. H., and Johns, E. W., *Eur. J. Biochem. 47,* 263 (1974).

107. Gilbert, W., and Müller-Hill, B., *Proc. Nat. Acad. Sci. U.S.A. 58,* 2415 (1967).

108. Riggs, A. D., Suzuki, H., and Bourgeois, S., *J. Mol. Biol. 48,* 67 (1970).

109. Lin, S. Y., and Riggs, A. D., *Nature 228,* 1184 (1970).

110. Wang, J. C., Barkley, M. D., and Bourgeois, S., *Nature 251,* 247 (1974).

111. Clement, R., and Daune, M. P., *Nucleic Acids Res. 2,* 303 (1975).

112. Felsenfeld, G., Sandeen, G., and von Hippel, P. H., *Proc. Nat. Acad. Sci. U.S.A. 50,* 644 (1963).

113. Rajn, E. V., and Davidson, N., *Biopolymers 8,* 743 (1969).

114. von Hippel, P. H., and McGhee, J. D., *Ann. Rev. Biochem. 41,* 231 (1972).

115. Gabbay, E. J., Sanford, K., and Baxter, C. S., *Biochemistry 11,* 3429 (1972).

116. Gabbay, E. J., Sanford, K., Baxter, C. S., and Kapicak, L., *Biochemistry 12,* 4021 (1973).

117. Novak, R. L., and Dohnal, J., *Nature New Biology 243,* 155 (1973).

118. Li, H. J., Rothman, R., and Pinkston, M. F., *J. Biol. Chem. (submitted).*

119. Chang, C., Weiskopf, M., and Li, H. J., *Biochemistry 12,* 3028 (1973).

120. Yu, S. S., Epstein, P., and Li, H. J., *Biochemistry 13,* 3713 (1974).

121. Yu, S. S., Li, H. J., and Shih, T. Y., *Biochemistry* (1976).

122. Marvin, D. A., Spencer, M., Wilkins, M. H. F., and Hamilton, L. D., *J. Mol. Biol. 3,* 547 (1961).

123. Tunis-Schneider, M. J. B., and Maestre, M. F., *J. Mol. Biol.* 53, 521 (1970).

124. Li, H. J., in *Chromatin Structure: A Model*, in *Molecular Biology of the Mammalian Genetic Apparatus - Its Relationship to Cancer, Aging and Medical Genetics, Part A.*, ed. P. O. P. Ts'o, Associated Scientific Publishers, Elsevier Medica, North Holland *(in press)* (1976).

125. Crick, F. H. C., and Klug, A., *Nature 255*, 530 (1975).

126. Gennis, R. B., and Cantor, C. R., *J. Mol. Biol.* 65, 381 (1972).

127. Studdert, D. S. Patronic, M., and Davis, R. C., *Biopolymers 11*, 761 (1972).

128. Usatyi, A. F., and Shlyvokhtenko, L. S., *Biopolymers 12*, 45 (1973).

Chapter 3

HISTONE-DNA INTERACTIONS:

CIRCULAR DICHROISM STUDIES

GERALD D. FASMAN

Graduate Department of Biochemistry
Brandeis University
Waltham, Massachusetts 02154

The genetic material of higher organisms appears to be much more complex than the genomes of viruses and bacteria. One aspect of this complexity is due to the interaction of DNA with histones, non-histone proteins and chromosomal RNA in the eukaryotic chromosomes. This close association of the family of basic proteins, the histones, with DNA in all higher organisms has led to two suggestions concerning their function. The first is their role in genetic regulation as first proposed by Stedman and Stedman (1,2); the second is a structural one: maintenance of the overall conformation of the nucleoproteins, or packaging of the DNA in the chromosome (3-5). However, to differentiate these roles, or even to demonstrate them has proven to be extremely difficult.

The DNA of the nucleus of each cell contains all the information necessary for duplication of the complete organism. However, this does not happen and only certain parts of the chromosome, certain genomes, are expressed in each cell; thus, liver cells reproduce liver cells, etc., and regeneration is carefully controlled. The mechanisms of such control are yet beyond our understanding.

Before delving into DNA-histone interactions, the aspects of circular dichroism spectroscopy, which are pertinent to these studies, will be briefly reviewed. Only the bare essentials will be discussed; for greater detail, the review of Brahms and Brahms is recommended (6).

The four main bases found in DNA are composed of two

Nomenclature: The following equivalents are used throughout: H1, F1, Ia: H2A, F2a2, IIb1: F2b, IIb2: H3, F3, III: H4, F21, IV.

purines, adenine and guanine, and two pyrimidines, cytosine and thymine (Fig. 1). These are the basic chromophores responsible for light absorption in the ultraviolet region. A typical absorption spectrum of these bases, that of adenosine

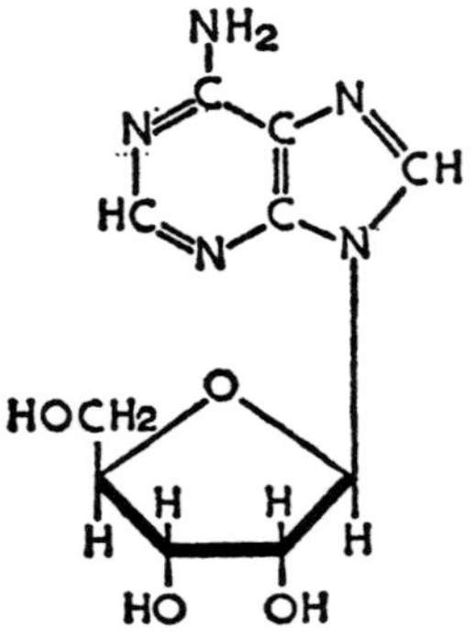

adenosine
9-β-ᴅ-ribofuranosyl
adenine

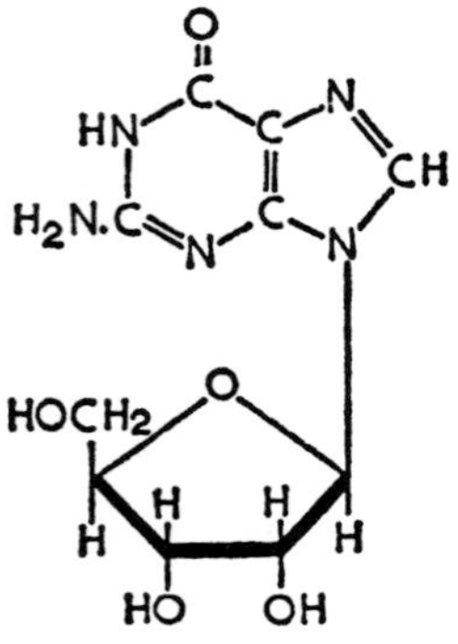

guanosine
9-β-ᴅ-ribofuranosyl
guanine

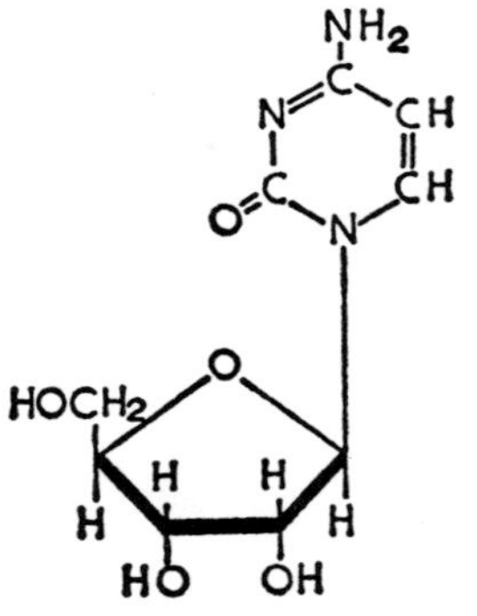

cytidine
1-β-ᴅ-ribofuranosyl
cytosine

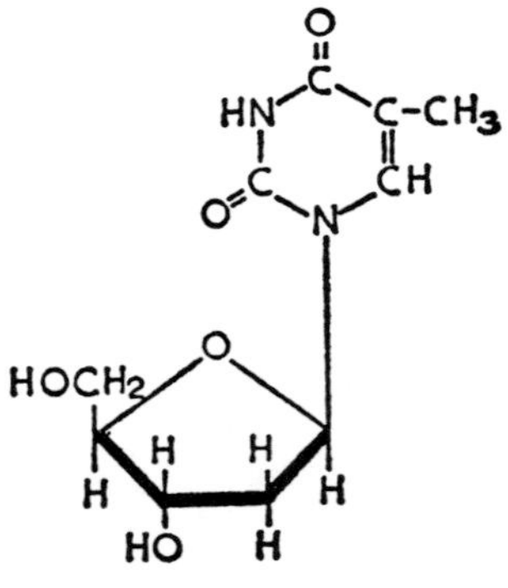

thymidine
1-β-ᴅ-2-deoxyribo-
furanosylthymine

Figure 1. The main bases found in DNA.

72

triphosphate is seen in Fig. 2. The main electronic transitions responsible for this absorption spectra are $\pi \rightarrow \pi^*$ and $n \rightarrow \pi^*$. The two requirements necessary for optical activity are: an absorption band and asymmetry. It is this combination which is responsible for circular dichrosim. Thus in the case of ATP, the main absorption band is found around 275 nm.

Circular dichroism (CD) has been employed to detect the asymmetry of DNA, i.e., a conformational probe. Although the exact conformation of DNA cannot be obtained from the circular dichroism spectra, despite the fact that theoretical studies along these lines are well advanced (6), it is an excellent probe to follow changes in asymmetry, i.e., conformational changes. This is the main application to be utilized in this discussion of the application of circular dichrosim spectra for the study of chromatin structure.

When two or more chromophores, which contain asymmetric centers, are stacked, e.g., the two DNA strands of the double helix, and their transition moments interact, several CD spectra are theoretically possible, as seen in Fig. 3 (7). The B-form of DNA has a spectra similar to curve 2 which has been called a conservative spectra. Once the double strand has formed, it is still possible to alter its dimensions (asymmetry) without destroying the base pairing, by tilting the bases either relative to the vertical axis or the horizontal axis. This is shown in Fig. 4, where the tilt and twist of the planar bases are shown. Changes of asymmetry of this type cause different transition moment interactions and consequently cause changes in the CD spectra. So, for instance, if by binding a protein onto DNA causes a tightening of the DNA helix, or causes a twisting of the bases, i.e., a conformational change, a different CD spectra will be produced.

X-ray analysis of DNA has demonstrated three canonical forms termed A, B and C (8,9,10). The dimensions of the various forms of DNA are seen in Table I. The conditions necessary to produce these forms, i.e., relative humidity and salt concentration, are also listed. One of the most significant differences in these forms is the angle between the perpendicular to the helix axis and the bases (tilt). The A form has a 20° tilt and the B form has a 2° tilt. Other differences, such as the number of bases per turn, are also noted. These data were obtained on oriented fibers under specific conditions of relative humidity and ionic strength. Utilizing the same set of conditions, Tunis-Schneider and Maestre (11) obtained films of DNA and examined their CD spectra and found that each of the A, B and C forms

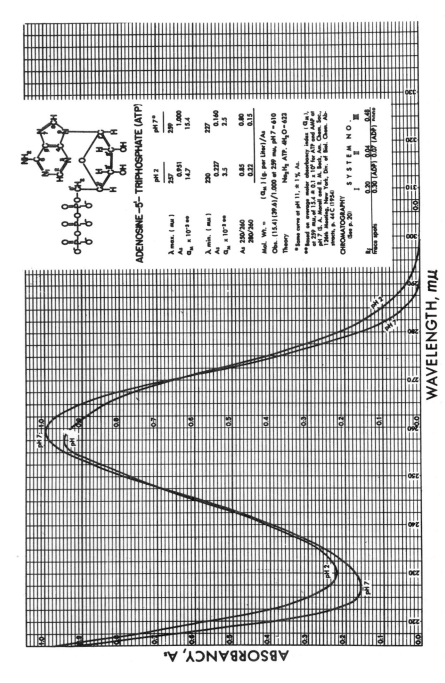

Figure 2. The absorption spectrum of adenosine-5'-triphosphate (ATP): Circular OR-10. Jan. 1956, P-L Biochemical, Inc., Milwaukee, Wisconsin 53205. Reprinted with permission.

74

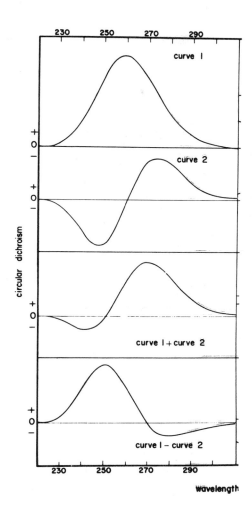

*Figure 3. The shape of circular dichroism curves. The com-
bination of the top two curves gives all others. Tinoco, I.,
Jr. (1968) J. Chim. Phys. 65, 91. Reprinted with permission
of Journal de Chimie Physique.*

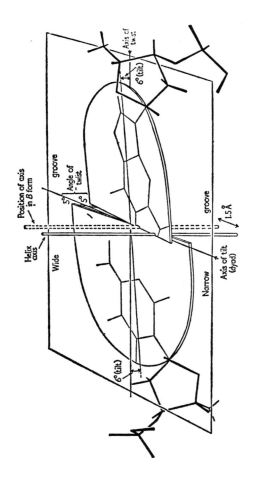

Figure 4. Perspective drawing showing one nucleotide pair in the structure. The bases are shown as if drawn on solid plates which are rotated around the axis of tilt and twist by 6° and 5°, respectively. The corresponding angles in the model for the B-form DNA are 2° tilt, and 5° twist. The base pair is moved 1.5 Å away from the helix axis relative to its position in B-form DNA. Marvin, D. A., Spencer, M., Wilkins, M. H. F., and Hamilton, L. D. (1961) J. Mol. Biol. 3, 547. Reprinted with permission of Academic Press.

Table I[*]

DIMENSIONS OF THE DIFFERENT FORMS OF DNA

DNA	Pitch	Residues per turn	Translation per residue	Rotation per residue	Angle between perpendicular to helix axis and bases	Dihedral angle between base planes	Furanose[a] out of plane atoms	φ_{CN}[b]	Ref.
A form, Na salt, 75% humidity	28.15±0.16	11	2.55	32.7°	20°	16°	C2′ = −0.13 C3′ = +0.53	−14.1°	130
B form, Na salt, 92% humidity	34.6	10	3.46	36°					131
B form, Li salt, 66% humidity	33.7±0.1	10	3.37	36°	2°	5°	C2′ = +0.19[c] C3′ = −0.10	−86.7°	132
C form, Li salt, 66% humidity	31.0	9.3	3.32	39°	6°	10°	C2′ = +0.41 C3′ = −0.05	−74.6°	133
DNA-RNA hybrid Na salt 75% humidity	28.8±0.5	11	2.62	32.7°	~20°				134
dAT, B form, Li salt, 66% humidity	33.4±0.2	10	3.34	36°					135
dABrU, B form, Li salt, 66% humidity	33.4±0.2	10	3.34	36°					135

[a,b] The values given here were computed by Langridge & MacEwan (personal communication). They are the displacements from the plane containing C1′, O1′, and C4′. The plus sign indicates that the displacement is on the same side as the C5′(endo).
[c] Haschemeyer & Rich (personal communication) calculate C2′ to be 0.26 Å endo displaced from the least-squares plane through the remaining four atoms.

[*] From D. Davies, Annual Review Biochemistry 36, Part 1, p. 351 (1967).

had characteristic spectra, shown in Fig. 5 (12). Thus, CD spectra can easily identify which of the three forms of DNA that exist under a particular set of conditions. The B form spectra is also that found in aqueous solution.

NUCLEIC ACID CONFORMATION

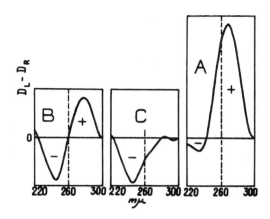

Figure 5. Schematized representation of the CD spectra for the A, B and C forms of DNA. The dotted lines are drawn through the absorption maximum. Ivanov, V. I., Minchenkova, L. E., Schyolkina, A. K., and Poletayev, A. I. (1973) Bio-polymers 12, 89. Reprinted with permission of John Wiley & Sons, Inc.

The other main component to be discussed herein is the proteins which associate with the DNA. Proteins and poly-peptides are capable of assuming three main conformations, the α-helix, the β-pleated sheet and the random coil (or irregular conformation). The CD spectra of these three con-formations are seen in Fig. 6 (13): Curve 1, is that observed for the α-helix, has two negative troughs, at 222 and 208 nm (-35,700 and -32,600, respectively) and a positive peak at 191 nm (76,900). Curve 2 is that found for the β-pleated sheet, which has a negative band at 217 nm (-18,400) and a positive peak at 195 nm (31,900) and curve 3 is that obser-ved for the random conformation, has a small positive peak at 217 nm (4,600) and a large negative band at 197 nm (-41,900). Thus, the basic conformations of proteins are easily distinguished by their CD spectra. Simplistically, if one now combines the above six CD spectra, one should potentially have all the CD contributions that could exist

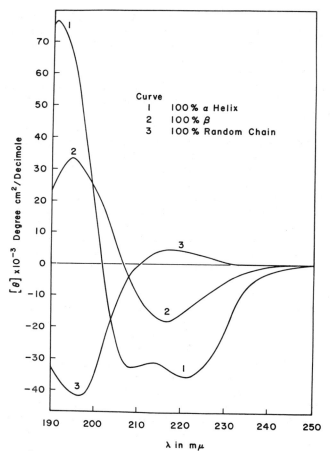

Figure 6. Circular dichroism spectra of poly-L-lysine in the α, β and random conformation. Greenfield, N., and Fasman, G. D. (1969) Biochemistry 8, 4108. *Reprinted with permission of the American Chemical Society.*

in chromatin. However, it will be evident in the discussion later that this is not the case. With this modest discussion of the background of the CD spectra of DNA and proteins, we now proceed to discuss DNA-histone interactions.

The conformation of DNA in chromatin has been the object of much research. Does the B form exist, or has it been altered? X-ray analysis has been extensively used to probe this question. Wilkins, Pardons and Richards (14) believe the X-ray patterns obtained of nucleoprotein can best be explained in terms of a super helix of DNA as seen in Fig. 7 (14). In this hypothesis, the DNA double helix maintains the B-form.

The CD spectrum of chromatin, (Fig. 8) shows that DNA in the presence of chromosomal proteins is considerably altered relative to DNA (also shown) (15). The positive band of DNA at 277 nm (8,500) is diminished to 3,990. The chromatin CD spectrum below 240 nm is attributable to proteins in the α-helical (40%) and random conformations. This alteration of the CD spectrum of DNA in chromatin can easily be shown to be attributable to the chromosomal proteins by the addition of sodium dodecylsulphate which has been reported to be able to dissociate proteins from DNA (16,17). The DNA portion of the CD spectrum, in the presence of sodium dodecylsulphate is restored to that of native DNA. The proteins which are dissociated contain considerable less α-helical content as viewed by the decrease in the 222 nm band. Thus, the conformation of both the DNA and the proteins appear to change upon dissociation. This evidence would suggest that the DNA in chromatin has been altered from the B-form, or the overall asymmetry of the DNA in chromatin is different from that found in DNA alone. Many other reports verify this CD spectrum of chromatin (e.g., 18-21). The conservative spectra of DNA (B-form) has been interpreted by Tinoco (22) to be the result of base stacking perpendicular to the helical axis. Furthermore, the nonconservative nature of the RNA spectrum, for example, is attributed to the large tilt of the bases with respect to the helical axis (7). Alterations in the secondary structure of DNA can produce significant CD spectral changes, as shown by theoretical studies (23-25). The supercoil model of Wilkins *et al.* would also introduce a new larger asymmetric unit capable of producing different CD spectra. Thus, there are several possible explanations for the altered DNA spectra found in chromatin.

If an altered DNA conformation, could be established, without question, it would have direct application to genetic regulation. An altered DNA structure would not be recognized by RNA polymerase and thus repression would be brought about by the proteins associated with the DNA.

The first histone to be considered is H1, the lysine-rich histone.

H1 was first suggested to possibly play a role in maintaining the structural integrity of chromatin (26); for this reason, it was first chosen for study in our laboratory. The

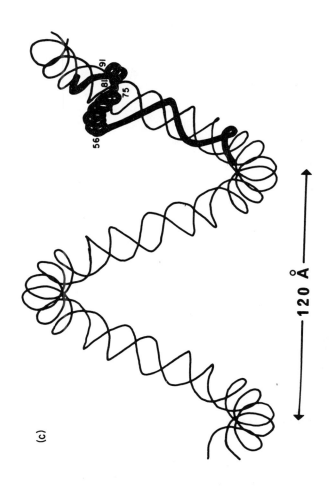

(c)

— 120 Å —

Figure 7. The proposed supercoil structure of nucleohistone. One possible manner that the histones (thick line) can attach to DNA. Some of the histone is presumed to be α-helical. Pardon, J., and Richards, B. (1973) Biol. Marcomol. 6, 53. Reprinted with permission of Marcel Dekker, Inc.

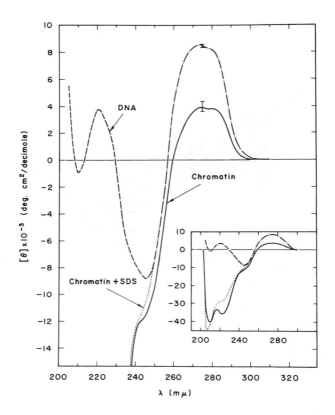

Figure 8.   CD spectra of calf thymus chromatin, chromatin in
0.1% sodium dodecylsulphate and DNA.   (———) calf thymus
chromatin; (·····), chromatin in the presence of 0.1% (W/W)
sodium dodecylsulphate; (----), pure calf thymus DNA.   Solutions in 0.14M NaF, 0.01M Tris-HCl, pH 8.0.   Conc. DNA =
1.6 x 10⁻⁴M (PO₄).   Mean residue ellipticity is based on DNA
residue concentration.   Shih, T., and Fasman, G. D. (1970)
J. Mol. Biol. 52, 125.   Reprinted with permission of Academic
Press.

examples discussed will be mainly from the laboratory of the author: the reason is a practical one, in that the diagrams were on hand. Equally illustrative examples are widely available in the literature. In Fig. 9 is seen the CD spectra of Hl and calf thymus DNA (27). The spectrum of DNA is that of the B-form, is a conservative spectrum. The CD spectrum of Hl is that of the random conformation for polypeptides. It is seen that in the region of the main ellipticity band of DNA, 275-280 nm, the histone has no contribution whatsoever. Consequently, if the conformation of DNA is altered, e.g., by the addition of a protein, this will cause a change in the ellipticity (≃280 nm) and one can monitor this conformational change without interference from the protein. Likewise, it is seen that in the main region of protein ellipticity, 220-200 nm, the ellipticity bands of DNA make a modest contribution. Therefore, if the conformation of a protein is altered upon binding to DNA, one can likewise monitor this change. Thus, the interdependence of the conformations of these two biological macromolecules can be monitored by CD. Hl is a histone which has a molecular weight of 21,000, is composed of 214 amino acids with 61 lysines, 3 arginines and a total of 16 acidic amino acids. Thus, it is a very basic protein (28). The basic groups are not evenly distributed, and 39 are found in the carboxyl end of the molecule. Cole and coworkers (28) have suggested that the basic C-terminus is the main binding region to DNA, while the N-terminus, the hydrophobic nonbasic moiety plays a special role.

The complexes of Hl and DNA were made by associating the components by means of gradient dialysis. The two components were mixed at high ionic strength and dialysed down, in steps, to the desired ionic strength, in the manner described by Bonner and coworkers (29).

Upon complexing Hl with DNA, by gradient dialysis, an altered CD DNA spectrum was obtained as seen in Fig. 10 (27). As increasing amounts of histone are added, the peak at 275 nm first decreases and red shifts until at a protein/DNA (moles amino acid residues/nucleotide residues, moles) ratio of 1, it disappears. The negative band at 245 nm increases in magnitude with increasing ratios and shifts to 251 nm at r = 1. At lower wavelengths, the spectrum becomes more negative, both due to the added histone, as well as causing the DNA band to become negative. Curve 3 of this series looks very similar to chromatin itself. Thus, by binding Hl, the CD spectrum of DNA becomes altered, indicating a change of asymmetry, i.e., a conformational change has occurred. The changes in CD spectra could also be

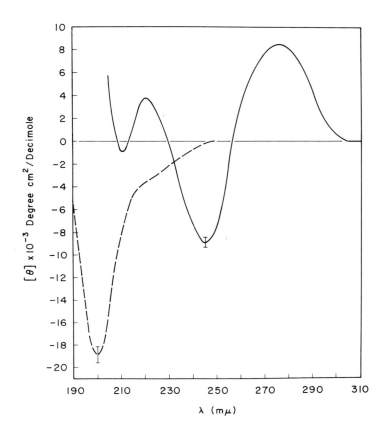

Figure 9. CD spectra of calf thymus DNA and H1 histone.
(———) DNA, 1.2 x 10⁻³M residue; (----) H1 histone, 1.18 x
10⁻³M residue. Solutions in 0.14M NaF, pH 7.0. Fasman, G.
D., Schaffhausen, B., Goldsmith, L., and Adler, A. (1970)
Biochemistry 9, 2814. *Reprinted with permission of the*
American Chemical Society.

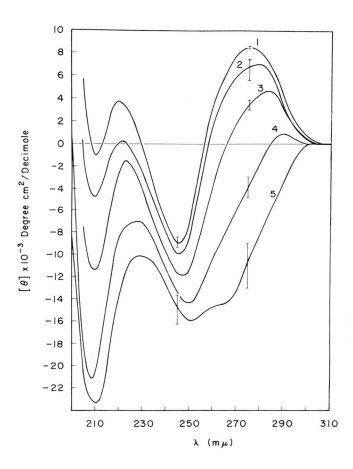

Figure 10. Circular dichroism spectra of H1:DNA complexes as a function of r, the histone (mole residue):DNA (mole phosphate) ratio at 10⁻³M DNA (mole phosphate) in 0.14M NaF, pH 7.0: Curve 1, native DNA (calf thymus); 2, r = 0.25; 3, r = 0.50; 4, r = 0.75 and 5, r = 1.0. Error bars represent reproducibility and noise dependence. Uncorrected for histone contribution. Fasman, G. D., Schaffhausen, B., Goldsmith, L., and Adler, A. (1970) Biochemistry 9, 2814. Reprinted with permission of the American Chemical Society.

brought about by varying the concentration of the complex at a fixed ratio. In Fig. 11 is seen the effect of varying the complex concentration at r = 0.75 in 0.14M NaF, where the complex is varied from 10^{-3} to 10^{-5}M (phosphate) DNA. Two plots are shown: (a) the ratio $[\theta]_{285}/[\theta]_{247}$, and (b) $[\theta]_{285}$, as a function of concentration. At 10^{-5}M DNA, the values approach that found for DNA alone. However, as the complex concentration is increased, between 5 x 10^{-5} and 4 x 10^{-4}M DNA, there is a large decrease in $[\theta]_{285}$. Thereafter, at r = 0.75, it appears that increased concentrations of complex do not cause further changes in the CD spectra. Thus, it appears that the interaction between complexes plays an important role in determining the final conformation. That is, formation of specific aggregates or super complexes is required for large changes in CD spectra.

In a similar fashion, there is a dependence of the CD spectra of the Hl:DNA complex on the ionic strength of the solution. The effect of variation in the ionic strength upon the CD spectra of the Hl:DNA complexes at r = 0.5 is seen in Fig. 12, for three salt concentrations (NaF = 0.01, 0.14, 0.20M). The spectrum at 0.01M is the same as that for native DNA. As the ionic strength is increased, the positive band at $[\theta]_{285}$ decreases and red shifts and the band at 245 nm becomes larger and red shifts, as observed for increasing the r ratio. Thus, the importance of the shielding of charge (by increasing the ionic strength) is seen, for the interaction between complexes which consequently cause the altered CD spectra. Native DNA is required for these altered CD spectra, as denatured DNA produces none of these effects. Thus, there is some specificity of Hl, binding to native DNA. Could this specificity reside in the recognition of certain nucleotide sequences with certain amino acid sequences? If this were true, then such studies would be of interest to the multiplicity of protein:DNA interactions. To further elaborate upon this theme, other histones-DNA interactions were studied.

Histone H4, the arginine-rich histone, with a molecular weight of 11,000, has about one-half the molecular weight of Hl. This histone has the opposite amino acid distribution, relative to Hl. Most of the basic amino acids are found in the N-terminus of the molecule, while in Hl, the basic portion is found in the C-terminus (30). Upon binding H4 to DNA, by gradient analysis, it is possible to obtain two types of CD spectra (31). It is necessary to add urea to the dialysis procedure to prevent H4 aggregation, and these two spectra depend upon when the urea is removed in the dialysis procedure. If the urea is removed at a salt

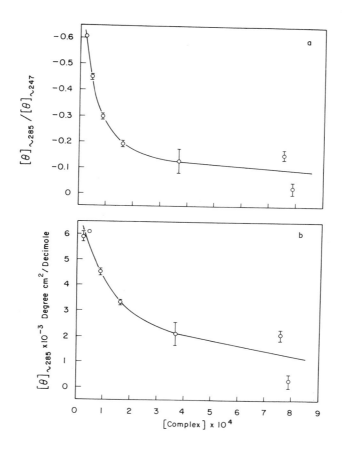

Figure 11. The dependence of the circular dichroism spectra upon Hl:DNA (calf thymus) complex concentration: Hl:DNA at r = 0.75 in 0.14M NaF at pH 7.0. (a) [θ]∼285/[θ]∼247 as a function of DNA concentration (M phosphate), termed complex concentration; (b) [θ]∼285 (ellipticity peak) vs. complex concentration. Fasman, G. D., Schaffhausen, B., Goldsmith, L., and Adler, A. (1970) Biochemistry 9, 2814. Reprinted with permission of the American Chemical Society.

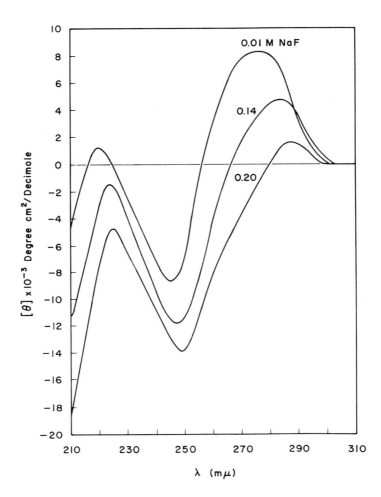

Figure 12. Circular dichroism spectra of Hl:DNA (calf thymus) complexes as a function of ionic strength (NaF concentration) for $10^{-3}M$ (phosphate) DNA at r = 0.5. Fasman, G. D., Schaffhausen, B., Goldsmith, L., and Adler, A. (1970) Biochemistry 9, 2814. Reprinted with permission of the American Chemical Society.

concentration of 0.14M NaCl, then altered CD spectra are obtained as seen in Fig. 13. As r increases from 0 to 0.5 (not shown), the CD spectra is essentially unchanged; from r = 1.0-1.5, a gradual blue shift and an increase in amplitude of the 275 nm band is observed and a maximum value of $[\theta]_{270} = 13,800$ is obtained at r = 1.5. The 245 nm negative

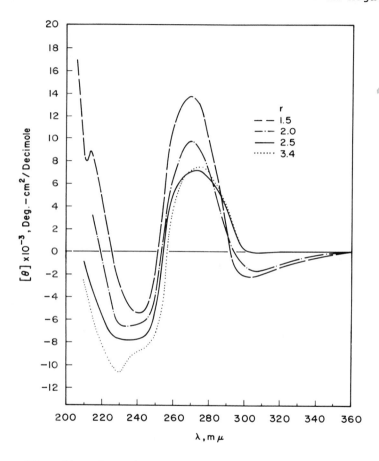

Figure 13. Circular dichroism spectra of H4:DNA (calf thymus) complexes as a function of r, the histone DNA ratio. Complexes were prepared by gradient dialysis by removal of urea at 0.15M NaCl. CD spectra was measured in 0.14M NaF - 0.001M Tris (pH 7.0). r = 1.5, ----; 2.0, -.-.-; 2.5,————; 3.4, ·····. Shih T., and Fasman, G. D. (1971) Biochemistry 10, 1675. Reprinted with permission of the American Chemical Society.

ellipticity band is concurrently decreased in amplitude and
blue shifted. A new negative band centered at 305 nm is
generated. When r becomes larger than 1.5, the magnitude of
the CD change now decreases and at r = 2.5, the CD spectrum
is again close to that of DNA. The CD spectra of H4:DNA
complexes show no concentration dependence, as was found for
H1:DNA complexes. Different CD results were obtained if
urea was removed at 0.015M NaCl. No changes in the CD spectra of the complex is observed under these conditions
(Fig. 14). Thus, the exact environmental media is important
in formation of specific complexes. The CD spectra of H4:DNA
at r = 1.5 (Fig. 13) looks similar to the CD spectra of RNA
(32), the theoretically calculated CD spectrum of the A form
of DNA (23) and the measured CD spectra of A-form DNA (11).
Thus, this complex differs from the H1:DNA complex by having
a specific stoichiometry, zero concentration dependence, and
yields an A-like DNA spectra. The spectra of the unaltered
CD spectra of the complex (Fig. 14) illustrates the fact

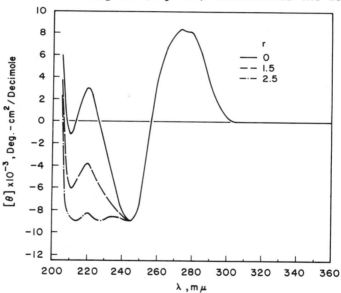

Figure 14. Circular dichroism spectra of H4:DNA (calf thymus) complexes as a function of r, the histone DNA ratio.
Complexes were prepared by gradient dialysis by removal of
urea at 0.015M NaCl. CD spectra was measured in 0.01M NaF -
0.001M Tris (pH 7.0). Shih, T., and Fasman, G. D. (1971)
Biochemistry 10, 1675. Reprinted with permission of the
American Chemical Society.

that CD is a very effective measure of the concentration of protein in DNA solutions. The ellipticity band near 200 nm can be used to evaluate the concentration present. Spectra of this type, where there is no DNA CD change can also be utilized to evaluate the conformation of the bound polypeptide. By subtracting out the CD contribution of the native DNA, one obtains Fig. 15, which indicates that the histone has assumed quite a different conformation with considerable β structure, when bound, to that obtained in solution alone. Free H4 responds significantly to the ionic strength of the solution. At low ionic strength (0.01M NaF) H4 yields a

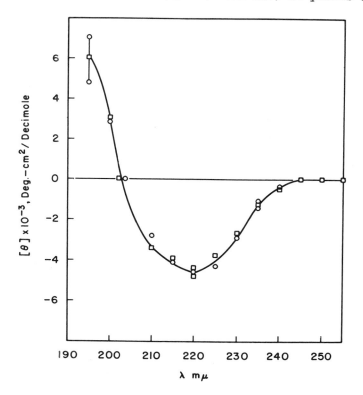

Figure 15. *Circular dichroism spectra of H4 bound to native DNA conformation. Complex prepared as described in Fig. 14, at r = 1.5. The CD spectra of DNA was subtracted from the CD spectra of the complex. [θ] based on amino acid residue concentration. Shih, T., and Fasman, G. D. (1971) Biochemistry 10, 1675. Reprinted with permission of the American Chemical Society.*

spectra (Fig. 16) similar to that found for a random coil.
In 0.14M NaF, the conformation changes drastically to that
of a protein with approximately 24% α-helix, 36% β and 40%
random coil. Perhaps this change of conformation with ionic
strength is the reason for the differences in CD spectra
when binding to DNA under different ionic conditions, as
discussed above. These experiments illustrate the interdependence of the conformation of both the DNA and protein.
One or the other can have its conformation altered upon
binding, or both can have changes in conformation upon interaction. H4 is capable of undergoing time-dependent conformational changes induced by changes in ionic strength (33).
The time-dependent change of the CD spectra of H4 as various

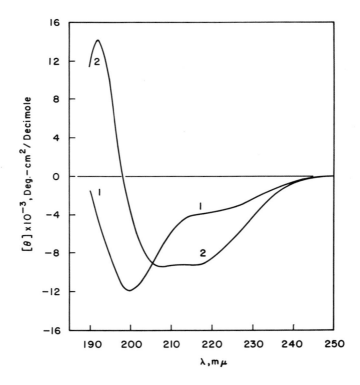

*Figure 16. Circular dichroism spectra of H4 as a function of
ionic strength. (1) 0.01M Tris (pH 7.0); (2) 0.14M NaF -
0.002M Tris (pH 7.0). Shih, T., and Fasman, G. D. (1971)
Biochemistry 10, 1675. Reprinted with permission of the
American Chemical Society.*

NaCl concentrations is seen in Fig. 17. There are both fast and slow conformational changes occurring. At low salt concentration only a fast change occurs, from a random coil to a partially α-helical structure. At high salt concentration, the same fast change occurs to the α-helical conformation, followed by a slow change to a β structure.

One can examine the CD spectra of the H4:DNA complex in more detail to evaluate the type of binding involved. By plotting the difference spectra, relative to DNA, one obtains the diagrams seen in Fig. 18. The sigmoid curve (the inset of Fig. 18) indicates that the structural transitions are cooperative relative to the ratio of H4 and DNA.

One must exercise caution in the interpretation of CD spectra obtained from scattering solutions. Two major artifacts of the optical activity of suspensions are recognized (34-37). First scattering of the light beam by particles is capable of distortion of the CD spectrum. Since CD is the measurement of the difference in absorption between left and right circularly polarized light, only asymmetric but not symmetric scattering of these two polarized components contribute to the measured CD. The direct experimental measurement of the magnitude of the asymmetric scattering in the present systems is still unavailable. But a theoretical consideration concludes for particles of polypeptides or polynucleotides with radius less than 1μ, that the scattering distortion of the CD spectrum is quite small (34). The second possible artifact is the flattening effect of the absorption and optical activity spectra near the absorption region, arising from the shadowing effect of chromophores in the particles, which is equivalent to the reduction of the effective concentration (38). These effects are considered to be minimal in the systems studied herein. Studies utilizing the fluorescat cell (39), which collects all scattered light, indicate that these possible artifacts are indeed small.

Chromatin of eukaryotic cells is a DNA complex with five major histone species, as well as other proteins. It is, therefore, of interest to examine reconstituted complexes formed from mixtures of different histones in order to evaluate their relative contributions to the final CD spectrum. Histones H1 and H4 induce different types of CD changes with DNA (Fig. 10, Fig. 13). Therefore, it is of interest to inquire whether these two histones act additively, or if not, synergistically or antagonistically in altering the CD spectrum of DNA. Histones H1:H4, mixed at a ratio of 1:4, and added to DNA at various ratios yield the CD spectra seen in Fig. 19 (32). Very little change of the DNA CD spectrum was

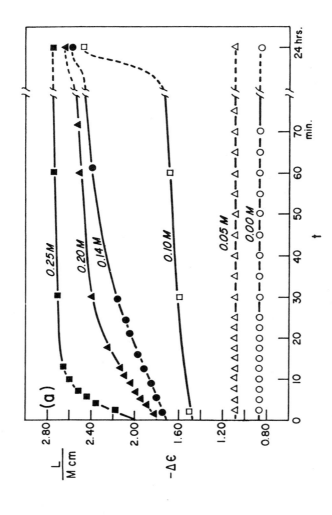

Figure 17. The circular dichroism time dependence of H4 at various NaCl concentrations in 5.0 x 10⁻³M cacodylate buffer, pH 6.5. Wickett, R. R., Li, H. J., and Isenberg, I. (1972) Biochemistry 11, 2952. Reprinted with permission of the American Chemical Society.

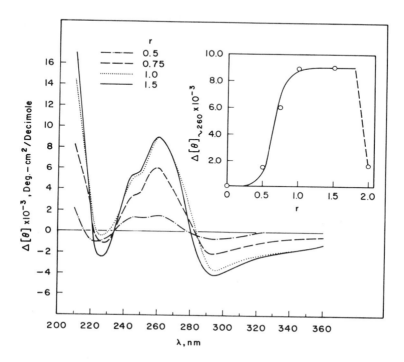

Figure 18. *Difference CD spectra of H1:DNA (calf thymus) complex and DNA. Difference spectra Δ[θ] = ([θ]complex – [θ]DNA) were obtained by subtracting control DNA, (r = 0) from complexes at different r ratios (amino acid residue/ nucleotide):r, 0.5,,–·–··–; 0.75, ––––; 1.0, ····; 1.5, ———. The inset Δ[θ] at the 260 nm peak of the difference spectra as a function of r. Solvent, 0.14M NaF – 1 mm Tris, pH 7.0. Shih, T., and Fasman, G. D. (1972) Biochemistry 11, 398. Reprinted with permission of the American Chemical Society.*

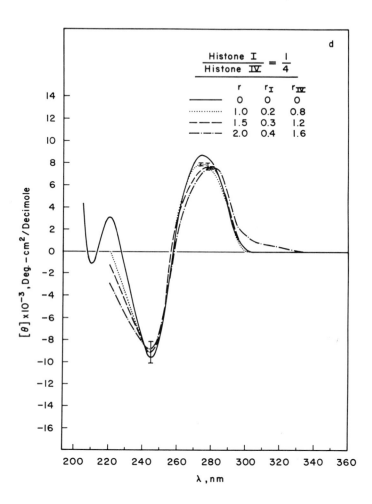

Figure 19. Circular dichroism spectra of DNA (calf thymus)
complexes with mixtures of H1 and H4 (1 to 4) at various
ratios. Histone DNA ratios: ———, 0; ····, 1.0; ----, 1.5;
and -·--·-, 2.0. Shih, T., and Fasman, G. D. (1972) Biochem-
istry 11, 398. Reprinted with permission of the American
Chemical Society.

observed. From these results and studies at various input ratios, one can conclude that these two histones do not act additively in inducing CD changes of DNA. Histone H1 appears antagonistic to and effectively blocks H4 at low relative ratios while H4 blocks H1 less effectively and requires far greater amounts to alter the CD spectrum of DNA.

Complexes of the other histones (H2A, H2B, H3), with DNA, have been studied in a similar fashion to that described for H1 and H4. These studies will be described briefly. H2B, the slightly lysine-rich histone, containing 125 amino acid residues, has a molecular weight of 13,800 and has a lysine:arginine ratio of 2.5. In 0.14M NaF, the CD curves indicate that H2B has about 30% α-helix (40). Upon complexing with DNA, at different ratios, the CD spectra obtained are seen in Fig. 20 (40). As increasing amounts of histone are added, the positive ellipticity band, in the 280 nm region, becomes larger and at r = 3.0 is greater than 40,000. This behavior is similar to that found for H4.

Histone H2A, a slightly lysine-rich protein, has 129 amino acids, with the more basic region residing in the N-terminus region. The CD spectra of complexes reconstituted from calf thymus DNA and H2A are shown in Fig. 21 (41). The effect of H2A is to greatly augment the positive CD band of DNA, in a manner similar to that of H4 and H2B. The arginine-rich histone H3 contains 135 residues, has a typical amino acid distribution with many basic residues clustered in the N-terminal part of the protein. Whem complexes are reconstituted with calf thymus DNA and H3 (oxidized or reduced), the CD spectra of the DNA is only slightly altered (41).

In summary, it can be stated that each histone binds in a specific manner to DNA and yields a characteristic CD spectra. H2A, H2B and H4 complexes with DNA cause an increase in the positive CD band of DNA, while H1 causes a large decrease in this CD band, yielding large negative values. H3 causes very slight alterations of the DNA CD spectrum. The reduced form causes a slight increase in the main DNA CD band, while the oxidized form produces a slightly reduced CD band. It is also evident that the histones are capable of undergoing conformational changes themselves which are dependent on the ionic strength (i.e., shielding) which may be considered equivalent to charge neutralization when the positively charged histone interacts with the negatively charged DNA. Thus, it appears that there is a conformational dependence of the DNA on bound histone and vice versa, the proteins undergo conformational changes upon binding to DNA. From the study of mixed histones on the CD spectra of DNA,

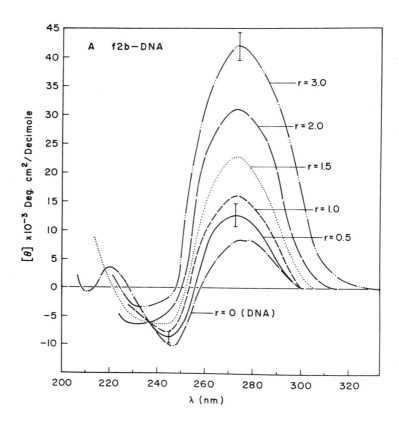

Figure 20. *Circular dichroism spectra of DNA (calf thymus)*
complexes with H2B at various ratios (r), as indicated.
Complex concentration = 1.2 - 1.4 x 10^{-4}M nucleotide residue;
solvent, 0.14M NaCl + 0.002M Tris (pH 7.0). Continuous Gdn.
HCl gradient dialysis used for complex formation. Adler,
A. J., Ross, D. G., Chen, K., Stafford, P. A., Woiszwillo,
M. J., and Fasman, G. D. (1974) Biochemistry 13, 616.
Reprinted with permission of the American Chemical Society.

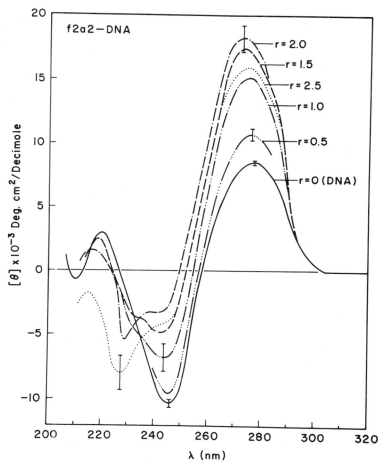

Figure 21. Circular dichroism spectra of complexes reconstituted from DNA (calf thymus) and H2A by means of Gdn. HCl
gradient dialysis. The concentration of histone (moles of
peptide residues) to DNA (moles of nucleotide residues) in
the complexes, r, is varied from r = 0 (DNA alone) to
r = 2.5. Solvent, 0.14M NaCl - 0.002M Tris (pH 7.0).
Adler, A. J., Moran, E. C., and Fasman, G. D. (1975) Biochemistry 14, 4179. Reprinted with permission of the American
Chemical Society.

it would appear that the sum of the independent histone contributions is not a valid model for the chromatin CD spectrum because the histones interact strongly with one another when bound to DNA.

What can be said about the conformational changes associated with the altered DNA CD spectra observed upon complexing of histones? The non-conservative CD spectra of H4:DNA complexes (Fig. 13) show an increase in the positive band and a new negative band at 305 nm, which could be classified as an A-type spectra, similar to that found for A-type DNA (11). When a protein is bound to DNA that is equivalent to placing it in a different ionic strength media. Neutralization of charge brought about by binding to DNA-phosphates, is similar to charge shielding at high ionic strength. Thus, it is not unexpected to observe conformational changes of histones upon binding to DNA.

When DNA CD curves lie midway between an A-form and B-form, one could suggest that a mixture of those two forms exists. Similarly, curves lying below that of B form DNA could be considered to be partially in the C form. Hanlon *et al.* (42) strongly support such a view. By isolating chromatin by various published procedures, the CD spectra shown in Fig. 22 were obtained (42). The curve marked (oooxooo) reflects a mixture of 47% B and 53% C and the curve marked (●●●▲●●●) is a mixture of 36% B and 64% C. These differences were attributed to removal of ions from the thymus nucleohistone in the early stages of isolation from the tissue. The proposed model to account for the above nucleohistone CD spectra and the meltout behavior of these complexes is shown in Fig. 23. The percent of B character of DNA in the nucleohistone corresponds to the fraction of bases melting out in transitions I, II and III. Conversely, the percent C character corresponds to the fraction of bases melting out in transitions IV and V. The major difference in conformation of areas bound with protein is due to the amount of ions present. Similar reasoning is frequently found to account for the altered DNA CD spectra found in the literature. Later in this article this point will be seriously questioned.

The CD changes induced in the DNA CD spectrum by complexing H4 are quite different from those of H1. From the primary sequences, it is known that the clustering of cationic residues also occurs in different regions in these two histones (30). In H4, these occur in the N-terminus region, while in H1, they are found in the C-terminus. The interesting question can be raised - are these different CD spectra the result of specific binding of certain amino acid

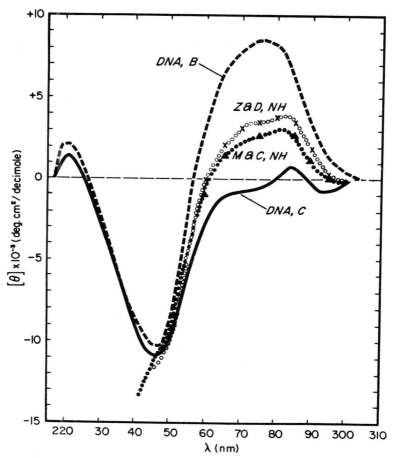

Figure 22. Comparison of the circular dichroic spectra of DNA and nucleohistone: (oooxooo) nucleohistone isolated by Zubay and Doty method and (●●●▲●●●) nucleohistone isolated by Maurer and Chalkey method. The B and C forms of DNA are also included. Johnson, R. S., Chan, A., and Hanlon, S. (1972) Biochemistry 11, 4347. Reprinted with permission of the American Chemical Society.

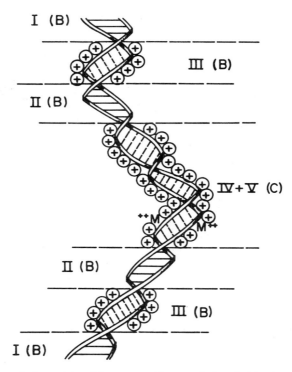

Figure 23. Schematic diagram of a model of thymus nucleo-histone. Johnson, R. S., Chan, A., and Hanlon, S. (1972) Biochemistry 11, 4347. Reprinted with permission of the American Chemical Society.

sequences to specific DNA base sequences? Are there specific coding regions of recognition on these macromolecules? At this stage of development, such ideas are highly speculative-but perhaps substantive enough to be taken seriously. Another interesting observation is the dependence of the plasticity of the histone-DNA complexes upon the annealing conditions. Eukaryotic chromosomes have long been observed to undergo very characteristic morphological changes during cell division and perhaps these various conformational changes observed with these histone-DNA complexes are models of these morphological changes in the cell. Thus, a beginning is being made to understand, on a molecular bases, these complex biological phenomena.

Recently, several interesting reports have appeared concerning the structure of DNA in concentrated polymer solutions. Lerman and co-workers (43,44) report that upon the addition of a neutral polymer, such as polyethylene oxide (PEO), in the presence of salt, to high molecular weight DNA, the DNA undergoes a cooperative structural change which results in a compact molecular conformation, a condensed state. The DNA is forced into a compact state by excluded volume interactions and undergoes a spontaneous rearrangement to an ordered tertiary structure characterized by a CD spectrum differing greatly from that found in aqueous solution. The compactness was inferred principally from changes in sedimentation velocity, at concentrations where the sedimentation is concentration independent and therefore presumably unimolecular. At higher concentrations, aggregation is visible. As a critical concentration of polymer (PEO) and salt are required for this induced transition, they have termed this PSI or Ψ-DNA. The CD spectra of Ψ-DNA are seen in Fig. 24, produced by various concentrations of PEO, after various times of mixing. Extremely large negative ellipticity bands are observed. Lerman (43,44,45) believes this spectra is caused by a superfolding of the DNA (i.e., ordered spaghetti) and may be comparable to the condensed form of DNA in the chromosome. X-ray scattering studies (45) have demonstrated that Ψ-DNA is still in the B-form. Thus, it is possible to supercoil the DNA into a new asymmetric tertiary structure without destroying the basic B-structure of the Watson-Crick double helix, and obtain a new type, Ψ-DNA, spectra. Lerman (44) has pointed out that this new spectra is similar to that obtained by other means (Fig. 25), such as by complexing with histone H1 or poly-L-lysine. More recently, Cheng and Mohr (46,47) have studied Ψ-DNA in greater detail. They have found that the GC content of the DNA plays a role in the type of CD spectra obtained in PEO (Fig. 26).

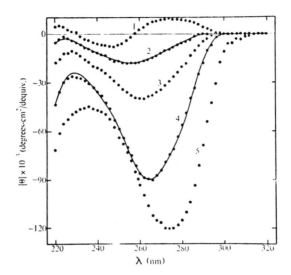

*Figure 24. Representative circular dichroism spectra of T7
DNA in different concentrations of polyethylene oxide (PEO)
at different times after mixing. The circles represent
measured spectra in PEO concentrations and at the time after
mixing as follows: 1, 80.9 mg/ml, 0.7 hr; 2, 98.1 mg/ml,
0.7 hr; 3, 126.5 mg/ml, 0.7 hr; 4, 90.1 mg/ml, 0.7 hr;
5, 126.5 mg/ml, 48 hr. DNA = 22 x 10⁻⁶M Na⁺. Jordon, C. F.,
Lerman, L. S., and Venable, J. H., Jr.* (1972) Nature New
Biol. 236, 67. *Reprinted with permission of Macmillian
Journals, Ltd.*

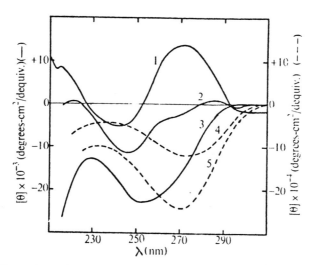

Figure 25. The circular dichroism spectra of DNA in various complexes or perturbing solvents. Curves are redrawn from published data. 1, DNA:H4 complex, r = 1.5; 2, DNA in ethylene glycol; 3, DNA:H1 complex, r = 1.0; 4, T7 DNA in 0.2M NaCl and 126.5 mg/ml PEO, 48 hr after mixing; 5, DNA:poly-L-lysine complex, r = 1.1. All data except curve 4 with calf thymus DNA. The ordinate calibration on left applies to solid curves, while the calibration on right, corresponding to a ten-fold larger value, applies to the dashed curves. Jordon, C. F., Lerman, L. S., and Venable, J. H., Jr. (1972) Nature New Biol. 236, 67. Reprinted with permission of Macmillian Journals, Ltd.

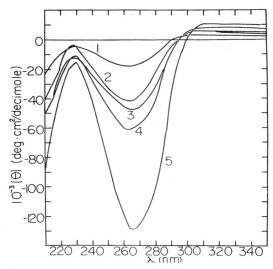

Figure 26. Circular dichroism spectra of Ψ-DNA samples of varying GC content. 1, C. perfringens, 31% GC; 2, human placenta, 40% GC; 3, calf thymus, 42% GC; 4, E. coli, 51% GC; 5, M. luteus, 72% GC. Solvent 0.35M KCl, 100 mg/ml PEO. Cheng, S. M., and Mohr, S. C. (1975) Biopolymers 14, 663. Reprinted with permission of John Wiley & Sons.

They also found it was possible to melt out the Ψ-DNA spectra, and at the elevated temperature obtain a B-type spectra. The meltout profiles for Ψ-DNA, of different GC contents is seen in Fig. 27. As expected, the higher GC content requires a higher temperature to melt out the Ψ-DNA. These authors also found that the dyes which intercalate, e.g., ethidium bromide and proflavin, into B form DNA, do not intercalate into Ψ-DNA and vice versa, a B-form DNA with an intercalated dye, cannot form Ψ-DNA. The meltout of Ψ-DNA in the presence of proflavin is seen in Fig. 28, upon cooling the spectra does not return to the Ψ-type but now gives the spectra obtained for the intercalated B form. This demonstrates that the meltout of Ψ-DNA does not destroy the B-form structure. Thus, the tertiary fold is melted out, but not the secondary structure (double helix).

Thus, the previous interpretation of the A⇌B⇌C transitions for either chromatin or for histone-DNA complexes must be carefully reexamined, for perhaps a tertiary fold, induced by the proteins, is causing the altered DNA-CD spectra.

There have been many studies using alcohols to study

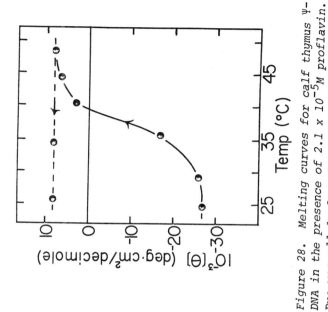

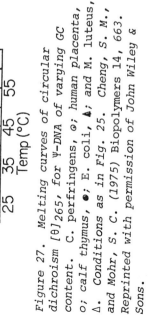

Figure 28. Melting curves for calf thymus Ψ-DNA in the presence of 2.1 x 10⁻⁵M proflavin. Dye was added after formation of Ψ-state. $[\theta]$ measured at 270 nm, at a DNA concentration of 8.5 x 10⁻⁵M. Conditions as in Fig. 25. Cheng, S. M., and Mohr, S. C. (1975) Biopolymers 14, 663. Reprinted with permission of John Wiley & Sons.

Figure 27. Melting curves of circular dichroism $[\theta]_{265}$ for Ψ-DNA of varying GC content. C. perfringens, ⊘; human placenta, ⊙; calf thymus, ●; E. coli, ▲; and M. luteus, △. Conditions as in Fig. 25. Cheng, S. M., and Mohr, S. C. (1975) Biopolymers 14, 663. Reprinted with permission of John Wiley & Sons.

conformational changes in DNA. Circular dichroism spectra
of DNA change drastically upon addition of alcohol, presumably indicating conformational changes. An example of such
studies is seen in Fig. 29 (12). At concentrations less

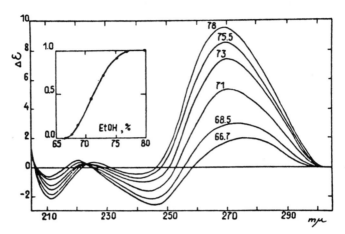

*Figure 29. Influence of ethanol upon the circular dichroism
of DNA (calf thymus). DNA concentration = 5 x 10^{-5}M (PO$_4$)
containing 5 x 10^{-4}M NaCl. Curves are labeled with the per
cent ethanol. The dependence of the relative CD change at
270 nm upon ethanol concentration is shown in the insert.
Ivanov, V. I., Minchenkova, L. E., Schyolkina, A. K., and
Poletayev, A. I. (1973) Biopolymers 12, 89. Reprinted with
permission of John Wiley & Sons.*

than 65% ETOH, the CD spectrum is that of B-DNA. As additional alcohol is added, the positive ellipticity band
increases to a maximum at 78% ETOH; the spectrum is now similar to that of A-type DNA. The authors interpret these
results in terms of a B → A transition, which is reversible
on dilution. However, it has been possible to crystallize
DNA from ethanol solutions, and in these crystals, by both
electron microscopy and X-ray crystallography (48), the DNA
has been shown to be extremely compact and still in the B
form. So there again by superfolding, or aligning of the DNA
in a super asymmetry, it appears possible to obtain altered
DNA CD spectra, <u>without</u> necessitating a change in the B form
of the DNA. The crystals so obtained could be redissolved,
have a B-form DNA CD spectrum and showed the usual hyper-
chromism upon melting.

Another approach to the study of chromatin, is to remove
the histones stepwise and look at the resulting nucleoprotein

to determine the effect of the deleted histone. The CD curves of Fig. 30 are illustrative of this approach (49). The 0.7M NaCl treatment removes H1 and the $[\theta]_{277}$ is slightly reduced. Upon treatment with 1.0M NaCl, causing removal of H2A, H2B and H3, the 277 nm band increases to the value found for native DNA. On the left of Fig. 30 is seen the effect in the CD curves of removing the proteins.

Recent studies on the digestion of chromatin with nucleases (50) has ellicited wide interest. These studies can elucidate the distribution of the histones along the DNA chain. Areas which are protected by protein will not be digested, while regions of DNA which are easily accessible, or on which protein is weakly bound, should be attacked by the nuclease. By such procedures approximately 50% of the DNA is digested, leaving a DNA-histone particle. Particles obtained by digestion with micrococcal nuclease and then further treated with trypsin have been studied by CD (51). The CD spectra (Fig. 31) of the whole chromatin shows a decrease $[\theta]$ in the 280 nm region compared to native DNA (also shown), while the partially digested chromatin (PS particle) shows an even greater depression of this ellipticity band. The DNA in the PS particle is of the order of 100 nucleotide pairs in average length. This altered DNA CD spectra must arise either directly from interaction of the DNA and protein, causing some modified DNA structure or causing a condensation of DNA into a third order asymmetric fold. Mild trypsin hydrolysis leads to a dramatic unfolding of these particles. After removal of 20% of the protein by trypsin, the hydrodynamic behavior approximates that expected for an extended rod-like DNA covered with the remaining protein and yields a CD spectra closer to that of DNA [Fig. 31 (51)]. Thus, this unfolding due to trypsin treatment also leads to reversion of the CD spectrum to a DNA like type. This PS particle is similar to the ν bodies visualized by electron microscopy studies (52) and is believed to consist of tetramers or octomers of histones dispersed along the DNA chain (53,54).

One of the original suggestions (1) for the function of histone H1 was the possible role in a regulatory mechanism in the transcription of DNA; more recently, this role has been assigned to the non-histone proteins of the nucleus. It has been demonstrated by Langan (55), and others, that in liver, histone phosphorylation occurs in response to the pancreatic hormone glucagon, upon initiation of protein synthesis. Thus, a mechanism is suggested whereby phosphorylation of H1 would cause a change in DNA-histone, or histone-histone interaction resulting in derepression of the template activity of the associated DNA, thereby allowing RNA synthesis.

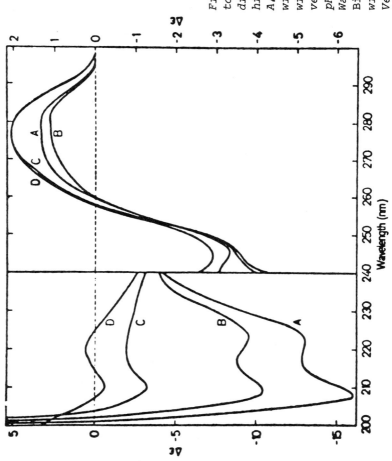

Figure 30. The effect of histone depletion on the circular dichroism of deoxyribonucleohistone (DNH) calf thymus. A, Whole DNH; B, DNH treated with 0.7M NaCl; C, DNH treated with 1.0M NaCl; D, DNA. Solvent 0.7 mM sodium phosphate, pH 6.8. Henson, P., and Walker, O. J. (1970) Eur. J. Biochem. 16, 524. Reprinted with permission of Springer-Verlog, New York, Inc.

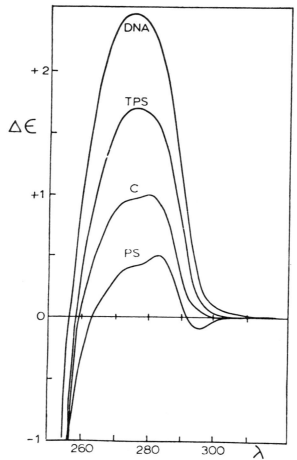

Figure 31. Circular dichroism spectra of whole chromatin (calf thymus) (C), Nuclease-resistant fragment (PS). Trypsin digested PS fractions (TPS) and DNA. Solvent is 10 mM Tris, pH 3.0, room temperature. Sahasrabuddhe, C. G., and Van Holde, K. E. (1974) J. Biol. Chem. 249, 152. Reprinted with permission of the American Society of Biological Chemists.

Circular dichroism has been utilized to demonstrate that such phosphorylation can potentially bring about a conformational change which might play a role in such depression (56). Enzymatically phosphorylated Hl (57) which has been mono-phosphorylated at Ser-37 (58), called Site A, was compared to normal Hl, alone and in association with DNA. The CD spectrum of Hl or phosphorylated Hl (A-PO$_4$-Hl) have identical CD spectra. However, when bound to DNA, in reconstituted complexes, significant differences were found (Fig. 32) (56). As the ratio of histone/DNA is increased, the positive ellipticity band $\sim$ 280 nm is slowly decreased. However, at equal ratios, the A-PO$_4$-Hl causes far less change than does Hl alone. Perhaps the clearest manner to view these progressive CD difference alterations is that the presence of PO$_4$ on Hl reduces its ability to cause deviation in the DNA CD spectra. Another phosphorylation site has been identified on Hl, at Ser 106 (59), site B, and it is possible to isolate monophosphorylated material at site A or B or at both sites simultaneously (57,60,61). The CD spectra of complexes with DNA of these three species is seen in Fig. 33 (62). The diphosphorylated Hl complex gives a spectra identical to DNA alone at r = 0.5. At r = 1.0, the Hl-P(B) is more effective in moderating the conformational change of the DNA upon binding. Perhaps these phosphorylations cause a modification of the histone-histone and/or histone:DNA interaction, thus reducing the state of aggregation in chromatin, thus making the DNA more accessible for transcription.

In addition to phosphorylation, histones are acetylated and methylated in the nucleus. The sequence of histone H4 is virtually identical in calf (63) and pea tissue (64). This conservation of primary sequence throughout evolution implies that H4 is an essential protein in the nucleus. Histone H4 is enzymatically acetylated (65) *in vivo* to form ε-N-acetyl lysine residues at specific sites. Di- and tri-acetylated specificities have also been found (66); the sites of acetylation in calf thymus H4 are at lysines 5, 8 and 12. CD studies of reconstituted complexes of these acetylated H4 species with DNA have been performed and are shown in Fig. 34 (67). The effect of nonacetylated H4, when bound to DNA, is to cause an increase in the positive ellipticity band (31,67). The monoacetylated H4 alters the interaction of the histone with DNA in reconstituted complexes, greatly diminishing the conformational change in DNA that can be induced by non-acetylated H4. This acetylation in the basic N-terminal end of the molecules, which removes only one positive change, apparently changes the ability of H4 to combine

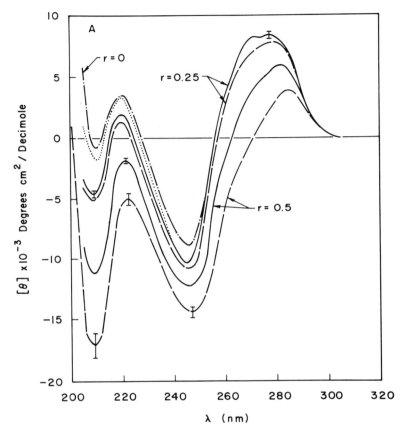

Figure 32. *Circular dichroism spectra of calf thymus H1:DNA complexes: effect of histone phosphorylation. The value of r, the histone (moles peptide)/DNA (moles of nucleotide residues) is varied. Complexes with phosphorylated H1, (———); Complexes with control (non-phosphorylated) H1, (----). A, r = 0.25 and 0.5. Native DNA (r = 0) is shown for comparison, (-·-·-); B, r = 0.6 and 0.75. Adler, A. J., Schaffhausen, B., Landan, T. A., and Fasman, G. D. (1971) Biochemistry 10, 909. Reprinted with permission of the American Chemical Society.*

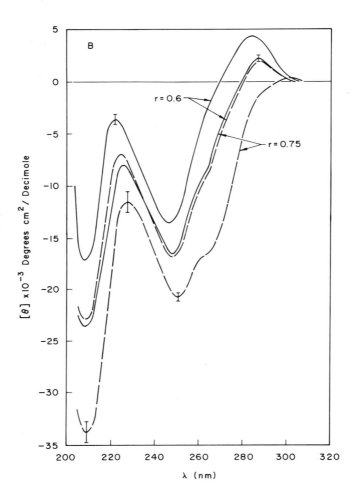

Figure 32B.

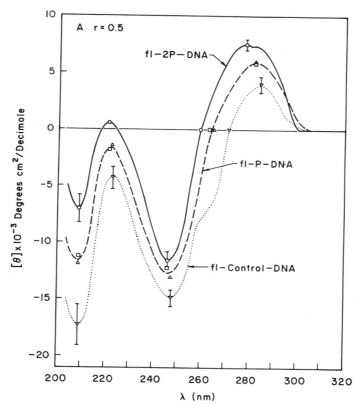

*Figure 33. Circular dichroism spectra of complexes recon-
stituted from DNA and Hl. Effect of histone phosphorylation.
Complexes with doubly phosphorylated, ○————○; complexes with
Hl phosphorylated only at site A, □----□ ; complexes with Hl
phosphorylated only at site B, △---△; complexes with control
(non-phosphorylated) Hl, ▽-- ▽. A. r = 0.5, spectra for
sites A and B are superimposable. B, r = 1.0. Adler, A. J.,
Langan, T. A., and Fasman, G. D. (1972)* Arch. Biochem. Biophys. 153, 769. *Reprinted with permission of Academic
Press.*

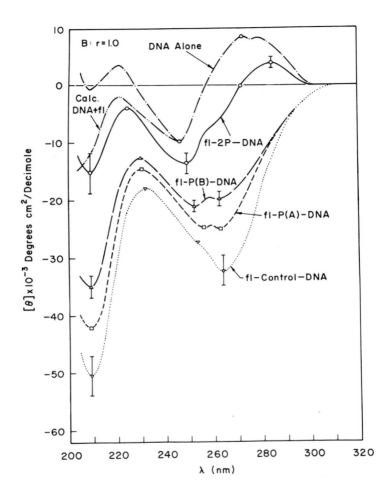

Figure 33 B.

with DNA in a specific manner leading to characteristic
H4:DNA complexes which cause a conformational change in DNA.
Although the physiological role of H4 acetylation is not
known, it has been shown that there is a correlation between
the RNA template activity of chromatin and acetylation. In
sea urchins sperm cells, which are not capable of RNA synthesis, the H4 histone fraction is completely nonacetylated,
although several acetylated forms exist in the sea urchin
embryo at a time of extensive gene activation (68). These
results imply a genetic regulatory function of H4 acetylation

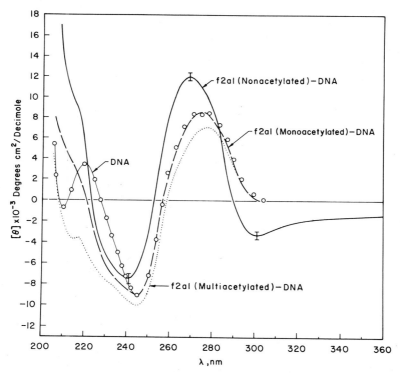

*Figure 34. Circular dichroism spectra of calf thymus DNA
alone and of complexes reconstituted from DNA with non-
acetylated, monoacetylated and multiacetylated H4: Effect of
histone acetylation. Complexes at r = 1.5, with non-acetylated (——); monoacetylated (----) and multiacetylated
(····). DNA alone (0——0). Adler, A. J., Fasman, G. D.,
Wangh, L. J., and Allfrey, V. G. (1974) J. Biol. Chem. 249,
2911. Reprinted with permission of the American Society of
Biological Chemists.*

which may operate through the mechanism demonstrated by the CD studies.

The individual histones appear to bind to DNA with independent specificities. To further investigate the specificity of the interaction of the histones, in their native conformation, with DNA, the cleaved products of histones where bound to DNA and their CD properties investigated. Histone H1 can be cleaved by N-bromosuccinimide (69) at the one tyrosyl residue, yielding two large fragments: the amino-terminal third of the protein (N) (Mol. wt. 6,000, 63 peptide residues, 19% lysine) and the carboxyl-terminal (C) remainder of the chain (mol. wt. 15,000, 152 residues, 33% lysine). Forty-five of the 59 lysine residues are clustered in the C-terminal half of the molecule, as are 16 of the 22 prolines. This distribution suggests that the C-fragment may be the primary site of electrostatic binding of H1 to DNA. The high cationic charge density, plus the lack of protein secondary structure (27,70) promoted by the high proline content, would be conducive to strong interaction with DNA (71). The CD curves of complexes of H1 with DNA and its fragments are seen in Fig. 35 (72). The intact H1 molecule causes a large decrease in the main ellipticity band at 275 nm, the N fragment causes little or no change, while the C-fragment causes even greater distortion of the DNA CD spectrum than does DNA attachment to the intact H1 molecule. The N-terminal, rich in hydrophobic residues, did however bind to the DNA. Of importance is the fact that the N segment of the intact H1 appears to cause a decrease in the conformational effect upon DNA when binding of the C-terminal segment occurs. This moderating effect of the N-segment of H1, when covalently bound to the C-segment, may be accounted for by the tertiary structure of H1, such that the N segment might be folded back on the C portion. CD spectra of various recombinations of intact and fragments of H1 with DNA are seen in Fig. 36 (72). The combined C and N fragments in complex with DNA cause greater CD distortion than H1 alone, as does H1 plus the N fragment. This again illustrates that the intact H1 molecule, due to its tertiary structure, binds in a different manner than the sum of its parts. Histone folding, i.e., tertiary structure, is consistent with the nuclease digestion studies of chromatin (73) which demonstrated that the histones are not evenly distributed along the DNA helix in chromatin.

Another example of histone binding to DNA which shows the whole histone is not merely the sum of its parts, is that of histone H2B (40). H2B can be cleaved with cyanogen bromide (74) to yield two fragments of nearly equal length, but

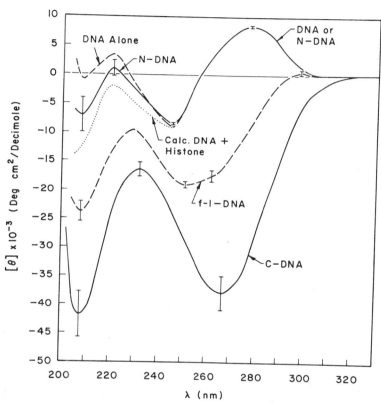

Figure 35. *Circular dichroism spectra of complexes reconstituted from calf thymus DNA and H1, or the N and C fragments, at r = 1.0. N and C indicate the N-terminal and C-
terminal H1 fragments. The CD of DNA alone is given for
comparison; the curve "calculated DNA and histone" is the
calculated sum of isolated DNA and H1 (or C or N fragment).
Fasman, G. D., Valenzuela, M. S., and Adler, A. J. (1971)
Biochemistry 10, 3795. Reprinted with permission of the
American Chemical Society.*

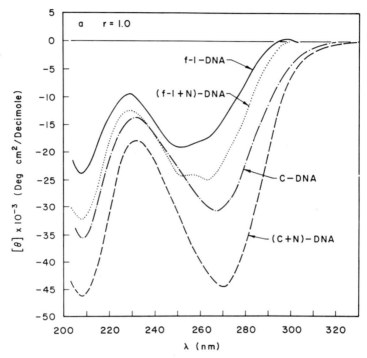

*Figure 36. Circular dichroism of calf thymus DNA complexes
with Hl and with mixtures of its components at various
ratios. For each set of curves the (C + N):DNA complex contains equal amounts of fragments C and N, sufficient to
reconstruct the Hl polypeptide chain at the Hl concentration
indicated. The C:DNA complex contains this same amount of C
peptide but no N. The (Hl + N):DNA complex includes complete
Hl chains at the indicated concentration plus an equal number
of N chains. Solvent 0.14M NaF + 0.002M Tris (pH 7.0).
Reference complex, Hl:DNA, r = 1.0. Values of r for components: 0.71 for C, 0.29 for N. Fasman, G. D., Valenzuela,
M. S., and Adler, A. J. (1971) Biochemistry 10, 3795.
Reprinted with permission of the American Chemical Society.*

with different charge density. The N-terminal half-molecule, N, contains residues 1-58 and 32% of its residues carry a positive charge. Half-molecule C has residues 63-125 and is 20% charged. When intact H2B is complexed to DNA, the resulting CD spectra (Fig. 37) shows an enhanced 275 nm band. The

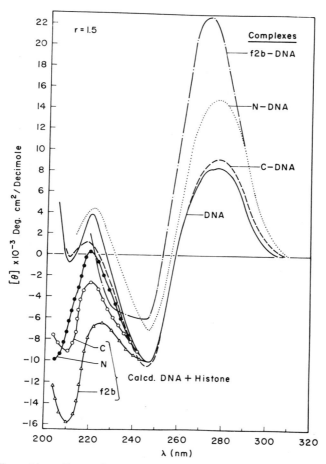

Figure 37. Circular dichroism spectra of complexes reconstituted from calf thymus DNA and H2B and its fragments, at r = 1.5. The CD spectra of DNA is shown for comparison (———). Adler, A. J., Ross, D. G., Chen, K., Stafford, P. A., Woiswillo, M. J., and Fasman, G. D. (1974) Biochemistry 13, 616. Reprinted with permission of the American Chemical Society.

N fragment, which is the more positively charged fragment, causes similar changes to that of the intact molecule, but of reduced magnitude, while the C fragment causes very little perturbation of the DNA CD spectra.

A technique that has been widely used in DNA studies is that of hyperchromicity (increase in optical density) to follow temperature induced transitions, e.g., denaturation. A similar approach has been developed to follow DNA denaturation by observing the CD changes during such transitions. The results of the use of this technique are shown in Fig. 38 (75). With DNA alone, as the temperature is increased from 10° to 40°, one observes a gradual increase in $[\theta]_{280}$, which has been termed a "premelt". Then a large decrease in $[\theta]_{280}$ is observed, with $T_m = 46.6^{\circ} \pm 1.5^{\circ}$, when denaturation occurs. The thermal denaturation curve of an annealed complex, poly(L-lysine$^{84.5}$ L-valine$^{15.5}$):DNA, $r = 0.25$ is also shown in Fig. 38. One observes a larger premelt, the first meltout is observed at $46.2^{\circ} \pm 1.5^{\circ}$, for the unbound DNA, and a second temperature meltout is found at $92.6^{\circ} \pm 2.0^{\circ}$ corresponding to the bound DNA. The temperature induced transitions can be seen more clearly by plotting derivatis, i.e., $d[\theta]/dT$ vs T, as shown in Fig. 39 (75). The various T_ms can be obtained with greater precision by the use of such plots.

Using the circular dichroism melting technique, the binding of H4, and its fragments, to DNA has been studied (76). H4 can be cleaved with cyanogen bromide, at methionine 84, into two fragments, an N-terminal 1-83 residue basic fragment and a C-terminal 18 residue (85-102) fragment. The reconstituted complex of intact H4:DNA yields a CD spectra (Fig. 13) with a large enhanced principal CD band; however, addition of the C or N peptide to DNA (Fig. 40) produces very little alteration of the DNA CD spectra. However, if the CD melting curves of these two complexes are examined, distinct differences are observed (Fig. 41,A,B). DNA and C-DNA complex exhibit only one major CD transition, at about 50° (Fig. 41A), showing that C-DNA consists mainly of free DNA. The small CD melt at $\sim 75^{\circ}$ may be due to GC-rich regions. The complexes H4:DNA and N-H4:DNA, show a major CD denaturation (Fig. 41B) in the T_{m2} region with the H4:DNA transition about 11° higher than the N-H4:DNA melt. Thus, despite a small change in the original CD spectra of the N-H4:DNA complex, one can clearly see the effect of binding in the CD meltout. Also evident is the fact that the intact molecule is necessary for the specific effect of H4 binding to DNA, for without the C-terminus, the remainder, the N-terminus, produces distinctly different effects upon binding. Examination of the complete CD spectra of the H4:DNA complex

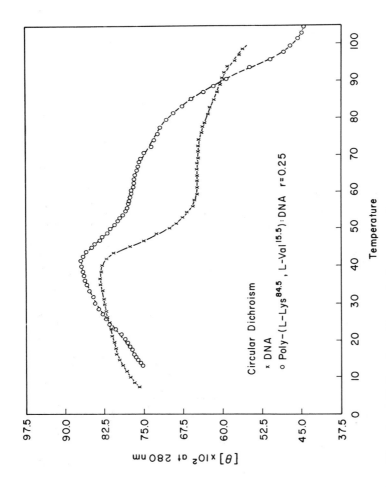

Figure 38. Circular dichroism thermal denaturation at 280 nm. DNA, (x–x–x); complex of poly-(L-lysine[84.5] L-valine[15.5]):DNA (o–o–o), r = 0.25; both in 2.5 x 10⁻⁴M EDTA, pH 7.0. Mandel, R., and Fasman, G. D. (1974) Biochem. Biophys. Res. Commun. 59, 672. Reprinted with permission of Academic Press.

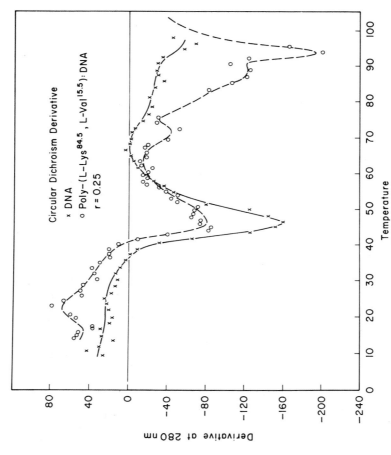

Figure 39. The derivative plot $d[\theta]/dT$ vs T of the circular dichroism denaturation curves shown in Fig. 38. DNA (xxxx); a complex of poly(L-lysine84.5 L-valine15.5):DNA, (oooo), $r = 0.25$. Mandel, R., and Fasman, G. D. (1974) Biochem. Biophys. Res. Commun. 59, 672. Reprinted with permission of Academic Press.

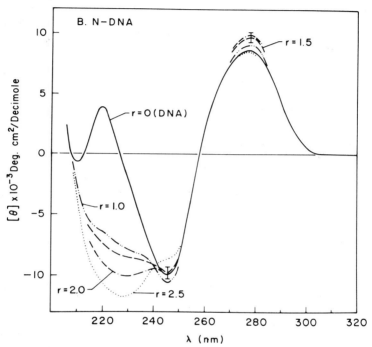

Figure 40. *Circular dichroism spectra of complexes reconsti-*
tuted from calf thymus DNA and the N-terminal fragment of H4,
at various r values. *Solvent 0.14M NaCl - 0.002M Tris*
(pH 7.0). *Adler, A. J., Fulmer, A. W., and Fasman, G. D.*
(1975) Biochemistry 14, 1445. *Reprinted with permission of*
the American Chemical Society.

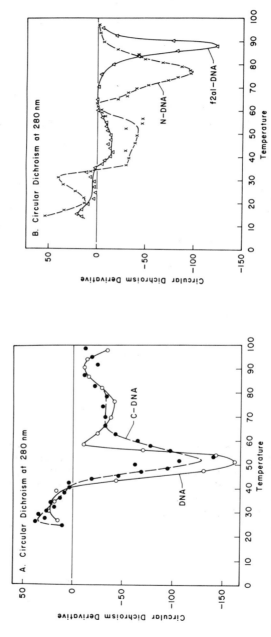

Figure 41. Derivative with respect to temperature d[θ]/dT vs temperature of circular dichroism thermal denaturation curves at 280 nm. (A) DNA and C-DNA complex, r = 1.5; (B) intact H4:DNA and N-DNA complexes, both at r = 1.5. Adler, A. J., Fulmer, A. W., and Fasman, G. D. (1975) Biochemistry 14, 1445. Reprinted with permission of the American Chemical Society.

during the melting profile yields the results in Fig. 42. At 23° is seen the "native complex", at 61°, the free DNA has melted, however, the DNA CD has changed slightly, and at 102° the bound DNA has melted and this spectra is that of denatured DNA (76).

CD temperature profiles of chromatin and histone depleted chromatin have also been published (77). In Fig. 43 is seen a derivative curve, plotted in a different manner, i.e., $\Delta[\theta]/[\theta]$ for nucleoprotein (DNP) obtained from chicken erythrocytes (77) and for complexes which have been extracted at various NaCl concentrations. One transition is seen for native DNA, at about 44°C. In the complete nucleoprotein complex, there are transitions at $\sim$ 40°, 60° and 72°C. As the histones are progressively removed, by extraction with higher NaCl concentrations, the two higher transitions slowly decrease, until after the 2.0M NaCl extraction, the CD plot is similar to that of native DNA. Thus, the importance of histone:DNA and histone:histone interactions is demonstrated for maintenance of the chromatin structural stability.

Numerous studies have been published utilizing synthetic poly-α-amino acids as models of histones, and their interactions with DNA. One such study has utilized the polypeptide poly(L-lysine^{40} L-alanine^{60}). The effect on the CD spectra of the complexes of poly(L-Lys^{40} L-Ala^{60}):DNA with different r ratios is seen in Fig. 44 (78). There was little change in the main ellipticity band at 275 nm; however, the increase in protein ratio can easily be seen in the 220 nm region. The binding was demonstrated by means of hyperchromicity meltout studies. The effect of varying the ionic strength on the CD of the complex was next investigated, and is shown in Fig. 45. As the ionic strength is increased, the main ellipticity band at $\sim$ 260-280 nm becomes larger and blue shifted. The authors suggest that a B $\rightarrow$ A transition has occurred. A study of the polymer, poly(L-Lys^{50} L-Tyr^{50}) binding to DNA, yielded CD spectra shown in Fig. 46 (79). Very little alteration of the DNA CD spectra occurred, however, binding was again demonstrated by means of hyperchomicity-temperature profiles. By observing changes in absorbance and fluorescence of these complexes, the authors suggest that tyrosine is not intercalating into the DNA.

The CD studies on the interaction of histones with homologous DNA have demonstrated that the interaction of the various histones appear to have a high degree of specificity; each capable of altering the conformation of the DNA in a unique fashion. The binding of a single histone to DNA

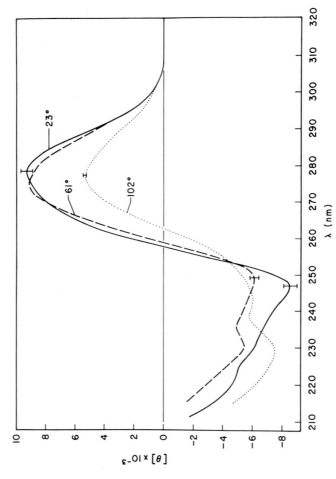

Figure 42. Circular dichroism spectra of the intact H4:DNA complex, r = 1.5 at various temperatures. Solvent 2.5 x 10⁻⁴M EDTA (pH 7.5). Adler, A. J., Fulmer, A. W., and Fasman, G. D. (1975) Biochemistry 14, 1445. Reprinted with permission of the American Chemical Society.

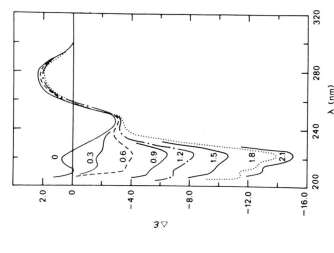

Figure 44. Circular dichroism spectra of poly
$(L-Lys^{40} Ala^{60})$:DNA complexes at various r
values. Solvent, 2.5 x 10^{-4}M EDTA, pH 8.0.
Direct mixed complexes. Pinkston, M. F., and
Li, H. J. (1974) Biochemistry 13, 5227. Reprinted with permission of the American Chemical Society.

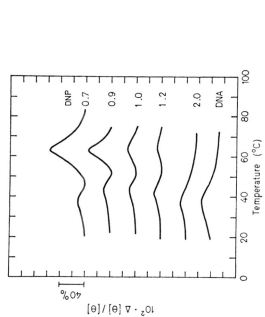

Figure 43. Circular dichroism melting curve
derivatives, $\Delta[\theta]/[\theta]$ vs temperature for
chicken erythrocyte nucleoprotein (DNP) and
for different nucleoprotein, extracted at the
concentration of NaCl indicated, as well as
purified chicken erythrocyte DNA. Solvent,
1 mM NaCl - 0.2 mM EDTA, pH 5.6. Wilhelm,
F. X., de Murcia, G. M., Champagne, M. H.,
and Duane, M. P. (1974) Eur. J. Biochem. 45,
431. Reprinted with permission of
Springer-Verlog, New York

129

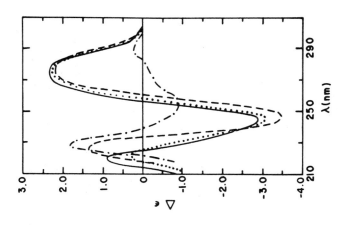

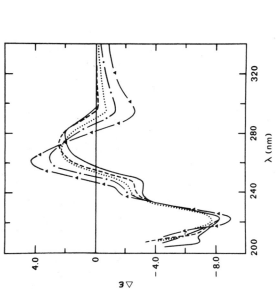

Figure 45. The circular dichroism of a poly (L-Lys⁴⁰ Ala⁶⁰):DNA complex at various ionic strengths at r = 1.2. NaCl concentrations: 0.00M, (——); 0.05M, (— — —); 0.10M (····); 0.15M (—·—·—); and 0.20M (—▲—▲—). Pinkston, M. F., and Li, H. J. (1974) Biochemistry 13, 5227. Reprinted with permission of the American Chemical Society.

Figure 46. Circular dichroism of poly (L-Lys⁵⁰ L-Tyr⁵⁰):DNA complexes at various r values. r = 0, (——); 0.31, (····); and the difference between r = 0 and r = 1.55, (—···—). Direct mixed complexes. Santella, R. M., and Li, H. J. (1974) Biopolymers 13, 1909. Reprinted with permission of John Wiley & Sons.

appears to be of a cooperative nature, and there is also a cooperativity between various sections within a single polypeptide chain. Thus, the conformation of the histone plays an important role. Therefore in chromatin histone:histone interactions, coupled with the individual specificity of histones plus the other constituents, play a united role in their overall interaction with DNA. These interactions can be modified by phosphorylation, acetylations, etc., causing conformational mobility of histone: DNA complexes, further implying a high degree of specificity of the interactions. These conformational modifications offer a mechanism for conformational changes in DNA, which may have significance in the depression of the template activity of DNA in chromatin. The specificity alluded to here might be of the same type observed in other biological systems such as in ribosomes, repressors, initiation factors and polymerases. This specificity may be due to defined base sequence and conformation interacting with defined amino acid sequence and conformation.

As electron microscopy has become more frequently used in chromatin studies, the interrelationship between electron microscopy and CD studies is of interest. The effect on the CD spectra of chromatin at various ionic strengths is seen in Fig. 47 (80). As the ionic strength is increased, the main DNA ellipticity band in the 280 nm region is slightly decreased while the protein CD band around 210-220 nm becomes more negative. Both display a 28% change. The electron micrographs of these samples are seen in Fig. 48 (80). Thus, as the ionic strength is decreased, the large masses disappear and what appears to be the fundamental strand becomes more obvious. Nodular elements are visible in all preparations, and these are probably the same structures now characterized as ν bodies (52). There appears to be a direct correlation between the state of condensation and the CD spectra obtained. When highly condensed, the DNA ellipticity band at $\simeq$ 250 nm is decreased and the protein band at 210-200 nm is more negative. These effects are reversible, both in EM and CD. Electron micrographs have also been obtained for single histone:DNA complexes. Micrographs for H1:DNA and H4:DNA complexes are seen in Fig. 49. Whereas the H4:DNA material consists largely of 0.3-0.5 μm globular aggregates, the aggregates of H1 complexes were substantially smaller (20-40 nm), occurring as small clusters. The CD spectra (Fig. 10 and Fig. 13) are distinctly different: H1 producing a large negative CD band, while H4 produces an increased positive ellipticity band. Thus, the morphological differences seen in micrographs are also distinguished by very specific

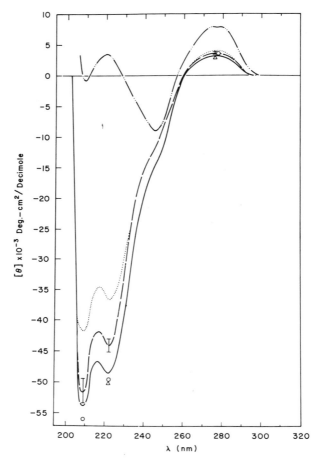

Figure 47. Circular dichroism spectra of native sheared calf thymus chromatin in various ionic strengths: 0.01M Tris with or without added 0.014M NaF, (····); 0.001M NH₄OAc, (----); 0.05M NH₄OAc, (o——o); 0.07M NH₄OAc, (———); 0.14M NH₄OAc, (△——△). The CD of isolated DNA in 0.14M NaF is given by (-·-·-). Slayter, H., Shih, T. Y., Adler, A. J., and Fasman, G. D. (1972) Biochemistry 11, 3044. Reprinted with permission of the American Chemical Society.

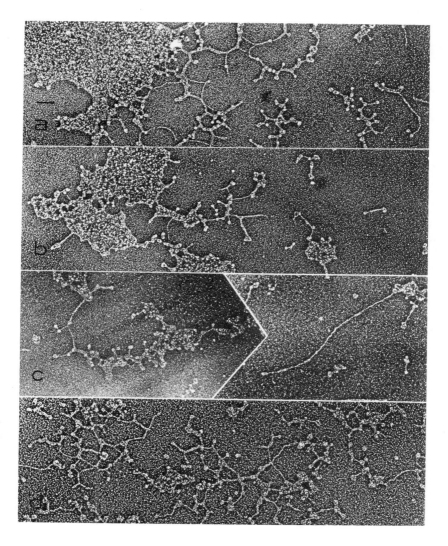

Figure 48. Electron micrographs of chromatin reflecting changes of ionic strength upon structure. Rotary shadow cast with platinum (75,000 x): (a) 0.14M NH₄OAc; (b) 0.07M NH₄OAc; (c) 0.005M NH₄OAc; (d) 0.001M NH₄OAc. Bar indicates 100 nm. Slayter, H., Shih, T. Y., Adler, A. J., and Fasman, G. D. (1972) Biochemistry 11, 3044. Reprinted with permission of the American Chemical Society.

133

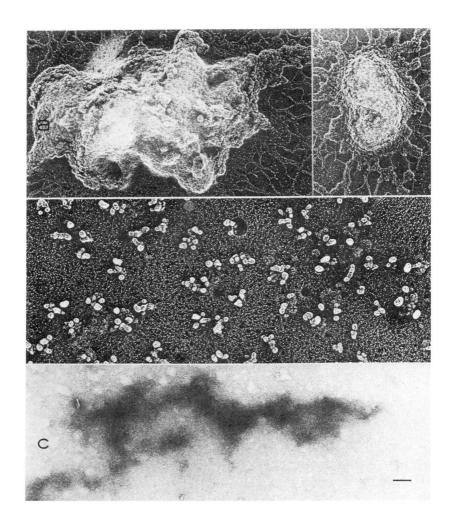

Figure 49. Comparison of H1:DNA and H4:DNA complexes at high ionic strength in 0.07M NH4OAc. (a) H4:DNA complex, rotary shadow cast (36,191 x); (b) H1:DNA complex, rotary shadow cast (36,191 x); (c) H1:DNA complex, negatively contrasted with uranyl acetate (36,191 x). Bar indicates 100 nm. Slayter, H., Shih, T. Y., Adler, A. J., and Fasman, G. D. (1972) Biochemistry 11, 3044. Reprinted with permission of the American Chemical Society.

CD spectra. It is clear that the CD spectra reflect not only secondary structure, but the tertiary structure of the aggregates.

The studies discussed herein have demonstrated the utility of using circular dichroism as a very sensitive probe for conformational changes associated with chromatin and model systems thereof. Thus, this spectroscopic tool, circular dichroism, can play a major role in our understanding of the biological interactions which occur in the cell nucleus, and which are central to understanding biological control mechanisms.

Added in Proof

Histone-histone interactions have been shown to be important facets of chromatin structure (81,82). Isenberg and coworkers have elegantly demonstrated the changes in CD spectra upon addition of salts (e.g., sodium phosphate) and upon various histone interactions. A typical CD spectra of histone H2B and the effect of sodium phosphate at pH 7.4, causing a conformational change is seen in Fig. 50 (83). The analysis of such curves is shown in Fig. 51. The change in conformation is that from the random coil to the α-helix, whereby ∿18 residues have become helical (83). Upon complexing H2B and H4 in a 1:1 ratio, the changes in CD at 220 nm are seen in Fig. 52 (83). H4 CD changes with time, while H2B does not. The CD spectra of the complex H2B:H4 is seen to alter with time, but the change is not the calculated average of the two histones. Thus, a conformational change has been induced upon their interaction. The increase in helical content has been calculated to be a minimum of eight residues from a total of 227 residues, giving a total helical content of 39 residues, or 42%.

ACKNOWLEDGEMENT: This is publication #1051 from the Graduate Department of Biochemistry, Brandeis University, Waltham, Mass. 02154. The writing of this article was generously supported by grants from the U. S. Public Health (GM17533), National Science Foundation (GB29204X) and the American Cancer Society (P-577).

REFERENCES

1. Stedman, E., and Stedman, E. Nature 166, 780 (1950).
2. Idem (1951) Phil. Trans. R. Soc. (London) B235, 565.
3. Elgin, S. C. R., Froehner, S. C., Smart, J. E., and Bonner, J. in Advances in Cell and Molecular Biology 1, Dupraw, E. J. (Ed.) p. 1, Academic Press, N.Y. (1971).

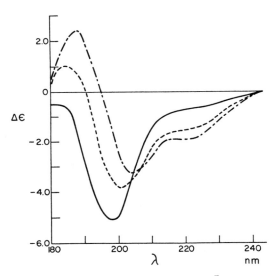

*Figure 50. CD spectra of H2B (1.2 x 10⁻⁵M) from 250 to 180
nm as measured in the vacuum CD spectrophotometer: in water
(pH 4.3) (———); in 0.0033M sodium phosphate (pH 7.4) (----);
and in 0.026M sodium phosphate (pH 7.4) (-·-·-). D'Anna,
J. A., Jr., and Isenberg, I. (1972) Biochemistry 11, 4017.
Reprinted with permission of the American Chemical Society.*

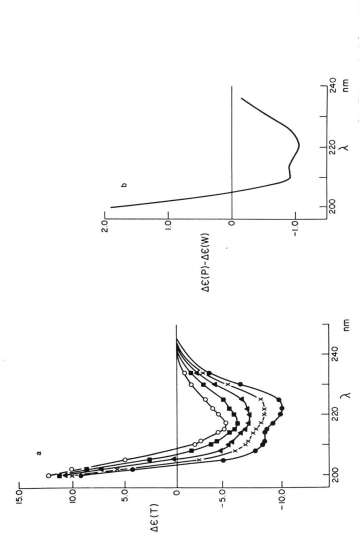

Figure 51. (a) Computed difference spectra for coil–100% α helix (●); coil–75% α helix, 25% β sheet (x); coil–50% α helix, 50% β sheet (▲); coil–25% α helix, 75% β sheet (■); and coil–100% β sheet (o). (b) Difference spectrum computed from CD spectra of histone H2B in 0.14M phosphate and in water (pH 7.4). D'Anna, J. A., Jr., and Isenberg, I. (1972) Biochemistry 11, 4017. Reprinted with permission of the American Chemical Society.

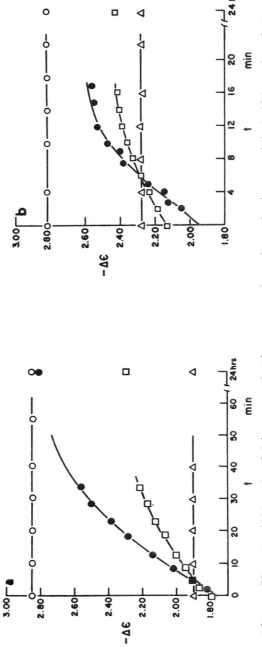

Figure 52. CD (220 nm) of histone solutions as a function of time at (a) 0.016 M phosphate, pH 7.0, and (b) 0.24M NaCl, pH 7.0: 0.90 x 10⁻⁵M H4 (●); 0.90 x 10⁻⁵M H2B (△); 0.90 x 10⁻⁵M H2B and H4, measured (o) and calculated (□) for a mixture of non-interacting histones. D'Anna, J. A., Jr., and Isenberg, I. (1973) Biochemistry 12, 1035. Reprinted with permission of the American Chemical Society.

4. Allfrey, V. G. in *Histones and Nucleohistones*, Phillips, D. M. P. (Ed.) p. 241, Plenum Press, N. Y. (1971).

5. ibid. Bradbury, E. M., and Crane-Robinson, C. p. 85.

6. Brahms, J., and Brahms, S. in *Biological Macromolecules, 4*, Fasman, G. D., and Timasheff, S. N. (Eds.) p. 191, Marcel Dekker, New York (1970).

7. Tinoco, I., Jr., *J. Chim. Phys. 65*, 91 (1968).

8. Marvin, D. A., Spencer, M., Wilkins, M. H. F., and Hamilton, L. D., *J. Mol. Biol. 3*, 547 (1961).

9. Langridge, R., Wilson, H. R., Hooper, C. W., Wilkins, M. H. F., and Hamilton, L. D., *J. Mol. Biol. 2*, 19, 38 (1960).

10. Fuller, W., Wilkins, M. H. F., Wilson, H. R., and Hamilton, L. D., *J. Mol. Biol. 12*, 60 (1965).

11. Tunis-Schneider, M. J., and Maestre, M. F., *J. Mol. Biol. 52*, 521 (1970).

12. Ivanov, V. I., Minchenkova, L. E., Schyolkina, A. K., and Poletayev, A. I., *Biopolymers 12*, 89 (1973).

13. Greenfield, N., and Fasman, G. D., *Biochemistry 8*, 4108 (1969).

14. See review Pardon, J., and Richards, B. in *Biological Macromolecules Vol. 6*, Fasman, G., and Timasheff, S., (Eds.) p. 1, Marcel Dekker, New York (1973).

15. Shih, T., and Fasman, G. D., *J. Mol. Biol. 52*, 125 (1970).

16. Marmur, J., *J. Mol. Biol. 3*, 208 (1961).

17. Shirey, T., and Huang, R. C. C., *Biochemistry 8*, 4138 (1969).

18. Permogorov, V. I., Debakov, V. G., Sladkova, I. A., and Rebentish, B. A., *Biochim. Biophys. Acta 199*, 556 (1970).

19. Simpson, R. T., and Sober, H. A., *Biochemistry 9*, 3103 (1970).

20. Wilhelm, F. X., Champagne, M. H., and Duane, M. P. *Eur. J. Biochem. 15*, 321 (1970).

21. Henson, P., and Walker, I. O., *Eur. J. Biochem. 16*, 524 (1970).

22. Tinoco, I., Jr., *J. Amer. Chem. Soc. 86*, 297 (1964).

23. Johnson, W. C., Jr., and Tinoco, I. Jr., *Biopolymers 7*, 727 (1969).

24. Moore, D. S., and Wagner, T. E., *Biopolymers 12*, 201 (1973).

25. Studdert, D. S., and Davis, R. C., *Biopolymers 13*, 1377 (1974).

26. Mirsky, A. E., Burdick, C. J., Davidson, E. H., and Littau, V. C., *Proc. Nat. Acad. Sci. U.S.A. 61*, 592 (1968).

27. Fasman, G. D., Schaffhausen, B., Goldsmith, L., and Adler, A., *Biochemistry 9,* 2814 (1970).
28. Rall, S. C., and Cole, R. D., *J. Biol. Chem. 246,* 7175 (1971).
29. Huang, R. C., Bonner, J., and Murray, K., *J. Mol. Biol. 8,* 54 (1964).
30. DeLange, R. J., and Smith, E. L., *Accounts of Chemical Research 5,* 368 (1972).
31. Shih, T., and Fasman, G. D., *Biochemistry 10,* 1675 (1971).
32. Shih, T., and Fasman, G. D., *Biochemistry 11,* 398 (1972).
33. Wickett, R. R., Li, H. J., and Isenberg, I., *Biochemistry 11,* 2952 (1972).
34. Ottaway, C. A., and Wetlaufer, D. B., *Arch. Biochem. Biophys. 139,* 257 (1970).
35. Urry, D. W., Hinners, T. A., and Masotti, L., *Arch. Biochem. Biophys. 137,* 214 (1970).
36. Gordon, D. J., and Holzworth, G., *Arch. Biochem. Biophys. 142,* 481 (1971).
37. Schneider, A. S., Schneider, M. T., and Rosenheck, K., *Proc. Nat. Acad. Sci. U.S.A. 66,* 793 (1970).
38. Duysens, L. N. M., *Biochim. Biophys. Acta. 19,* 1 (1956).
39. Dorman, B. P., Hearst, J. E., and Maestre, M. F., *Methods in Enzymology XXVII,* p. 767 (1973).
40. Adler, A. J., Ross, D. G., Chen, K., Stafford, P. A., Woiszwillo, M. J., and Fasman, G. D., *Biochemistry 13,* 616 (1974).
41. Adler, A. J., Moran, E. C., and Fasman, G. D., *Biochemistry 14,* 4179 (1975).
42. Johnson, R. S., Chan, A., and Hanlon, S., *Biochemistry 11,* 4347 (1972).
43. Lerman, L. S., *Proc. Nat. Acad. Sci. U.S.A. 8,* 1886 (1971).
44. Jordon, C. F., Lerman, L. S., and Venable, J. H., Jr., *Nature New Biol. 236,* 67 (1972).
45. Mantiatis, T., Venable, J. H., Jr., and Lerman, L. S., *J. Mol. Biol. 84,* 37 (1974).
46. Cheng, S. M., and Mohr, S. C., *FEBS Lett. 49,* 37 (1974).
47. Cheng, S. M., and Mohr, S. C., *Biopolymers 14,* 663 (1975).
48. Giannoni, G., Padden, F. J., Jr., and Keith, H. D., *Proc. Nat. Acad. Sci. U.S.A. 62,* 964 (1969).
49. Henson, P., and Walker, O. J., *Eur. J. Biochem. 16,* 524 (1970).
50. Hewish, D. R., and Burgoyne, C. A., *Biochem. Biophys. Res. Commun. 52,* 504 (1973).

51. Sahasrabuddhe, C. G., and Van Holde, K. E., *J. Biol. Chem. 249,* 152 (1974).

52. Olins, A. L., and Olins, D. E., *Science 183,* 330 (1974).

53. Kornberg, R. D., *Science 184,* 868 (1974); Kornberg, R. D., and Thomas, J. O., *Science 184,* 865 (1974).

54. Thomas, J. O., and Kornberg, R. D., *Proc. Nat. Acad. Sci. U.S.A. 72,* 2626 (1975).

55. Langan, T. A., *J. Biol. Chem. 244,* 5763 (1969).

56. Adler, A. J., Schaffhausen, B., Langan, T. A., and Fasman, G. D., *Biochemistry 10,* 909 (1971).

57. Meisler, M. H., and Langan, T. A., *J. Biol. Chem. 244,* 4961 (1969).

58. Langan, T. A., Rall, S. C., and Cole, R. D., *J. Biol. Chem. 244,* 4961 (1971).

59. Bustin, M., and Cole, R. D., *J. Biol. Chem. 245,* 1458 (1970).

60. Langan, T. A., *Fed. Proc. 30,* 1089 (1971).

61. Langan, T. A., *Science 162,* 579 (1968).

62. Adler, A. J., Langan, T. A., and Fasman, G. D., *Arch. Biochem. Biophys. 153,* 769 (1972).

63. DeLange, R. J., Fambrough, D. M., Smith, E. L., and Bonner, J., *J. Biol. Chem. 244,* 319 (1969).

64. DeLange, R. J., Fambrough, D. M., Smith, E. L., and Bonner, J., *J. Biol. Chem. 244,* 5669 (1969).

65. Pogo, B. G. T., Pogo, A. O., Allfrey, V. G., and Mirsky, A. E., *Proc. Nat. Acad. Sci.U.S.A. 59,* 1337 (1968).

66. Wangh, L., Ruiz-Carrillo, A., and Allfrey, V. G., *Arch. Biochem. Biophys. 150,* 44 (1972).

67. Adler, A. J., Fasman, G. D., Wangh, L. J., and Allfrey, V. G., *J. Biol. Chem. 249,* 2911 (1974).

68. Wangh, L., Ruiz-Carrillo, A., and Allfrey, V. G., *Arch. Biochem. Biophys. 150,* 44 (1972).

69. Bustin, M., and Cole, R. D., *J. Biol. Chem. 244,* 5291 (1969).

70. Boublik, M., Bradbury, E. M., and Crane-Robinson, C., *Eur. J. Biochem. 14,* 486 (1970).

71. Bustin, M., and Cole, R. D., *J. Biol. Chem. 245,* 1458 (1970).

72. Fasman, G. D., Valenzuela, M. S., and Adler, A. J., *Biochemistry 10,* 3795 (1971).

73. Clark, R. J., and Felsenfeld, G., *Nature New Biol. 229,* 101 (1971).

74. Iwai, K., Hayashi, H., and Ishikawa, K., *J. Biochem. (Tokyo) 72,* 357 (1972).

75. Mandel, R., and Fasman, G. D., *Biochem. Biophys. Res. Commun. 59,* 672 (1974).

76. Adler, A. J., Fulmer, A. W., and Fasman, G. D., *Biochemistry 14*, 1445 (1975).
77. Wilhelm, F. X., de Murcia, G. M., Champagne, M. H., and Daune, M. P., *Eur. J. Biochem. 45*, 431 (1974).
78. Pinkston, M. F., and Li, H. J., *Biochemistry 13*, 5227 (1974).
79. Santella, R. M., and Li, H. J., *Biopolymers 13*, 1909 (1974).
80. Slayter, H., Shih, T. Y., Adler, A. J., and Fasman, G. D., *Biochemistry 11*, 3044 (1972).
81. Kornberg, R. D., and Thomas, J. O., *Science 184*, 865 (1974).
82. Van Holde, K. E., and Isenberg, I., *Acct. Chem. Res. 8*, 327 (1975).
83. D'Anna, J. A., Jr., and Isenberg, I., *Biochemistry 11*, 4017 (1975).

Chapter 4

CHROMATIN SUBUNITS

HSUEH JEI LI

Division of Cell and Molecular Biology
State University of New York at Buffalo
Buffalo, New York 14214

I.   INTRODUCTION

One of the major advances in the research of chromatin
structure during the past five years is the discovery, con-
firmation and elaboration of chromatin subunits and their
structures, an advance which has been made by numerous
scientists from various laboratories. Conformational studies
of histones (See Chapter 1) and studies of histone-histone
association using sedimentation, electron microscopy or electrophoresis combined with cross-linking technique (1-16) have
contributed greatly to the evolution and understanding of the
concept of chromatin subunits. The subject of histone-histone
interaction has been reviewed elsewhere by Isenberg (17) and
also by Li (18). The present chapter will be devoted primarily to those experiments directly related to the subunits
isolated either from chromatin or from nuclei.

II.   CHROMATIN SUBUNITS ISOLATED BY NUCLEASE DIGESTION

Clark and Felsenfeld (19) first extensively examined the
protection of chromatin DNA against staphylococcal nuclease
digestion, although nuclease was used to study chromatin
earlier by Murray (20) and about the same time by Itzhaki
(21). Clark and Felsenfeld (19) observed that only about 50%
of chromatin DNA was digested by the nuclease; the DNA isolated from the protected fragments of chromatin was double-
stranded and about 170 base pairs long on the average. Using
nuclease digestion as a criterion, chromatin does not appear
to have uniform distribution of histones. Instead, it is a
complex composed of different regions, some susceptible, some
resistant to nuclease. The latter regions are rather homogeneous, which implies the existence of some kind of subunit in

chromatin.

In 1973, Hewish and Burgoyne (22a) isolated nuclei which were autodigested by endogenous nuclease and the DNA was then isolated from the nuclei. These authors found a series of repeating fragments corresponding to pieces of DNA which were multiples of a fundamental unit about 200 base pairs in length. Therefore, proteins, primarily histones, bound on DNA in chromatin must be distributed in a regular way, so that only certain restricted regions are accessible to the nuclease. The results of Hewish and Burgoyne (22a) were confirmed by Noll (22b) when an exogenous nuclease, staphylococcal nuclease, was added to the nuclei from the outside. Independently, Rill and Van Holde (23) isolated a low-salt soluble fraction of chromatin subunits, PS particles, after the chromatin had been treated by staphylococcal nuclease. These PS particles contained 1.85 mg of protein per mg of DNA and exhibited circular dichroism (23) and thermal denaturation (24) properties different from those of isolated chromatin. It should be noted, however, that this fraction of chromatin subunits contained only 0 to 20% of total DNA in chromatin, depending upon the length of time it was exposed to nuclease digestion (23), so the properties of these PS particles (23,24) cannot represent the average of all the 50% nuclease-resistant fragments of chromatin reported by Clark and Felsenfeld (19).

The above results of both exogenous and endogenous nuclease digestion of DNA in either isolated chromatin or nuclei (19-24) suggest the existence of some kind of chromatin subunits. Are these subunits isolated either from chromatin or nuclei related to each other? In order to answer this question, Sollner-Webb and Felsenfeld (25) studied the kinetics of staphylococcal nuclease digestion of duck reticulocyte nuclei and chromatin. Fig. 1a shows that the kinetics of nuclease digestion of DNA in nuclei is identical to that of DNA in chromatin (25,26), implying that similar chromatin subunits can be obtained from either source. Fig. 1b shows the distribution of DNA fragments isolated from nuclease-resistant fractions of nuclei subjected to various degrees of digestion by staphylococcal nuclease. When only 2% of the DNA is digested, there is a characteristic pattern comprising DNA fragments of 185 base pairs (monomer) and multiples of this unit (oligomers or multimers). As the digestion proceeds, the oligomers are replaced by monomers with the appearance of a new 140 base pair fragment. Finally at the limit of digestion (47-48% acid soluble), the 185 base pair monomer is nearly fully cleaved to form the 140 base pair fragment. Apparently, proteins in chromatin provide at

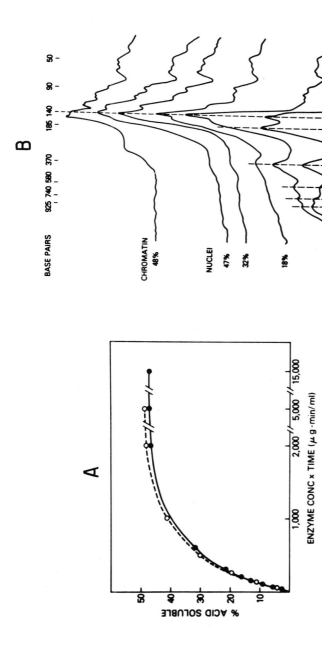

Figure 1. Kinetics of nuclear digestion.
(a) Release of acid-soluble products. Duck
reticulocyte nuclei (●) and duck reticulocyte
chromatin (○) were digested by staphylococcal
nuclease. (b) Densitometric scans of polyacrylamide gels of protected DNA after
electrophoresis. From B. Sollner-Webb, and G. Felsenfeld (1975). Biochemistry 14, 2915. Reprinted with permission from the American Chemical Society.

least two different levels of protection of DNA against nuclease digestion: one covers a length of 185 base pairs of DNA
and the other 140 base pairs. Furthermore, the proteins
responsible for these two levels of protection must be next
to each other along chromatin DNA since the 185 base pair
monomer is the precursor of the 140 base pair fragment. Similar results were reported by Axel (27) and Shaw et al. (28).
Further digestion of the core fragments by nuclease yields
smaller fragments of DNA (25).

The results of Fig. 1 (25) would be in accord with the
earlier suggestion of Li et al. (29) that histone H1 might
protect DNA less well against nuclease digestion than other
histones, if the core segment of 140 base pairs represents
that portion bound by a subunit of histones other than H1 (an
octamer of H2A, H2B, H3 and H4, for example) while the adjacent fragment of 45 base pairs (digestable from the 185 base
pair monomer) is covered by histone H1. If this is true, it
would explain the correlation between the 50% nuclease-
resistant fragments first reported by Clark and Felsenfeld
(19) and also shown in Fig. 1a (25) with the 50-60% base
pairs covered by histones in histone H1-depleted chromatin as
measured by thermal denaturation (29-31). Also, calculations
from thermal denaturation data of chromatin and NaCl-treated
partially dehistonized chromatin indicate that an octamer of
H2A, H2B, H3 and H4 would bind 130-150 base pairs while one
H1 molecule would cover 30-40 base pairs (31,32).

The above results obtained from nuclease-resistant chromatin (25,27) and the conclusion as to the length of DNA
bound by histone H1 and the octamer of other histones (32)
were recently confirmed by the independent reports of Simpson
and Whitlock (33), Shaw et al. (28) and Varshavsky et al.
(34). By digesting rat liver nuclei with an endogenous endo-
nuclease, Simpson and Whitlock (32) observed DNA fragments of
about 205 base pairs which were reduced to fragments of about
160 base pairs if digestion were allowed to continue. Shaw
et al. (28) obtained various fractions of chromatin subunits
after a mild treatment of chicken erythrocyte chromatin by
staphylococcal nuclease (Fig. 2a). When histones from the
different fractions of the monomer pool were examined by gel
electrophoresis, lysine-rich histones H1 (I or f1) and H5
(V or f5) were specifically lost in the smaller monomer (Fig.
2b). These authors (28) further determined that the smaller
monomers were comprised of only 140 base pairs of DNA, while
the larger ones contained 180 base pairs. The latter was
recently corrected to be 200 base pairs (K. E. Van Holde,
private communication). Using sucrose gradient centrifugation, Varshavsky et al. (34) independently fractionated mouse

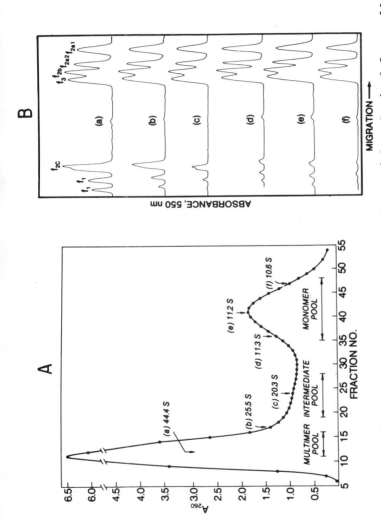

Figure 2. (a) Separation on Biogel A-5m of nucleoprotein particles obtained from a 10-min digestion of chicken erythrocyte nuclei. The numbers given next to a, b, c, etc. are the sedimentation coefficients of individual fractions. (b) Polyacrylamide-Na Dod SO₄ gel patterns (15%) of proteins from fractions (a) through (f) in Fig. 2a. From B. R. Shaw, T. M. Herman, R. T. Kovacic, G. S. Beaudreau, and K. E. Van Holde (1976). Proc. Nat. Acad. Sci. U.S.A. 73, 505. Reprinted with permission from the National Academy of Sciences, U.S.A.

Ehrlich ascites tumor chromatin into mono-, di-, and oligo-
nucleosomes after mild nuclease treatment (Fig. 3a). Purified
mononucleosomes appear as two separate bands in polyacrylamide
gel (Fig. 3b). The mononucleosome with greater mobility in
Fig. 3b lacks histone H1, while that of slower mobility contains all species of histones (Fig. 3c). Further electrophoretic analysis of mononucleosomal DNA by these authors
indicated that the DNA segment from the H1-depleted mononucleosome contained 170 base pairs while that of the mononucleosome with all five histones had 200 base pairs.

Among various authors (25,27,28,33,34), there is a general agreement that the main monomers of chromatin subunits
prepared by mild nuclease digestion contain a DNA segment of
180-205 base pairs which are bound by all species of histones.
The segment can be further digested into the smaller monomers
composed of 140-170 base pairs bound by whole histones minus
the lysine-rich histone H1 or minus H1 + H5, depending upon
the source of chromatin. The implication from these studies
is that one histone H1 molecule binds 30-45 base pairs adjacent to a segment of 130-170 base pairs bound by the core
histone subunit of H2A, H2B, H3 and H4, as predicted by an
earlier model (29,32). It should be emphasized, however,
that thermal denaturation results (29-31) simply predict that
30-40 base pairs of DNA are bound directly by some portions
of one histone H1 molecule but do not predict the exact manner of binding. Therefore, this would not contradict any
hypothesis which required that portions of histone H1 be
linked to other histone molecules in the neighboring histone
subunits, or that H1 binds to nonhistone proteins as proposed
by Cole (35), or even that it binds two crossed DNA segments
in a superhelical DNA (36,37).

III. CHROMATIN SUBUNITS VISUALIZED UNDER ELECTRON MICROSCOPE

Anderson and Moudrianakis in 1969 (38) observed a series
of nodules when chromatin was examined in the electron microscope. Also in the electron microscope, Olins and Olins (39,
40) reported that chromatin from rat thymus fixed by formaldehyde appeared as a string of beads as shown in Fig. 4. The
string of beads was also reported by Woodcock (41). The
average diameter of these beads was reported to be 83 ± 23Å
and that of the connecting filaments as 15Å. If chromatin
from rat liver rather than thymus was examined, the average
diameter of the beads was seen to be 60 ± 16Å (40).

Oudet et al. (42) observed (Fig. 5) that, in the absence
of formaldehyde fixation, chromatin obtained from nuclei
directly lysed on the grids, appeared as a series of spherical beads with a diameter of 130 ± 11Å; some beads of nucleosomes were closely packed, whereas others were connected by

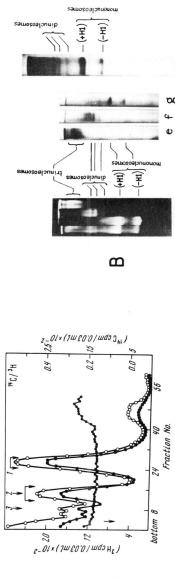

Figure 3. (a) Sucrose gradient centrifugation of nuclease digest of chromatin. ^3H (DNA) (●), ^{14}C (protein) (o) and $^{14}C/^3H$ (·). Arrows in the upper left of the graph indicate peaks of mono-, di- and trinucleosomes. (b) Polyacrylamide gel electrophoresis of DNP. Total nuclease digests of the chromatin (a and h), purified mononucleosomes (b and g), purified dinucleosomes (c and f) and trinucleosomes and higher oligomers of the nucleosomes (d and e). (c) Protein composition of mono- and dinucleosomes. Proteins from more slowly migrating mononucleosome (a), rapidly migrating mononucleosome (b), most rapidly migrating dinucleosome (c), dinucleosome with an intermediate mobility (d), and third dinucleosome with the lowest mobility (e). From A. J. Varshavsky, V. V. Bakayev, and G. P. Georgiev (1976). *Nucleic Acids Res.* 3, 477. Reprinted with permission from Information Retrieval Limited.

149

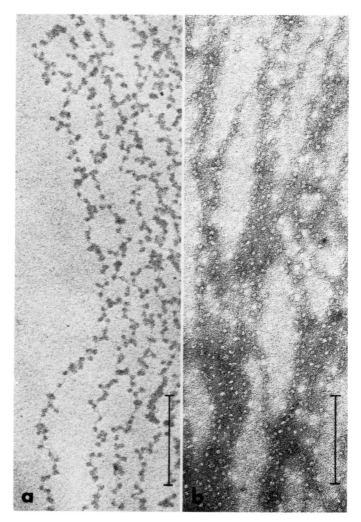

Figure 4. A string of ν-bodies of rat thymus chromatin. (a) Positively stained; (b) negatively stained. From A. L. Olins and D. E. Olins (1974). Science 183, *330. Reprinted with permission from the American Association for the Advancement of Science.*

Figure 5. High magnification view of chromatin of chicken liver nuclei lysed directly on an electron microscope grid. The bar indicates 0.25 μm. The arrows point to short DNA connecting strands. From P. Oudet, M. Gross-Bellard, and P. Chambon (1975). Cell 4, 281. Reprinted with permission from the Massachusetts Institute of Technology.

filaments. If H1-depleted chromatin was examined instead, a series of nucleosomes, similar to that of a string of beads, became apparent (Fig. 6). However, only a small portion of this H1-depleted chromatin possesses such regular beaded structure as shown in Fig. 6a (42). In fact, the main portion of this H1-depleted chromatin appears like naked DNA with a few nucleosomes scattered irregularly along the molecule. Some questions with regard to electron micrographs of chromatin perhaps should be raised here:

(a) With what frequency does the string of beads manifest itself in an ordinary preparation of chromatin, or what is the probability that a chromatin would appear like a string of regular beads? The hypothesis of a basic structure similar to a string of beads would be acceptable only if such probability is high.

(b) What are those nucleosomes scattered in the regions of naked DNA (the right half of Fig. 6a)? Are they chromatin subunits of histones or nonhistone proteins? Currently only histones are considered to be parts of chromatin subunits while nonhistone proteins are commonly excluded. Nevertheless, there is no solid ground for believing either that nonhistone proteins, when complexed with DNA in chromatin, cannot generate beaded structure visualized under the electron microscope, or that nonhistone proteins cannot protect DNA segments against nuclease digestion.

Beaded structure of chromatin in SV40 minichromosomes have been reported by Griffith (43) and also by Germond *et al.* (44) and in calf thymus chromatin by Langmore and Wooley (45). The diameter of these beads, however, varies greatly, ranging from $60 \pm 16\overset{\circ}{A}$ (40) to $134 \pm 14\overset{\circ}{A}$ (45) as summarized in Table I. The diameter of beads is an important parameter since it can set a theoretical limit on size and exclude some hypothetical models. For instance, Crick and Klug (46) proposed a kinky helix to describe the DNA within a chromatin subunit partly because a spherical body with a diameter of about $60\overset{\circ}{A}$ would be too small for 200 base pairs of DNA to bend smoothly within the subunit. However, such an event would not be unreasonable if the diameter of the nucleosomes were about $130\overset{\circ}{A}$ rather than $60\overset{\circ}{A}$ (47).

Chromatin subunits isolated by nuclease digestion have been visualized as spherical particles under the electron microscope (42,48-51). These particles of chromatin subunits supported the hypothesis of a string of beads (ν-bodies, PS particles or nucleosomes) as a basic structure of chromatin. However, nuclease digestion of chromatin would likely modify

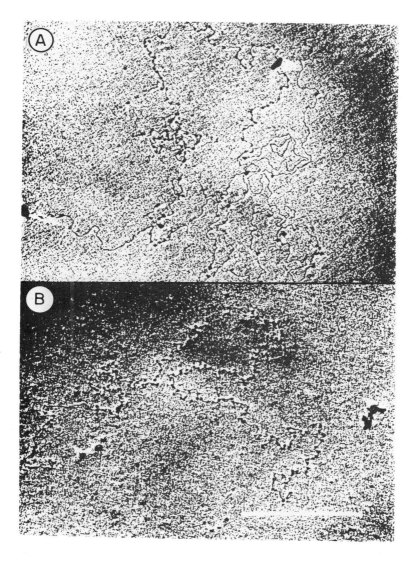

Figure 6. Electron micrographs of chicken liver chromatin after trypsin digestion of histone H1. From P. Oudet, M. Gross-Bellard, and P. Chambon (1975). Cell 4, 281. Reprinted with permission from the Massachusetts Institute of Technology.

Table I. Diameters of chromatin particles determined from
the electron microscopy.

Source	Formaldehyde Fixation	Diameters (Å)	Ref.
rat liver	yes	60 ± 16	40
rat thymus	yes	83 ± 23	40
SV-40 minichromosome	yes	109 ± 7	43
chicken liver (H1 depleted)	yes	96 ± 9	42
chicken liver (H1 depleted)	no	128 ± 12	42
SV40 Virions	no	131 ± 10	44
calf thymus	no	134 ± 14	45

the structure of histone-DNA complexes (31). Indeed, both melting (24) and circular dichroism (23) properties of PS particles differ from those of native chromatin. Histones in nuclease-resistant fractions of chromatin can be dissociated either at lower salt or at higher pH than can those in whole chromatin (9).

Most of the reports published in the last few years dealing with chromatin subunits seem to equate "beads, ν-bodies, or nucleosomes" from electron micrographs with chromatin subunits isolated from nuclease digestion. However, the hypothesis of subunits suggests simply a regular composition of histones and a given length of DNA without implying any particular morphology, while beads, ν-bodies or nucleosomes describe the general structure of these subunits (47). Such a distinction was elaborated by the following studies of chromatin in urea.

Carlson *et al.* (53) observed no beaded structure when chromatin was fixed by formaldehyde in the presence of urea. In other words, ν-bodies are destroyed by a high concentration of urea. However, urea previously has been shown to change only the physical properties of the chromatin, melting, circular dichroism and viscosity, but not the histone association with DNA (54-57). The addition of urea to chromatin solution therefore could favor the transition of chromatin subunit structure from a closed, beaded state to an open, coiled state without destroying the subunits themselves (31). Indeed, Jackson and Chalkley (52) reported that chromatin subunits isolated from nuclease digestion of chromatin in the presence of urea were similar to those isolated without urea. A similar observation was also reported by Oosterhof *et al.* (58).

IV. HETEROGENEITY OF CHROMATIN SUBUNITS

Although the data presented in Section II seem to indicate homogeneity among chromatin subunits whether with or without lysine-rich histones, DNA fragments of various lengths ranging from 50 to 200 base pairs have been isolated from nuclease-treated chromatin or nuclei (58-60). Among the products of nuclease-treated chromatin, Doenecke (61) isolated fractions with a low protein/DNA ratio and even with pure DNA. The length of DNA from these fractions is about half that of the monomer in chromatin subunits. Thus, when nuclease digestion is the means for preparing chromatin subunits, even when chromatins containing all five histones are used, one still has to consider not only the fractions containing DNA and histones alone, but all other fractions as well.

If nonhistone proteins, particularly the sequence-specific DNA binding nonhistone proteins, can bind tightly to DNA and protect it against nuclease digestion (47), then heterogeneity among chromatin subunits should be considered. Indeed, nonhistone proteins do exist in some chromatin subunits obtained by nuclease digestion of chromatin isolated from rat liver (62) or from calf thymus (Hu and Li, unpublished results).

Chromatin subunits containing the common histone composition (two each of H2A, H2B, H3 and H4 with or without H1) certainly could not exist if the chromatin lacked one or more species of histones. For instance, Fig. 7a shows that histones H1 and H3 exist in macronuclear chromatin isolated from *Tetrahymena pyriformis* whereas they are not found in micronuclear chromatin. Despite these differences in histone composition, after treatment with staphylococcal nuclease, the electrophoretic patterns of DNA isolated from these two types of nuclei are quite similar to each other (Fig. 7b) (63).

Lohr and Van Holde (64a) found chromatin subunits isolated from yeast although yeast chromatin was considered by the authors with a possible lack of histones H1 and H3. This possibility, however, is not supported by the presence of histones H1 and H3 in *Neurospora* which is closely related to yeast (64b). Despite the uncertainty about the presence of H1 and H3 in yeast, the results shown in Fig. 7 suggest that potentially histones can assume various types of assembly with one another. Once formed, these combinations of histones can bind DNA and protect it against nuclease digestion, the extent of protection depending upon the composition of histones present in each nucleus.

V. MECHANISM OF PROTECTION OF DNA AGAINST NUCLEASE DIGESTION
    BY PROTEIN BINDING

Since the initial report from Clark and Felsenfeld (19) that about 50% of chromatin DNA were resistant to nuclease digestion, this method has been widely used for preparation of chromatin subunits. However, the mechanism of protection of DNA against nuclease digestion by protein binding remained unclear.

With respect to the nuclease activity of cutting across both strands of DNA, research on chromatin indicates (a) that the DNA covered by histone H1 is less protected than that bound by other histones (25-28,32-34), (b) that the nuclease-resistant fragments become insoluble at the end of digestion

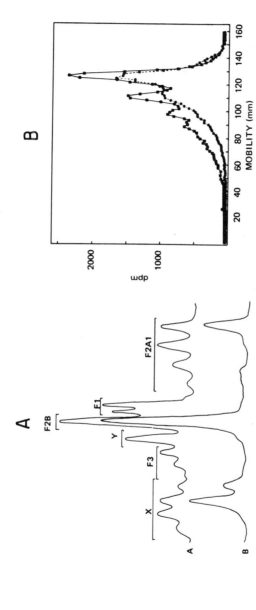

Figure 7. Densitometer tracing of long polyacrylamide gels containing (A) macronuclear and (B) micronuclear histones of strain HSM from Tetrahymena pyriformis. *F3, F2B, F1 and F2A1 are, respectively, histones H3, H2B, H1 and H4. (b) Electrophoretic analysis of partially digested [³H] thymidine-labeled micronuclear DNA (■) and [¹⁴C] thymidine-labeled macronuclear DNA (●).*

(19) and (c) that a larger percentage of chromatin DNA is protected if polylysine is added to chromatin before digestion (19).

Previously, Li *et al.* (28) suggested that H1 offered a lesser degree of protection for DNA than did the other histones. The reason offered was the lower α-helical content and less globular structure of histone H1 which might allow the nuclease to approach the DNA more easily than is possible with the other histones which possess more α-helices and more globular structures. Also, if histone H1 binds DNA as a monomer while each type of other histone binds DNA as a dimer, the width of each dimer might be able to shield DNA from nuclease attack more effectively than that of the monomer. Furthermore, it is also possible that H1-bound regions are more hydrated and more exposed to the medium than are the core regions. Under these circumstances, after a mild digestion, the H1-bound regions (30-45 base pairs) are well exposed to the medium and are more vulnerable to nuclease attack than the core regions, which appear to be condensed into beaded structure and therefore not well dissolved in the medium. If this is the case, nuclease could continue its digestion of the DNA originally covered by histone H1. Since histone H1 is known to be not strongly associated with other histones, it might be gradually released from DNA as the digestion proceeds until the H1-covered DNA is fully digested. At this point, the core subunits might become more condensed, insoluble and resistant to the nuclease attack.

Insolubility of the protein-DNA complex is certainly a major factor in protection of DNA against further nuclease digestion with chromatin wherein the end of digestion is accompanied by appearance of insoluble products (19). Perhaps the additional protection of chromatin DNA against nuclease digestion in a polylysine-chromatin complex, as observed by Clark and Felsenfeld (19),could result from the earlier appearance of insoluble products during the process of digestion of this complex than occurs in chromatin alone. To test the validity of this hypothesis, both polylysine-DNA and protamine-DNA complexes with various degrees of coverage were digested by staphylococcal nuclease. Insoluble materials were observed at the limit of digestion. The digested DNA in the supernatant was shown to be proportional to the fraction of free DNA in the original complex, whether bound by polylysine or by protamine (Shaw and Li, unpublished results). Therefore, the protection of DNA against nuclease digestion is not a property unique to histones, but rather a general one shared by many other types of proteins, including nonhistones and model proteins.

VI. STRUCTURE OF CHROMATIN SUBUNITS

Using DNase I, Noll (65) first demonstrated the presence of single-stranded polynucleotides containing multiples of ten bases (i.e., 40,50,60,etc.) when nuclei were digested by this enzyme. Noll's observations were later confirmed by Simpson and Whitlock (33) and Shaw *et al.* (28). Using DNase digestion as a criterion, it is apparent that there exists some kind of regularity in arrangement of histones among themselves and in association with DNA within chromatin subunits.

Many models dealing with the structure of chromatin have been proposed in the past few years by Kornberg (66), Van Holde *et al.* (67), Baldwin *et al.* (68), Hyde and Walker (69), Li (31), Weintraub (60), Bram *et al.* (70), Pardon *et al.* (71), Cole (35), Jackson *et al.* (9), Olins (51), Camerini-Otero *et al.* (72) and others. Most of the models depict the chromatin subunits as compact (51,60,67-72), whereas others suggest a supercoil as a possible alternative (9,31).

Despite different views on the secondary structures of chromatin subunits, either a compact bead or a relaxed coil, the internal regularity within the subunit first observed by Noll (65) does not seem to depend upon the secondary structure. For instance, Yaneva and Dessev (73) recently reported that the 10 nucleotide repeat could be obtained by DNase I digestion of chromatin either in the presence or absence of 5M urea, although v-bodies of chromatin treated by urea disappeared in the electron microscope (53). The experiments on chromatin in urea (53-58,73) indicate that the binding characteristics between histones and DNA are not substantially perturbed by urea despite the fact that this denaturant could destroy the secondary structure of both histones (55-57) and chromatin subunits (53).

The observed multiples of 10 nucleotides within chromatin subunits after DNase I digestion have been explained as due to specific folding of DNA within v-bodies or nucleosomes (46,65). In particular, Crick and Klug (46) proposed a model of kinky helix that the DNA within a nucleosome is kinked every 20 base pairs. The suggestions (46,65) of dependence of the internal regularity of 10 nucleotides upon specific secondary structures of chromatin subunits, however, are not supported by the observations of Yaneva and Dessev (73) in urea. An alternative explanation was recently offered (18) based upon the earlier model of chromatin strucstructure (32).

As shown in Fig. 8, if every dimer of histones H2A, H2B, H3 and H4 binds 30-40 base pairs of DNA as previously

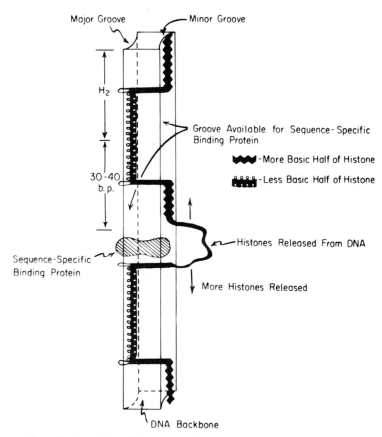

Figure 8. Schematic diagram showing internal regularity of histone binding to DNA within a chromatin subunit and a possible mechanism of gene activation through the binding of a sequence-specific binding protein in the empty groove of histone-bound regions.

calculated (31,32) and if the more basic regions of each
dimer bind primarily in the minor groove while the less basic
regions in the major groove as suggested (32), there will be
a regular spacing of 30-40 base pairs with respect to (a)
empty major or minor groove and (b) uncovered DNA backbone.
If each action of DNase I requires a binding site of 8-10
nucleotides along the backbone or groove, as implied from the
report of Hoard and Goad (74), each stretch of uncovered
backbone or groove of 30-40 base pairs in Fig. 8 would accom-
modate at most four DNase I molecules. Polynucleotides
resulted from DNase I digestion of each available stretch
would contain 10, 20, 30 and 40 nucleotides. Combined with
the adjacent stretches, products with multiples of 10 nucleotides could be generated (18).

The model shown in Fig. 8 has one additional feature in
that it offers a possible mechanism of specific histone
removal and gene activation. According to the model (Fig. 8)
each empty groove (major or minor) within a chromatin subunit
covers 30-40 base pairs. It might be long enough for specific
recognition by a sequence-specific binding protein. If such
binding does occur and the binding affinity is stronger than
that of histones, histones could be forced off and be released from the DNA. More histones in the vicinity could possibly be released through the mechanism of cooperative histone removal (a physical phenomenon which is the reverse of
cooperative binding of histones to DNA), enzymatic modification or digestion of released histones, or transcription,
after the activation of this particular gene. Thus, although
the model shown in Fig. 8 has not been proved by experiments,
it does have the capacity to explain the internal structure
within chromatin subunits and offers an interesting mechanism for histone removal from a specific gene and its activation for transcription.

To date, the experimental data dealing with structure of
chromatin subunits are very limited. Most discussions on the
subject are speculative, and have been treated at length in a
recent review of chromatin structure (18). One of the major
efforts currently needed in the research of chromatin structure is a determination of the internal structure of chromatin subunits.

ACKNOWLEDGEMENT: The author wishes to acknowledge Dr. Gary
Felsenfeld for the use of part of his lecture transcript in
this chapter and his comments on the manuscript.

REFERENCES

1. Fambrough, D. M., and Bonner, J., *J. Biol. Chem. 243*,
   4434 (1968).

2. Panyim, S., Sommer, K. R., and Chalkley, R., *Biochemistry 10,* 3911 (1971).
3. Kornberg, R. D., and Thomas, J., *Science 184,* 865 (1974).
4. Martinson, H. G., and McCarthy, B. J., *Biochemistry 14,* 1073 (1975).
5. Hardinson, R. C., Eichner, M. E., and Chalkley, R., *Nucl. Acids Res. 2,* 1751 (1975).
6. Hyde, J. E., and Walker, I. O., *FEBS Lett. 50,* 150 (1975).
7. Weintraub, H., Palter, K., and Van Lente, F., *Cell 6,* 85 (1975).
8. Thomas, J. O., and Kornberg, R. D., *Proc. Nat. Acad. Sci. U.S.A. 72,* 2626 (1975).
9. Jackson, V., Hoffman, P., Hardinson, R., Murphy, J., Eichner, M. E., and Chalkley, R., in *Molecular Biology of the Mammalian Genetic Apparatus - Its Relationship to Cancer, Aging and Medical Genetics,* Ed. P. O. P. Ts'o, Elsevier, North Holland (1976).
10. Edwards, P. A., and Shooter, R. V., *Biochem. J. 114,* 227 (1969).
11. Kelley, R. I., *Biochem. Biophys. Res. Commun. 54,* 1588 (1973).
12. Roark, D. E., Geohegan, T. E., and Keller, G. H., *Biochem. Biophys. Res. Commun. 59,* 542 (1974).
13. D'Anna, J. A., and Isenberg, I., *Biochem. Biophys. Res. Commun. 61,* 343 (1974).
14. Sperling, R., and Bustin, M., *Proc. Nat. Acad. Sci. U.S.A. 71,* 4625 (1974).
15. Sperling, R., and Bustin, M., *Biochemistry 14,* 3322 (1975).
16. Sperling, R., and Bustin, M., *Nucl. Acids Res. 3,* 1263 (1976).
17. Isenberg, I., in *Search and Discovery - A Volume Dedicated to Albert Szent-Gyorgi,* Ed. B. Kaminer, Academic Press (1976).
18. Li, H. J., *Chromatin Structure.* In *Hormone Action I. Steroid Hormone Receptors,* Eds. B. W. O'Malley, and L. Birnbaumer, Academic Press (*in press*).
19. Clark, R. J., and Felsenfeld, G., *Nature New Biology 229,* 101 (1971).
20. Murray, K., *J. Mol. Biol. 39,* 628 (1969).
21. Itzhaki, R. F., *Biochem. J. 125,* 221 (1971).
22a. Hewish, D. R., and Burgoyne, L. A., *Biochem. Biophys. Res. Commun. 52,* 504 (1973).
22b. Noll, M., *Nature 251,* 249 (1974).
23. Rill, R., and Van·Holde, K. E., *J. Biol. Chem. 248,* 1080 (1973).
24. Sahasrabuddhe, C. G., and Van Holde, K. E., *J. Biol.*

Chem. 249, 152 (1974).

25. Sollner-Webb, B., and Felsenfeld, G., *Biochemistry 14,* 2915 (1975).

26. Clark, R. J., and Felsenfeld, G., *Biochemistry 13,* 3622 (1974).

27. Axel, R., *Biochemistry 14,* 2921 (1975).

28. Shaw, B. R., Herman, T. M., Kovacic, R. T., Beaudreau, G. S., and Van Holde, K. E., *Proc. Nat. Acad. Sci. U.S.A. 73,* 505 (1976).

29. Li, H. J., Chang, C., Evagelinou, Z., and Weiskopf, M., *Biopolymers 14,* 211 (1975).

30. Li, H. J., and Bonner, J., *Biochemistry 10,* 1461 (1971).

31. Li, H. J., Chang, C., and Weiskopf, M., *Biochemistry 12,* 1763 (1973).

32. Li, H. J., *Nucl. Acids Res. 2,* 1275 (1975).

33. Simpson, R. T., and Whitlock, J. P., Jr., *Nucl. Acids Res. 3,* 117 (1976).

34. Varshavsky, A. J., Bakayev, V. V., and Georgiev, G. P., *Nucl. Acids Res. 3,* 477 (1976).

35. Cole, R. D., in *Molecular Biology of the Mammalian Genetic Apparatus - Its Relationship to Cancer, Aging and Medical Genetics, Part A.* Ed. P. O. P. Ts'o, Elsevier, North Holland (1976).

36. Vogel, T., and Singer, M. F., *J. Biol. Chem. 250,* 796 (1975).

37. Vogel, T., and Singer, M. F., *Proc. Nat. Acad. Sci. U.S.A. 72,* 2597 (1975).

38. Anderson, P. L., and Moudrianakis, E. N., *Biophys. J. 9,* A-54 (1969).

39. Olins, A. L., and Olins, D. E., *J. Cell. Biol. 59,* 252 (1973).

40. Olins, A. L., and Olins, D. E., *Science 183,* 330 (1974).

41. Woodcock, C. L. F., *J. Cell. Biol. 59,* 368 (1973).

42. Oudet, P., Cross-Bellard, M., and Chambon, P., *Cell 4,* 281 (1975).

43. Griffith, J. D., *Science 187,* 1202 (1975).

44. Germond, J. E., Hirt, B., Oudet, P., Cross-Bellard, M., and Chambon, P., *Proc. Nat. Acad. Sci. U.S.A. 72,* 1843 (1975).

45. Langmore, J. P., and Wooley, J. C., *Proc. Nat. Acad. Sci. U.S.A. 72,* 2691 (1975).

46. Crick, F. H. C., and Klug, A., *Nature 255,* 530 (1975).

47. Li, H. J., in *Molecular Biology of the Mammalian Genetic Apparatus - Its Relationship to Cancer, Aging and Medical Genetics,* Ed. P. O. P. Ts'o, Elsevier, North Holland (1976).

48. Van Holde, K. E., Sahasrabuddhe, C. G., Shaw, B. R.,

Van Bruggen, E. F. J., and Annberg, A. C., *Biochem. Biophys. Res. Commun. 60,* 1365 (1974).

49. Finch, J. T., Noll, M., and Kornberg, R. D., *Proc. Nat. Acad. Sci. U.S.A. 72,* 3320 (1975).

50. Bakayev, V. V., Melnickov, A. A., Osicka, V. D., and Varshavsky, A. J., *Nucl. Acids Res. 2,* 1401 (1975).

51. Olins, D. E., in *Molecular Biology of the Mammalian Genetic Apparatus - Its Relationship to Cancer, Aging and Medical Genetics, Part A,* Ed. P. O. P. Ts'o, Elsevier, North Holland (1976).

52. Jackson, V., and Chalkley, R., *Biochem. Biophys. Res. Commun. 67,* 1391 (1975).

53. Carlson, R. D., Olins, A. L., and Olins, D. E., *Biochemistry 14,* 3122 (1975).

54. Bartley, J. A., and Chalkley, R., *Biochim. Biophys. Acta 160,* 224 (1968).

55. Shih, T. Y., and Lake, R. S., *Biochemistry 11,* 4811 (1972).

56. Bartley, J. A., and Chalkley, *Biochemistry 12,* 468 (1973).

57. Chang, C., and Li, H. J., *Nucl. Acids Res. 1,* 945 (1974).

58. Oosterhof, D. K., Hozier, J. C., and Rill, R. L., *Proc. Nat. Acad. Sci. U.S.A. 72,* 633 (1975).

59. Rill, R. L., Oosterhof, D. K., Hozier, J. C., and Nelson, D. A., *Nucl. Acids Res. 2,* 1525 (1975).

60. Weintraub, H., *Proc. Nat. Acad. Sci. U.S.A. 72,* 1212 (1975).

61. Doenecke, D., *Cell 8,* 59 (1976).

62. Liews, C. C., and Chan, P. K., *Proc. Nat. Acad. Sci. U.S.A.* (in press).

63. Gorovsky, M. A., and Keevert, J. B., *Proc. Nat. Acad. Sci. U.S.A. 72,* 3536 (1975).

64a. Lohr, D., and Van Holde, K. E., *Science 188,* 165 (1975).

64b. Goff, C. G., *J. Biol. Chem.* (in press).

65. Noll, M., *Nucl. Acids Res. 1,* 1573 (1974).

66. Kornberg, R. D., *Science 184,* 868 (1974).

67. Van Holde, K. E., Sahasrabuddhe, C. G., and Shaw, B. R., *Nucl. Acids Res. 1,* 1579 (1974).

68. Baldwin, J. P., Boseley, P. G., Bradbury, E. M., and Ibel, K., *Nature 253,* 245 (1975).

69. Hyde, J. E., and Walker, I. O., *Nucl. Acids Res. 2,* 405 (1975).

70. Bram, S., Butler-Browne, G., Brady, P., and Ibel, K., *Proc. Nat. Acad. Sci. U.S.A. 72,* 1043 (1975).

71. Pardon, J. F., Worcester, D. L., Wooley, J. C., Tatchell, K., Van Holde, K. E., and Richards, B. M., *Nucl. Acids Res. 2,* 2163 (1975).

72. Camerini-Otero, R. D., Sollner-Webb, B., and Felsenfeld, G., *Cell* (in press).
73. Yaneva, M., and Dessev, G., *Nucl. Acids Res. 3,* 1761 (1976).
74. Hoard, D. E., and Goad, W., *J. Mol. Biol. 31,* 595 (1968).

Chapter 5

POST-SYNTHETIC MODIFICATIONS OF HISTONE STRUCTURE:

A Mechanism for the Control of Chromosome Structure
by the Modulation of Histone-DNA Interactions

VINCENT G. ALLFREY

The Rockefeller University
New York, New York 10021

The discovery of histone acetylation (1) and methylation (2) in 1964 and of histone phosphorylation in 1966 (3,4) opened a field which has completely altered earlier conceptions that histones are essentially passive and metabolically inert components of the chromatin. It is now clear that all of the major histone classes are subject to one or more modifications of their structure, and that most of the modifications occur in regions of the polypeptide chain that are most likely to interact with DNA. The proposition that post-synthetic, enzymatically-catalyzed modifications of DNA-associated proteins provide a mechanism for the control of chromatin structure and function (1,5-15) has been investigated for over a decade. It has been tested in dividing and non-dividing cell populations of a wide variety of species, in systems responding to hormones, mitogens, carcinogens and drugs, and in cells infected with transforming viruses (16, 17). A remarkable set of temporal and spatial correlations have emerged to support the view that alterations in histone structure are key events in the cell cycle and in chromosomal activation. The aim of this review is to reexamine and up-date some of the evidence, and to indicate new levels of complexity in this important and expanding area of chromatin research.

HISTONE ACETYLATION

Background -

The presence of acetyl groups in histones was first detected by Phillips in 1963 (18), and it was subsequently

established that N-acetylserine is the amino-terminal residue
of three of the major histone classes - Hl, H2a and H4 (18,
19). These observations prompted experiments on the relationship between histone acetylation and synthesis, comparing
the uptakes of radioactive acetate and isotopically-labeled
amino acids into the histones of calf thymus lymphocytes (1,
5,6). The results led to the surprising conclusions that
acetate uptake could take place in isolated nuclei in the
absence of histone synthesis, and with the production of a
previously-unknown amino acid, epsilon-N-acetyllysine (20,
21). Thus, it becomes necessary to discriminate between two
types of histone acetylation, one involving the α-amino group
of the terminal serine residue and the other involving the
epsilon-amino group of lysine residues within the polypeptide
chain. These two reactions will be considered separately in
this review.

NH_2-Terminal Acetylation of Histones: A Cytoplasmic Modifi-

cation of Nascent Histone Chains -

Despite the fact that the majority of histone molecules
contain NH_2-terminal acetyl groups, measurements of radioactive acetate incorporation into the histones of non-
dividing cells have shown little if any incorporation of
acetate into amino-terminal peptides; the greater part, by
far, of the isotopic acetate is recovered as epsilon-N-
acetylllysine (20-22). Thus, there is a clear distinction
between the metabolism of acetyl groups on the same histone
molecule - e.g., histone H4. Although NH_2-terminal acetylation is an uncommon event in cells which are not synthesizing
histones, it is readily detectable in dividing cell populations, as in regenerating liver (22), or in avian erythroblasts (23,24).

In the first cycle of cell division in regenerating rat
liver, histone synthesis becomes appreciable at 16 hours and
reaches a peak at about 24 hours post-partial hepatectomy.
In double-labeling experiments using ^{3}H-acetate as a precursor of the histone acetyl groups, and ^{14}C-serine as a marker
for newly-synthesized histones, it was observed that [3]H-
acetyl-N-^{14}C-serine could be recovered from proteolytic
digests of histones obtained at 16 hours but not before (22).
In the non-dividing hepatocytes of adult rat livers, acetate
uptake into the histones is largely due to the formation of
epsilon-N-acetyllysine; this reaction is not puromycin-
sensitive. In regenerating rat liver at the peak of histone
synthesis, ^{3}H-acetyl-N-^{14}C-serine also makes its appearance,
but not if puromycin is administered first (22). This is

the result expected if the acetylation of the NH_2-terminal serine residue is tightly-coupled to histone synthesis; puromycin, by blocking *de novo* synthesis, would prevent the production of histone molecules bearing ^3H-acetyl-N-^{14}C-serine at their amino termini. A similar puromycin block of amino-terminal acetylation was also observed in the histones of mammary gland cells (25).

In regenerating rat liver, the synthesis of histones H3 and H4 takes place on small polysomes in the cytoplasm (22), and, after short-term double-labeling experiments, the cytoplasmic histones yield ^3H-acetyl-N-^{14}C-serine upon proteolytic digestion. This doubly-labeled amino-terminal peptide was not recovered from the polysomes of puromycin-treated animals, as would be expected if amino-terminal acetylation were tightly coupled to histone synthesis (22). Further evidence for coupling was obtained by the use of puromycin to displace nascent histone chains from the isolated small polysomes of regenerating rat liver; the released peptides were shown to contain an acetylated NH_2-terminal serine residue. This result led to the suggestion that the initiation of histone synthesis in hepatocytes might involve the participation of an initiator tRNA bearing acetyl-serine (instead of methionyl-tRNA) (22). This is probably not the case; because more recent studies of histone synthesis in reticulocyte lysates, using histone mRNAs from ascites tumor cells, show that it is possible to initiate histone H1 and H4 chains with N-formyl-methionyl tRNA from yeast (26). An incorporation of N-formylmethionine into nascent histone chains has also been observed in an ascites cell-free protein synthesizing system (27). These results indicate the presence of the methionine initiation-codon in the histone messenger RNAs and strongly suggest that the natural initiation of histone chains proceeds from an amino-terminal methionine residue which is subsequently cleaved from the histone precursor molecule and does not appear in the final product. (None of the major histone classes has an amino-terminal methionine residue.) Whether other amino acids are cleaved from the amino-terminal region of the presumptive histone precursor is not known at this time. Yet, the rapid appearance of acetyl-N-serine on nascent histone chains released by puromycin (22) indicates that the methionine-containing sequence must be short-lived, and that the generation of acetyl-serine at the amino-terminus can occur before the nascent histone chain leaves the polysome.

What is the mechanism of NH_2-terminal acetylation? One possibility is that cleavage of the histone precursor generates a free amino group on a serine residue; this could then be acetylated by an enzymatically-catalyzed transfer of acetyl groups from acetyl-Coenzyme A to the serine-nitrogen. Enzymes capable of acetylating histones are known to occur in the cytoplasm and to modify both lysine-rich (H1) and argi- nine-rich (H4) histones (28). Whether the enzymatic modifi- cation actually involves formation of an amino-terminal acetylserine has not been clarified; one such cytoplasmic enzyme actually acetylates the lysine residues of histone H4 (29).

An alternative mechanism invokes the participation of an acetyl-N-seryl-transfer RNA. Acetylseryl-tRNA has been detected in regenerating rat liver after the simultaneous injection of ^{14}C-serine and ^{3}H-acetate. Treatment of the purified, doubly-labeled tRNA with ribonuclease released the terminal nucleoside which was identified as ^{3}H-acetyl-N-^{14}C- seryl-adenosine (22). The formation of acetylseryl-tRNA from acetyl CoA and "charged" seryl-tRNA has also been observed in cell-free systems from regenerating rat liver (22). The results raise the possibility that acetylseryl-tRNA could play a role in a terminal addition reaction in which methio- nine is cleaved from a histone precursor molecule and replaced by acetyl-serine. The utilization of tRNAs for the amino-terminal modification of preexisting polypeptide chains has been noted before; for example, in the puromycin-sensitive transfer of arginine from arginyl-tRNA to the NH$_2$-terminal position of preformed protein acceptors in rat liver extracts (30,31), and in the transfer of leucine or phenylalanine to bacterial proteins from the appropriate tRNAs (32). A simi- lar mechanism involving acetylseryl-tRNA could explain the puromycin-sensitivity of amino-terminal acetylation in his- tones (22).

In any case, it is clear that the details of the amino- terminal acetylation of three major histone classes remain to be worked out. The reaction takes place in the cytoplasm, and, judging by the stability of the acetylserine residues in histones, this reaction is essentially irreversible. This is in strong contrast to the acetylation of lysine residues in histones, which is often a rapidly reversible modification of histone structure which occurs mainly (but not exclusively) within the cell nucleus.

Acetylation of Histone Lysine Residues: Structures of Modi- fied Histones -

Since the discovery of epsilon-N-acetyllysine as a natural component of histones (20,21,33), careful amino acid sequence studies have established that a limited and specific set of lysine residues is subject to acetylation in each

histone class. In histone fraction H4, the sites of internal
acetylation have been identified as lysine residues at positions 5, 8, 12 and 16 of the polypeptide chain (8,9,33-38).
In histone fraction H3, acetylation may occur at lysine residues in positions 9, 14, 18 and 23 (9,39-41). There are also
multiple sites of lysine acetylation in histone fraction H2B
and at least one site in H2A (9,39,42). Structures indicating the sites of modification of the different histone
classes are presented in Figures 1-5.

It is important to stress that the potential for acetylation at multiple sites is not realized in every histone
molecule; i.e., particular lysine residues may or may not be
acetylated. Because this type of histone modification is a
rapid and reversible process (1,5,6) which can proceed in
the absence of histone synthesis or degradation, acetate
"turnover" on stable histone molecules is readily detectable
in non-dividing cells (7,23,43-46) and in isolated cell
nuclei (1,5,6,16). The result is that preparations of purified histones such as H3 and H4 are internally heterogeneous.
Each histone class comprises a mixture of polypeptide chains
of identical amino acid sequence, some of which are internally
acetylated to different degrees, while others are not acetylated at all. Each of these subfractions may then, in addition, differ with regard to other forms of substitution,
such as methylation or phosphorylation. (Some sequence
heterogeneity in two different forms of H3 adds additional
complexity (47,71).) Clearly, the relative proportions of
the various acetylated and non-acetylated forms of a particular histone constitute an important set of variables in
describing the composition of nuclei, chromatin fractions,
or nucleosomes from different tissues, or from the same
tissue at different stages of development.

Enzymatic Basis of Histone Acetylation -

The acetylation of lysine residues in histones is catalyzed by acetyltransferases which utilize acetyl-Coenzyme A
as the acetyl group donor (5,28,29,48-57). It is now clear
that the enzymes catalyzing this type of histone modification
are complex. Multiple forms of histone-specific acetyltransferases have been isolated from rat thymus nuclei and shown
to differ in molecular weights, isoelectric points and substrate specificities. One acetyltransferase showed preference for histone H1, while two others preferentially acetylated histone H4 and failed to acetylate H3 (56). The histone acetyltransferases from different organs differ in their
patterns of elution from DEAE-cellulose columns, suggesting
differences in the enzyme complements of different somatic

AC-SER-GLU-ALA-PRO-ALA-GLU-THR-ALA-ALA-PRO-ALA-ALA-PRO-ALA-GLU-PRO-ALA-LYS-LYS-
1   10   20

-LYS-LYS-ALA-ALA-LYS-LYS-PRO-GLY-ALA-GLY-ALA-ALA-LYS-ARG-LYS-ALA-ALA-GLY-PRO-PRO-VAL-
SER-P   40
30

-SER-GLU-LEU-ILE-THR-LYS-ALA-VAL-ALA-ALA-SER-LYS-GLU-ARG-ASN-GLY-LEU-SER-LEU-ALA-ALA-
50   60

-LEU-LYS-LYS-ALA-LEU-ALA-ALA-GLY-GLY-TYR-ASP-VAL-GLU-LYS-ASN-SER-ARG-ILE-LYS-LEU-GLY-
70   80

-LEU-LYS-SER-LEU-VAL-SER-LYS-GLY-THR-LEU-VAL-GLU-THR-LYS-GLY-THR-GLY-ALA-SER-GLY-SER-
90   100

-PHE-LYS-LEU-ASP-LYS-LYS-ALA-ALA-SER-GLY-GLU-ALA-LYS-PRO-LYS-PRO-LYS-LYS-ALA-GLY-ALA-
110   120

-ALA-LYS-PRO-LYS-LYS-PRO-ALA-GLY-ALA-ALA-LYS-LYS-PRO-ALA-GLY-ALA,ALA,LYS,ALA,PRO (THR,
130   140

-PRO,LYS)(VAL-ALA-LYS)(LYS-ALA-VAL-LYS)(ALA-LYS-LYS)(SER-PRO-LYS)(LYS-ALA-LYS)(LYS-PRO-
150   160

-LYS)(ALA-PRO-LYS)(SER-ALA-ALA-LYS)(SER-PRO-ALA-LYS-PRO-LYS)(ALA-ALA-LYS-PRO-LYS-ALA-
170   180

-PRO-LYS-PRO-LYS)(ALA-ALA-LYS)(LYS)(ALA-ALA-LYS)(SER-PRO-ALA-LYS)(ALA-VAL-LYS-PRO-LYS)
190   200

(ALA-ALA-LYS-PRO-LYS)(ALA-ALA-GLY-ALA-LYS)(LYS-LYS-COOH)
210   220

Figure 1. Amino acid sequence of histone H1 (fraction 3) from rabbit thymus, indicating the presence of acetyl-N-serine at the amino-terminal; and a site of phosphorylation (cAMP-dependent) at serine-37. Other sites of phosphorylation are not shown.

```
     P              Ac                                                            P
     |              |                                                            |
    AC-SER-GLY-ARG-GLY-LYS-GLN-GLY-GLY-LYS-ALA-ARG-ALA-LYS-ALA-LYS-THR-ARG-SER-SER-ARG-
     1                       10                                      20

    -ALA-GLY-LEU-GLN-PHE-PRO-VAL-GLY-ARG-VAL-HIS-ARG-LEU-LEU-ARG-LYS-GLY-ASN-TYR-ALA-GLU-
                             30                                      40

    -ARG-VAL-GLY-ALA-GLY-ALA-PRO-VAL-TYR-LEU-ALA-ALA-VAL-LEU-GLU-TYR-LEU-THR-ALA-GLU-ILE-
                             50                                      60

    -LEU-GLU-LEU-ALA-GLY-ASN-ALA-ALA-ARG-ASP-ASN-LYS-LYS-THR-ARG-ILE-ILE-PRO-ARG-HIS-LEU-
                             70                                      80

    -GLN-LEU-ALA-ILE-ARG-ASN-ASP-GLU-GLU-LEU-ASN-LYS-LEU-LEU-GLY-LYS-VAL-THR-ILE-ALA-GLN-
                             90                                      100

    -GLY-GLY-VAL-LEU-PRO-ASN-ILE-GLN-ALA-VAL-LEU-LEU-PRO-LYS-LYS-THR-GLU-SER-HIS-HIS-LYS-
                             110                                     120

    -ALA-LYS-GLY-LYS-COOH
                             129
```

Figure 2. Amino acid sequence of histone H2A, indicating the presence of acetyl-N-serine at the amino-terminal, and of epsilon-N-acetyllysine at lysine-5. Serine residues at positions 1 and 19 are potential sites of phosphorylation.

173

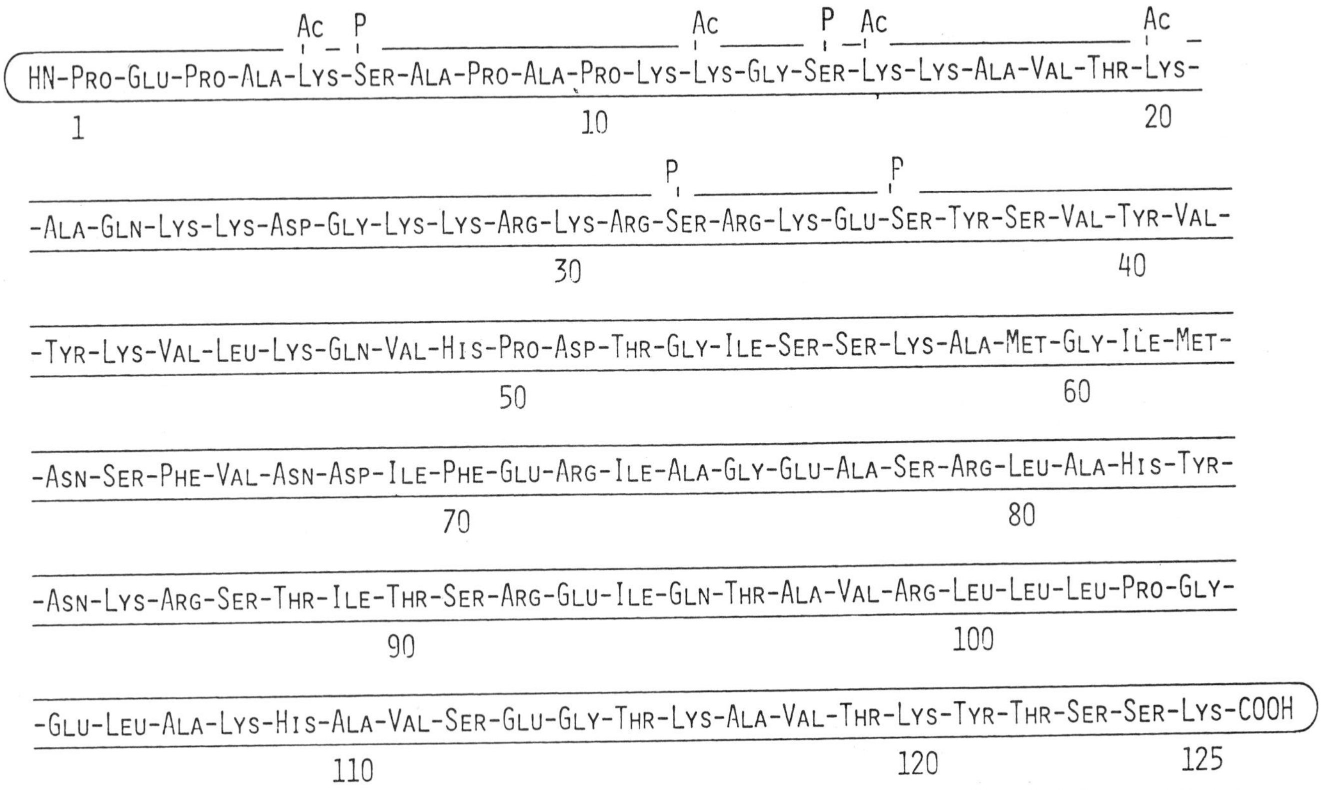

Ac P   Ac   P Ac   Ac

HN-Pro-Glu-Pro-Ala-Lys-Ser-Ala-Pro-Ala-Pro-Lys-Lys-Gly-Ser-Lys-Lys-Ala-Val-Thr-Lys-
1 _____ 10 _____ 20

                        P   P

-Ala-Gln-Lys-Lys-Asp-Gly-Lys-Lys-Arg-Lys-Arg-Ser-Arg-Lys-Glu-Ser-Tyr-Ser-Val-Tyr-Val-
30 _____ 40

-Tyr-Lys-Val-Leu-Lys-Gln-Val-His-Pro-Asp-Thr-Gly-Ile-Ser-Ser-Lys-Ala-Met-Gly-Ile-Met-
50 _____ 60

-Asn-Ser-Phe-Val-Asn-Asp-Ile-Phe-Glu-Arg-Ile-Ala-Gly-Glu-Ala-Ser-Arg-Leu-Ala-His-Tyr-
70 _____ 80

-Asn-Lys-Arg-Ser-Thr-Ile-Thr-Ser-Arg-Glu-Ile-Gln-Thr-Ala-Val-Arg-Leu-Leu-Leu-Pro-Gly-
90 _____ 100

-Glu-Leu-Ala-Lys-His-Ala-Val-Ser-Glu-Gly-Thr-Lys-Ala-Val-Thr-Lys-Tyr-Thr-Ser-Ser-Lys-COOH
110 _____ 120 _____ 125

Figure 3. Amino acid sequence of histone H2B, indicating the presence of multiple epsilon-N-acetyllysine residues at positions 5, 12, 15 and 20. Sites of phosphorylation at serine residues 6, 14, 32 and 36 are also shown.

174

Ac _____ Me/P ___ Ac _____ Ac
H₂N-Ala-Arg-Thr-Lys-Gln-Thr-Ala-Arg-Lys-Ser-Thr-Gly-Gly-Lys-Ala-Pro-Arg-Lys-Gln-Leu-
1 .... 10 20

Ac _____ Me/P
-Ala-Thr-Lys-Ala-Ala-Arg-Lys-Ser-Ala-Pro-Ala-Thr-Gly-Gly-Val-Lys-Lys-Pro-His-Arg-Tyr-
30 40

-Arg-Pro-Gly-Thr-Val-Ala-Leu-Arg-Glu-Ile-Arg-Arg-Tyr-Gln-Lys-Ser-Thr-Glu-Leu-Leu-Ile-
50 60

-Arg-Lys-Leu-Pro-Phe-Gln-Arg-Leu-Val-Arg-Glu-Ile-Ala-Gln-Asp-Phe-Lys-Thr-Asp-Leu-Arg-
70 80

-Phe-Gln-Ser-Ser-Ala-Val-Met-Ala-Leu-Gln-Glu-Ala-Cys-Glu-Ala-Tyr-Leu-Val-Gly-Leu-Phe-
90 100

-Glu-Asp-Thr-Asn-Leu-Cys-Ala-Ile-His-Ala-Lys-Arg-Val-Thr-Ile-Met-Pro-Lys-Asp-Ile-Gln-
110 120

-Leu-Ala-Arg-Arg-Ile-Arg-Gly-Glu-Arg-Ala-COOH
130 135

Figure 4. Amino acid sequence of histone H3, indicating sites of acetylation at lysine residues 9 (which is also a site of methylation in some H3 molecules), 14, 18 and 23. Phosphorylation may occur at serine residues in positions 10 and 28. Lysine-27 is subject of methylation.

```
        P         Ac        Ac        Ac                  Ac              Me
        |          |         |         |                   |               |
       (Ac-Ser-Gly-Arg-Gly-Lys-Gly-Gly-Lys-Gly-Leu-Gly-Lys-Gly-Gly-Ala-Lys-Arg-His-Arg-Lys-
        1                             10                                  20

       -Val-Leu-Arg-Asp-Asn-Ile-Gln-Gly-Ile-Thr-Lys-Pro-Ala-Ile-Arg-Arg-Leu-Ala-Arg-Arg-Gly-
                              30                                  40

       -Gly-Val-Lys-Arg-Ile-Ser-Gly-Leu-Ile-Tyr-Glu-Glu-Thr-Arg-Gly-Val-Leu-Lys-Val-Phe-Leu-
                       50                                  60

       -Glu-Asn-Val-Ile-Arg-Asp-Ala-Val-Thr-Tyr-Thr-Glu-His-Ala-Lys-Arg-Lys-Thr-Val-Thr-Ala-
                       70                                  80

       -Met-Asp-Val-Val-Tyr-Ala-Leu-Lys-Arg-Gln-Gly-Arg-Thr-Leu-Tyr-Gly-Phe-Gly-Gly-COOH)
                       90                          100       102
```

Figure 5. Amino acid sequence of histone H4, indicating multiple sites of acetylation at lysine residues 5, 8, 12 and 16. The amino-terminal serine residue is stably acetylated and transiently phosphorylated. Lysine-20 is a site of methylation.

tissues (54). However, the chromatographic behavior of the acetyltransferases is complex and can be altered by the presence of associated histones; e.g., two chromatographically separable forms of histone acetyltransferase from rat liver can be interconverted by removal of histones H2A and H3 (57).

The major approach in previous work on the purification and properties of histone acetylating enzymes has been to concentrate on enzymes present in isolated cell nuclei, where most of the acetylation of lysine residues takes place. Acetyltransferases have been described in nuclei from calf (5,6) and rat (56) thymus, from pigeon liver (48), and from rat kidney (54), liver (50,51,53,54,57) and brain (53) cells. From the salt concentrations required for the extraction of acetyltransferases from isolated nuclei or chromatin fractions, one can surmise that much of the activity is tightly associated with the chromatin (56,57). This conclusion is supported by autoradiographic evidence that histone acetylation occurs along the polytene chromosomes of *Chironomus thummi* (58) and in the maternal (but not paternal) chromosomes of *Planococcus citrii* (59).

There are also good indications that at least one form of acetyltransferase occurs in the cytoplasm (24,29,60). This enzyme preferentially acetylates histone H4 and it appears to play a special role in the early processing of nascent H4 chains (24,29).

In considering the specificity of the histone acetyltransferases, it is significant that histone H1, which does contain an acetyl-N-serine residue at its amino-terminal, does not contain detectable amounts of epsilon-N-acetyllysine. All of the other histone classes (H2A, H2B, H3 and H4) are subject to enzymatic acetylation of their lysine residues, but differ in their sites of acetylation and in the surrounding amino acid sequences (Figures 2-5). From the variety of modification sites in different histones, and the obvious preferences of different acetyltransferases for particular histones as substrates, it can be anticipated that many more enzymes of this type will be discovered. What will be needed, apart from purification of such enzymes to homogeneity, is a careful analysis of their substrate specificity to define the type of histone being modified, and the particular lysine residues in the sequence of that histone which are subject to acetylation. In such studies, it will be helpful to employ natural or synthetic peptides of known amino acid sequence, such as have been employed in the analysis of the substrate specificities and mechanisms of action of histone de-acetylating enzymes (61-63).

Moreover, it is very likely that enzymes which modify

the structure of histones and other DNA-associated proteins
are under close coordinate physiological controls. The
mechanisms by which histone H4 acetylation, for example, is
suddenly increased during gene activation by mitogens (7,64)
and hormones (6,65-70), or selectively activated in one
chromosome set but not in another (59), are not understood,
but they are likely to involve allosteric and other regulatory modifications of the acetyltransferases, themselves, as
well as the action of inhibitory factors.

Histone Deacetylation: Enzymatic Aspects –

    Early experiments on the incorporation of radioactive
acetate into the histones of isolated cell nuclei established
that the isotopically-labeled acetyl groups, once incorporated, were not stable, but were subject to "turnover" without
degradation of the polypeptide chain to which they were
attached (1,5,6). The release of previously-incorporated
acetyl groups is catalyzed by enzymes that attack the amide
linkage between the epsilon-amino group of the modified
lysine residue and the carboxy-carbon of the acetyl group.
The release of isotopically-labeled acetic acid from histones
which had previously incorporated radioactive acetate in vivo
or in vitro is the basis for most of the assays for enzyme
activity (61-63,72-80). A variety of techniques for the
separation and recovery of the released isotopically-labeled
acetate have been described. Those which depend upon extraction of the acetic acid from an acidic aqueous phase into
ethyl acetate are not, in our experience, quantitative (63,
81). Alternative procedures, such as the separation of
released acetate from histones by exclusion chromatography
(63,81), or by adsorption of the histones to phosphocellulose
(80) or nitrocellulose (81) filters permit more accurate
assessments of histone deacetylase activity. The deacetylation of histones can also be monitored without recourse to
isotopic-labeling techniques, by using high-resolution polyacrylamide gel electrophoresis for the separation of histone
H4 (and H3) subfractions which differ in their degree of
acetylation (23,24,82). The loss of acetyl groups leads to
an increase in positive charge and electrophoretic mobility;
and the slow-moving (acetylated) histone bands diminish in
amount after treatment with deacetylases – with a corresponding increase in the fast-moving (unmodified) band. The
changing amounts of histone in the various acetylated and
non-acetylated subfractions can be followed by quantitative
densitometry of the histone bands in the stained gels (77,82).
    The specificity of the histone deacetylases has been
studied using intact histones, and histone fragments obtained

by proteolytic digestion, labeled with radioactive acetate. Early studies by Inoue and Fujimoto established that a calf thymus deacetylase could distinguish between the natural sites of histone modification and random sites acetylated by ^{14}C-acetic anhydride (72,73). Vidali et al. (77) have described the purification and properties of a histone deacetylase from calf thymus nuclei. The enzyme has a molecular weight of 150-160,000 and an optimum activity in the pH range 7.5-8.5. Its amino acid composition shows a predominance of aspartic and glutamic acids over the basic amino acids, lysine, arginine and histidine, in keeping with the low isoelectric point (pH 4.5) of the enzyme. An additional factor contributing to the acidic nature of the deacetylase is a high degree of phosphorylation; the purified enzyme contains 1.3% phosphorus by weight, mainly in the form of phosphoserine. The enzyme occurs mainly, if not exclusively, in the cell nucleus. When calf thymus nuclei are isolated in non-aqueous media under conditions which preclude an exchange of water-soluble components between the nucleus and the cytoplasm, the nuclear fraction retains virtually all of the deacetylase activity for the enzymatic hydrolysis of epsilon-N-acetyllysine residues in histones H3 and H4 (77).

One of the most interesting properties of this enzyme is its selectivity for histones H3 and H4; when the deacetylase is added to a mixture of calf thymus histones in solution, H3 and H4 are selectively precipitated and deacetylated, leaving the other histone classes (H1, H2A and H2B) in the supernatant.

The substrate-specificity requirements of the calf thymus deacetylase have been examined using ^{3}H-acetylated peptides derived from histone H4 by thermolysin digestion (61), and synthetic peptides prepared by the Merrifield solid-phase procedure (83) labeled with ^{3}H-acetate and ^{14}C-acetate at lysine residues 16 and 12, respectively (62,63, 81). By studying the release of [3]H- and ^{14}C-acetate from these well-defined substrates, it was concluded that:

1.   the entire histone molecule is not required for recognition by the deacetylase; the amino-terminal fragment corresponding to residues 1-37 of histone H4 is deacetylated as rapidly as is the intact histone H4 molecule.

2.   very small peptide sequences, such as residues 10-21 (obtained by thermolysin digestion (61)), or residues 15-21 (synthesized by the solid-phase method (62)), which contain a radioactive epsilon-N-acetyllysine residue at position-16 are not deacetylated by the thymus nuclear enzyme.

3.   doubly-labeled synthetic peptides corresponding to H4

sequence 1-37 and containing epsilon-N-^{3}H-acetyl-lysine at
position -16 and epsilon-N-^{14}C-acetyl-lysine at position-12
release ^{3}H-acetate and ^{14}C-acetate simultaneously and at comparable rates. This suggests that the deacetylase can attack
both of these sites with equal probability (63,81).

The specificity requirements for the deacetylation of
the various forms of histone H3 (Figure 4) have not yet been
analyzed in comparable detail. Comparisons of the known
sites of lysine acetylation in histones H3 and H4 (Figures 4
and 5) do not reveal any obvious clues as to the sequence
requirements for recognition by deacetylating enzymes. It
seems unlikely that such varied sites are all deacetylated
by a single enzyme, although a partially-purified thymus
nuclear deacetylase does remove acetyl groups from positions-
12 and -16 of a histone H4 sequence (81). However, such a
preparation may comprise a mixture of enzymes of similar
physical properties but differing specificities. Some
support for the view that histone deacetylases are complex
derives from observations that some activities are chromatographically separable (74,76), and that some preparations are
unable to deacetylate chromatin-bound histones while others
have this activity (76).

A further complication is that histone deacetylases
appear to have naturally-occurring inhibitors of their function (73,77). There are numerous indications that the rate
of deacetylation of histones is under physiological control.
It can be varied by hormonal stimulation of the appropriate
target tissues (49) and by exposure of granulocytes to plant
lectins (84). Deacetylation is suppressed in the course of
liver regeneration (43) and increased after treatment with
the hepatocarcinogen, Aflatoxin Bl (44). How such control is
achieved remains an open question.

Correlations between Histone Acetylation and Transcriptional
Activity -

The first experiments on histone acetylation included
evidence that the inhibition of RNA polymerase by histones
is progressively suppressed by increasing degrees of histone
acetylation (1,5,6), and it was surmised that histone
acetylation *in vivo* might influence both the structure and
function of chromatin (1). Early studies of gene activation
in lymphocytes treated with phytohemagglutinin confirmed
this view, in that the acetylation of histones was augmented
within minutes after exposure to the mitogen, before any
increase in the rate of RNA synthesis in the activated lymphocyte (7).

In the past decade there have been many other indications

that histone acetylation is enhanced at early stages of gene activation - as induced by hormones, drugs, mitogenic agents, polyamines, or other stimuli (16) - while acetylation is suppressed in inactive regions of the chromatin (1,5,6,59, 85). Much of the earlier evidence relating histone acetylation to nuclear activity in RNA synthesis has been reviewed (16,86), and the major arguments for a control function of histone acetylation can be restated and updated briefly:

1. histone acetylation is increased in a variety of target tissues responding to an appropriate hormonal stimulus by an increase in transcriptional activity. For example, histone acetylation precedes RNA synthesis in polycythemic mouse spleen cells stimulated by erythropoietin (65), and in hepatocytes during the induction of tyrosine aminotransferase by insulin (69). Acetylation of the arginine-rich histones (H3 and H4) is sharply increased in the uterus after estradiol-17β administration (66), in the liver after cortisol injections (87), in the aldosterone-stimulated kidney (67,68) and heart muscle cells (68). Such responses are not limited to higher organisms; the fungus *Achlya ambisexualis* responding to sexual steroid acetylates its histones prior to an increase in RNA-synthetic capacity (70).

2. Rates of histone acetylation are altered as cells in culture change their capacity for RNA synthesis. For example, when HeLa cell cultures are exposed to thymine riboside, RNA synthesis is depressed, but it resumes within 60 minutes after removal of the anti-metabolite. Acetylation of the arginine-rich histones increases sharply at that time (88). Similar effects were seen in minimal-deviation Reuber hepatoma cells (88).

3. The rate of acetylation of histones is increased, and the loss of previously incorporated acetyl groups is curtailed during the early stages of gene activation after partial hepatectomy. The acetylation of liver histones reaches a peak at 3 - 4 hours and then declines. This is a period of extensive gene activation during which DNA template activity for RNA synthesis increases dramatically, reaching a plateau at about 6 hours. The increase in net acetyl content of the histones appears to precede the increased capacity for transcription by 1 - 2 hours (43).

4. Another example of an increase in histone acetylation before an increase in RNA synthesis is provided by comparisons of the rates of histone acetylation and RNA synthesis in lymphocytes at different times after the addition of a mitogen, such as phytohemagglutinin (PHA). The response of

the lymphocyte to PHA involves the synthesis of "new" RNA molecules and subsequent increases in the rates of protein synthesis. This type of transformation may be regarded as a case of gene activation in which the chromatin alters its patterns of transcription to provide the products needed for subsequent growth and division. The kinetics of acetate incorporation into the histones of PHA-treated lymphocytes are profoundly altered early in transformation; within a few minutes after the addition of PHA to the culture medium, ^{14}C-acetate uptake into the arginine-rich histones increased sharply. (Acetylation was not affected by puromycin, indicating that it was not coupled to histone synthesis, but represented the post-synthetic modification of previously-existing histone molecules (7).) These changes preceded the increase in rate of ^{3}H-uridine incorporation into the RNA of the transformed lymphocyte (7). An important point is that these results, obtained on a mixed population of lymphocytes responding to PHA in an unsynchronized manner, were confirmed by autoradiography of single cells. Autoradiographically, ^{14}C-acetate uptake was observed in the nucleus within 15 minutes of PHA stimulation, whereas enhanced RNA synthesis was seen only after 3 - 6 hours in culture (89). Thus, it cannot be argued that histone acetylation happens in one set of cells and RNA synthesis in another.

The question arises as to the relationship between such changes in the acetyl content of the histones and changes in chromosomal structure. There is good evidence that the reactivity of DNA is increased during the earliest stages of lymphocyte transformation. Killander and Rigler have observed that the amount of acridine orange binding to DNA of PHA-treated lymphocytes increases rapidly in the first 30 minutes (90,91). An increased "availability" of DNA is also evident by enhanced binding of ^{3}H-actinomycin D (92). Of particular interest was the observation that a chemical acetylation of the proteins of control lymphocytes increases the acridine orange binding to DNA, but acetylation of the PHA-treated cells does not lead to any further increase in the DNA dye-binding capacity. The results are consistent with the view that enzymatic acetylation of the histones provides a mechanism for increasing the availability of DNA for transcription.

In considering the significance of histone acetylation, it is important to point out that acetylation is not, in itself, a sufficient cause for the induction of RNA synthesis at previously repressed gene loci. For example, the transformation of PHA-treated lymphocytes can be blocked by the addition of cortisol to the culture medium, and no increase in RNA synthesis is observed (64). Under these conditions,

an increase in histone acetylation still occurs (64), and the
DNA of the chromatin becomes more accessible to binding by
acridine orange or actinomycin D, yet there is no obvious
stimulation of transcription. The results suggest that
changing the physical state of the chromatin is merely a
prelude to other, more specific reactions which are needed to
initiate RNA synthesis at particular gene loci. In this
view, the acetylation of histones provides a mechanism for
"releasing" the DNA template- the first step in a complex
chain of events which must be set into motion to modify the
patterns of transcription in the cells of higher organisms.

5.   Increases in histone acetylation have also been noted in
cell transformation by viruses; e.g., in SV 40-transformed
WI-38 cells (93) and in embryonic kidney cells infected with
adenovirus 2 (94). Of particular interest is the high level
of acetylation of the histones associated with the viral
DNA. Histones H3 and H4 derived from transforming wild-type
polyoma and SV 40 virus particles are much more acetylated
than are the corresponding histones of the host cells. The
same histone fractions prepared from seven non-transforming
mutants of polyoma virus did not show a high degree of
acetylation (17).

6.   In comparisons of different cell types from the same
organ, histone acetylation is more active in the more
actively transcribing cells. For example, the acetylation
of brain histones H3 and H4 is much greater in isolated
neuronal nuclei than in nuclei from glial cells (95). In
addition, a positive correlation was found to exist between
the extent of acetylation of chromatin-bound histones and the
extent of chromatin-templated UMP incorporation by endogenous
RNA polymerases (95).

7.   Drugs which modify RNA synthesis also influence histone
acetylation. For example, RNA synthesis in cardiac muscle
cells is enhanced by the administration of polyamines (96).
The rate of acetylation of histone H4 was found to increase
by 200% within 5 minutes after the addition of spermine to
a perfusion medium. Moreover, drugs which inhibit polyamine
and RNA synthesis, such as methylglyoxal bis (guanylhydra-
zone), cause a decrease in the acetylation of the cardiac
histones (96). Spermine is able to reverse this inhibition.
        Lysergic acid diethylamide stimulates the acetylation
of the histones of rabbit cerebral hemispheres and midbrain
within 30 minutes after administration of the drug. A
stimulation of RNA synthesis in brain nuclei has been
detected 2½ hours after administration of LSD *in vivo* (97).
This is another example of the programming of acetylation

VINCENT G. ALLFREY

before an increase in RNA synthesis.

8. The acetylation of histones is altered during the development and differentiation of different cell types. Different stages in chick embryonic muscle development are associated with increases in the rate of histone acetylation (98) and with decreasing amounts of a histone deacetylase (99).

Very striking changes in histone structural modifications are seen during spermatogenesis in trout. These have been analyzed with exemplary precision, defining the nature of the histone, the sites of modification by acetyl or phosphoryl groups, and the stage in sperm differentiation (8-10, 42,100). Some of these changes are related to histone synthesis in the spermatocyte preparing for meiosis (8,100), while others may be involved in the removal of histones from the DNA of spermatids, and their replacement by protamines (9,100). The variety and extent of such modifications has suggested a model in which the degree of acetylation of the histones acts to control their binding to DNA in the assembly of chromatin strands (8,9).

A similar program of histone acetylation has been observed at different stages in spermatogenesis in the rat (101).

Overall levels of histone acetylation reflect the transcriptional activity of avian red cells at successive stages in maturation. The maturation of the nucleated erythrocyte involves a programmed series of nuclear and cytoplasmic events that eventually lead to an almost complete cessation of RNA synthesis. The average RNA synthetic capacity of isolated erythroblasts and early polychromatic erythrocytes is at least 8 times higher than that of mature erythrocytes (82). There is a comparable decline in the rate of histone acetylation (23) and a definite decline in the degree of acetylation of histones H3 and H4 with increasing age of the cell (82). In addition, the histone deacetylase activities are significantly higher in the mature erythrocyte than in immature reticulocytes (46). However, as in the lymphocyte, the acetylation of histones and the act of transcription are not tightly coupled; inhibitors of RNA synthesis, such as Rifamycin AF/013 and Actinomycin D, do not simultaneously block ^{3}H-acetate incorporation into the histones of avian erythroid cells (23).

Correlations between histone acetylation and transcription become obscured in dividing cells engaged in histone synthesis, partly because acetylation of amino-terminal serine residues takes place at that time (22). Histone acetylation varies during the cell cycle of synchronized

mammalian cells (102,103) and some of it represents the terminal acetylation of newly-synthesized histone chains (104).
Acetylation of the terminal serine residues of histones Hl,
H2A and H4 (24) and of the erythrocyte-specific histone H5
(24,105) is restricted to the dividing erythroblast, However, there also appears to be a number of <u>internal</u> modifications of the nascent chains of histone H4 (24,60). The
nature of these modifications has been carefully analyzed in
the avian erythroblast where there are three distinct cytoplasmic modifications of nascent H4 chains; these include
acetylation and phosphorylation of the terminal serine residue, and acetylation of a lysine residue (24). The acetate
attached to the serine residue is stable, but the other
modifications are transient, and are removed within minutes
after the newly-synthesized histone enters the cell nucleus.
It is a matter of some interest that, in erythroblasts, such
cytoplasmic modifications involving serine phosphorylation
and lysine acetylation are limited to histone H4 (24), and
it has been suggested that the modified H4 molecules may be
involved in a "nucleation" step during the assembly of
chromatin particles. Intranuclear enzymes then further
alter the level of acetylation of the polypeptide chain (24,
60).

9.   In non-dividing cells (where the above complications are
absent, or less complicating), the acetylation of histone
lysine residues appears to be restricted to the template-
active regions of the chromatin. For example, subfractions
of thymus lymphocyte chromatin which differ in transcriptional activity *in vivo* and *in vitro* have been compared with
regard to their content of radioactive histones after incorporation of isotopic acetate; the histones of the transcriptionally-active fraction were much more radioactive than
those of the relatively inert heterochromatin fraction (5,
85). More recent studies extend the spatial correlation
between acetylation and sites of transcription. For example,
in the mealy bug, *Planococcus citri,* males preferentially
utilize the maternal chromosome set and sequester the paternal chromosomes in a heterochromatic mass. The maternal,
genetically-active euchromatic chromosome set incorporates
about seven times more ^{3}H-acetate than does the heterochromatic, largely-inactive paternal set (59). Since this type
of acetylation was not suppressed significantly by puromycin,
it can be concluded that epsilon-N-acetyllysine formation was
probably involved (59).

A remarkable correlation has been noted in *Tetrahymena
pyriformis,* where histones have been compared from the macro-
and micro-nuclei (106). The transcriptionally-active macro-

nucleus contains acetylated forms of histone H4, while the
genetically less-active micro-nucleus contains the corresponding histone in its non-acetylated form (106). The
absence of the acetylated forms of histone H4 in the inert
micronucleus of *Tetrahymena* is reminiscent of other observations that the histones of mature *Arbacia* sperm (cells which
are incapable of RNA synthesis) occur entirely in their non-
acetylated forms (45,107). Resumption of RNA synthesis in
the *Arbacia* embryo is accompanied by a reappearance of the
acetylated forms of histones H3 and H4 (45). There are other
instances of a parallel suppression of histone acetylation
and RNA synthesis. For example, the administration of the
potent hepatocarcinogen, Aflatoxin B1, leads to a sudden
increase in the rate at which acetyl groups are released from
histones in the liver. Aflatoxin B1 is known to suppress
DNA-dependent RNA polymerase activities of the liver within
15 minutes after administration. In that time a large proportion of previously-incorporated acetyl groups is lost (44).
The level of histone acetylation rises slowly after reaching
a minimum at 15-30 minutes, and the synthesis of RNA slowly
recovers.

Histone Acetylation and the Conformation of Histone-DNA Complexes

All of the previous correlations suggest that histone
acetylation modifies the structure and functional activity of
the chromatin, but they do not prove that a cause-and-effect
relationship exists. In fact, the general observation that
changes in histone acetylation precede changes in transcriptional activity argues strongly for a structural role of this
modification. Direct tests of the hypothesis that histone
acetylation regulates chromatin structure have become possible
due to the development of methods for the bulk separation of
histone subfractions which differ in their degree of acetylation (45). The separation of otherwise identical polypeptide
chains differing in their content of epsilon-N-acetyllysine
is made possible by the net decrease in positive charge of
the histone molecule as the free amino groups on lysine
residues are neutralized by acetylation. The charge difference is sufficient to allow separation of the modified and
unmodified histone molecules by ion-exchange chromatography
(45) and by gel electrophoretic techniques (e.g., 23,45).
Histone H4 from calf thymus nuclei has been separated into
its naturally-occurring non-acetylated, mono-acetylated, and
di-acetylated subfractions (45). Each of these subfractions
was compared with regard to its interactions with DNA, using
circular dichroism to monitor changes in DNA conformation.

186

The results indicate that the mono-acetylated form of histone H4 is far less effective than the non-acetylated form in producing conformational distortions of DNA, although both forms bind to the nucleic acid (108). This is the first direct evidence that the acetylation of a histone can affect its interactions with the DNA helix. However, it should be pointed out that the interaction of histone H4 with DNA *in vivo* involves a complex set of interactions between pairs of four different histones (H2A, H2B, H3 and H4) with each other as well as with DNA – to form the particles which appear in chromatin as nucleosomes (nu bodies) (109). The role of histone acetylation in the asembly of such particles and in the release of histone complexes from DNA remains to be investigated. It should be stressed, however, that the acetylation of the nucleosomal histones involves those regions of the polypeptide chain which are most likely to react with the enveloping DNA strand (110).

REFERENCES

1. Allfrey, V. G., Faulkner, R., and Mirsky, A. E., *Proc. Nat. Acad. Sci. U.S.A. 51,* 786 (1964).
2. Murray, K., *Biochemistry 3,* 10 (1966).
3. Kleinsmith, L. J., Allfrey, V. G., and Mirsky, A. E., *Proc. Nat. Acad. Sci. U.S.A. 55,* 1182 (1966).
4. Ord, M. G., and Stocken, L. A., *Biochem. J. 98,* 888 (1966)
5. Allfrey, V. G., *Can. Cancer Conf. 6,* 313 (1964).
6. Allfrey, V. G., *Cancer Res. 26,* 2026 (1966).
7. Pogo, B. G. T., Allfrey, V. G., and Mirsky, A. E., *Proc. Nat. Acad. Sci. U.S.A. 55,* 805 (1966).
8. Sung, M. T., and Dixon, G. H., *Proc. Nat. Acad. Sci. U.S.A. 67,* 1616 (1970).
9. Louie, A. J., Candido, E. P. M., and Dixon, G. H., *Cold Spring Harbor Symp. Quant. Biol. 38,* 803 (1974).
10. Louie, A. J., and Dixon, G. H., *Nature New Biol. 243,* 164 (1973).
11. Langan, T. A., *Proc. Nat. Acad. Sci. U.S.A. 64,* 1276 (1969).
12. Adler, A. J., Langan, T. A., and Fasman, G. D., *Arch. Biochem. Biophys. 153,* 769 (1972).
13. Balhorn, R., Bordwell, J., Sellers, L., Granner, D., and Chalkley, R., *Biochem. Biophys. Res. Commun. 46,* 1326 (1972).
14. Marks, D. B., Paik, W. K., and Borun, T. W., *J. Biol. Chem. 248,* 5660 (1973).
15. Bradbury, E. M., Inglis, R. J., and Matthews, H. R., *Nature 247,* 257 (1974).

16. Allfrey, V. G., in *Histones and Nucleohistones*, Phillips, D. M. P., ed., Plenum Publishing Company, Ltd., London, pp. 241-294 (1971).
17. Schaffhausen, B. S., and Benjamin, T. L., *Proc. Nat. Acad. Sci. U.S.A. 73,* 1092 (1976).
18. Phillips, D. M. P., *Biochem. J. 87,* 258 (1963).
19. Phillips, D. M. P., *Biochem. J. 107,* 135 (1968).
20. Gershey, E. L., Vidali, G., and Allfrey, V. G., *J. Biol. Chem. 243,* 5018 (1968).
21. Vidali, G., Gershey, E. L., and Allfrey, V. G., *J. Biol. Chem. 243,* 6361 (1968).
22. Liew, C. C., Haslett, G. W., and Allfrey, V. G., *Nature 226,* 414 (1970).
23. Ruiz-Carrillo, A., Wangh, L. J., and Allfrey, V. G., *Arch. Biochem. Biophys. 174,* 273 (1976).
24. Ruiz-Carrillo, A., Wangh, L. J., and Allfrey, V. G., *Science 190,* 117 (1975).
25. Marzluff, W. F., and McCarty, K. S., *J. Biol. Chem. 245,* 5635 (1970).
26. Keskes, E., Sures, I., and Gallwitz, D., *Biochemistry 15,* 2541 (1976).
27. Jacobs-Lorena, M., and Baglioni, C., *Mol. Biol. Rep. 1,* 113 (1973).
28. Pestanna, A., Sudilovsky, O., and Pitot, H. C., *FEBS Lett. 19,* 83 (1971).
29. Horiuchi, K., and Fujimoto, D., *J. Biochem. (Tokyo) 72,* 433 (1972).
30. Kaji, H., *Biochemistry 7,* 3844 (1968).
31. Leibowitz, M. J., and Soffer, R. L., *J. Biol. Chem. 246,* 5207 (1971).
32. Soffer, R. L., *J. Biol. Chem. 248,* 8424 (1973).
33. DeLange, R. J., Fambrough, D. M., Smith, E. L., and Bonner, J., *J. Biol. Chem. 244,* 319 (1969).
34. DeLange, R. J., Fambrough, D. M., Smith, E. L., and Bonner, J., *J. Biol. Chem. 244,* 5669 (1969).
35. Ogawa, Y., Quagliarotti, G., Jordan, J., Taylor, C. W., Starbuck, W. C., and Busch, H., *J. Biol. Chem. 244,* 4387 (1969).
36. Candido, E. P. M., and Dixon, G. H., *J. Biol. Chem. 246,* 3182 (1971).
37. Sautiere, P., Lambelin-Breynaert, M.-D., Moschetto, Y., and Biserte, G., *Biochimie 53,* 711 (1971).
38. Wilson, R. K., Starbuck, W. C., Taylor, C. W., Jordan, J., and Busch, H., *Cancer Res. 30,* 2942 (1970).
39. Candido, E. P. M., and Dixon, G. H., *Proc. Nat. Acad. Sci. U.S.A. 69,* 2015 (1972).
40. DeLange, R. J., Hooper, J. A., and Smith, E. L., *J. Biol. Chem. 248,* 3275 (1973).

41. Hooper, J. A., Smith, E. L., Sommer, K. R., and Chalkley, R., *J. Biol. Chem. 248,* 3275 (1973).
42. Candido, E. P. M., and Dixon, G. H., *J. Biol. Chem. 247,* 3868 (1972).
43. Pogo, B. G. T., Pogo, A. O., Allfrey, V. G., and Mirsky, A. E., *Proc. Nat. Acad. Sci. U.S.A. 59,* 1137 (1968).
44. Edwards, G. S., and Allfrey, V. G., *Biochim. Biophys. Acta 299,* 354 (1973).
45. Wangh, L. J., Ruiz-Carrillo, A., and Allfrey, V. G., *Arch. Biochem. Biophys. 150,* 44 (1972).
46. Sanders, L. A., Schechter, N. M., and McCarty, K. S., *Biochemistry 12,* 783 (1970).
47. Pathy, L., and Smith, E. L., *J. Biol. Chem. 250,* 1919 (1975).
48. Nohara, H., Takahashi, T., and Ogata, K., *Biochim. Biophys. Acta 127,* 282 (1966).
49. Libby, P. R., *Biochem. Biophys. Res. Commun. 31,* 59 (1968).
50. Gallwitz, D., *Biochem. Biophys. Res. Commun. 32,* 117 (1968).
51. Gallwitz, D., and Sekeris, C. E., *Hoppe-Seyler's Z. Physiol. Chem. 350,* 150 (1969).
52. Gallwitz, D., *Biochem. Biophys. Res. Commun. 40,* 236 (1970).
53. Bondy, S. C., Roberts, S., and Morelos, S., *Biochem. J. 119,* 665 (1970).
54. Gallwitz, D., *FEBS Lett. 13,* 306 (1971).
55. Racey, L. A., and Byvoet, P., *Exp. Cell Res. 64,* 366 (1971).
56. Gallwitz, D., and Sures, I., *Biochim. Biophys. Acta 263,* 315 (1972).
57. Lue, P. F., Gornall, A. G., and Liew, C. C., *Can. J. Biochem. 51,* 1177 (1973).
58. Allfrey, V. G., Pogo, B. G. T., Littau, V. C., Gershey, E. L., and Mirsky, A. E., *Science 159,* 314 (1968).
59. Berlowitz, L., and Pallotta, D., *Exp. Cell Res. 71,* 45 (1972).
60. Jackson, V., Shires, A., Tanphaichitr, N., and Chalkley, R., *J. Mol. Biol. 104,* 471 (1976).
61. Horiuchi, K., and Fujimoto, D., *J. Biochem. (Japan) 73,* 117 (1973).
62. Krieger, D. E., Levine, R. B., Merrifield, R. B., Vidali, G., and Allfrey, V. G., *J. Biol. Chem. 249,* 322 (1974).
63. Krieger, D., Doctoral Thesis, The Rockefeller University, New York (1976).
64. Ono, T., Terayama, H., Takaku, F., and Nakao, K., *Biochim. Biophys. Acta 179,* 214 (1969).

65. Takaku, F., Nakao, K., Ono, T., and Terayama, H., *Biochim. Biophys. Acta 195*, 396 (1969).
66. Libby, P. R., *Biochem. J. 130*, 663 (1972).
67. Libby, P. R., *Biochem. J. 134*, 907 (1973).
68. Liew, C. C., Suria, D., and Gornall, A. G., *Endocrinol. 93*, 1025 (1973).
69. DeVilliers-Graaff, G., and Von Holt, C., *Biochim. Biophys. Acta 299*, 480 (1973).
70. Horgen, P. A., and Ball, S. F., *Cytobios 10*, 181 (1974).
71. Marzluff, W. F., Sanders, L. A., Miller, D. M., and McCarty, K. S., *J. Biol. Chem. 247*, 2026 (1972).
72. Inoue, A., and Fujimoto, D., *Biochem. Biophys. Res. Commun. 36*, 146 (1969).
73. Inoue, A., and Fujimoto, D., *Biochim. Biophys. Acta 220*, 307 (1970).
74. Inoue, A., and Fujimoto, D., *J. Biochem. (Japan) 72*, 427 (1972).
75. Fujimoto, D., and Segawa, K., *FEBS Lett. 32*, 59 (1973).
76. Kaneta, H., and Fujimoto, D., *J. Biochem. (Japan) 76*, 905 (1974).
77. Vidali, G., Boffa, L. C., and Allfrey, V. G., *J. Biol. Chem. 247*, 7365 (1972).
78. Boffa, L. C., Gershey, E. L., and Vidali, G., *Biochim. Biophys. Acta 254*, 135 (1971).
79. Libby, P. R., *Biochim. Biophys. Acta 213*, 234 (1970).
80. Horiuchi, K., and Fujimoto, D., *Anal. Biochem. 69*, 491 (1975).
81. Krieger, D. E., Merrifield, R. B., Vidali, G., and Allfrey, V. G., (Manuscript in preparation, 1976).
82. Ruiz-Carrillo, A., Wangh, L. J., Littau, V. C., and Allfrey, V. G., *J. Biol. Chem. 249*, 7358 (1974).
83. Merrifield, R. B., *Advan. Enzymol. 32*, 221 (1969).
84. Pogo, B. G. T., Allfrey, V. G., and Mirsky, A. E., *J. Cell Biol. 35*, 477 (1967).
85. Allfrey, V. G., *Fed. Proc. (Symp.) 29*, 1447 (1970).
86. Hnilica, L. S., in *The Structure and Biological Functions of Histones*, CRC Press, Cleveland, Ohio, 79 (1972).
87. Allfrey, V. G., Pogo, B. G. T., Pogo, A. O., Kleinsmith, L. J., and Mirsky, A. E., *CIBA Foundation Study Group 24*, 42 (1966).
88. Wilhelm, J. A., and McCarty, K. S., *Proc. Amer. Assoc. Cancer Res. 9*, 77 (1968).
89. Mukherjee, A. B., and Cohen, M. M., *Exp. Cell Res. 54*, 257 (1968).
90. Killander, D., and Rigler, R., *Exp. Cell Res. 39*, 710 (1965).

91. Killander, D., and Rigler, R., *Exp. Cell Res. 54,* 163 (1969).
92. Darzynkiewicz, Z., Bolund, L., and Ringertz, N. P., *Exp. Cell Res. 55,* 120 (1969).
93. Krause, M. V., and Stein, G. S., *Exp. Cell Res. 92,* 175 (1975).
94. Ledinko, N., *J. Virol. 6,* 58 (1970).
95. Sarkander, H. I., Fleischer-Lambropoulos, H., and Brade, W. P., *FEBS Lett. 52,* 40 (1975).
96. Caldarera, C. M., Casti, A., Guarnieri, C., and Moruzzi, G., *Biochem. J. 152,* 91 (1975).
97. Brown, I. R., and Liew, C. C., *Science 188,* 1122 (1975).
98. Boffa, L. C., and Vidali, G., *Biochim. Biophys. Acta* 236, 259 (1971).
99. Boffa, L. C., Gershey, E. I., and Vidali, G., *Biochim. Biophys. Acta 254,* 135 (1971).
100. Candido, E. P. M., and Dixon, G. H., *J. Biol. Chem. 247,* 5506 (1972).
101. Grimes, S. R., Chae, C.-B., and Irvin, J. L., *Arch. Biochem. Biophys. 168,* 425 (1975).
102. Shepherd, G. R., Noland, B. J., and Hardin, J. M., *Biochim. Biophys. Acta 228,* 544 (1971).
103. Shepherd, G. R., *Biochim. Biophys. Acta 299,* 485 (1973).
104. Jackson, V., Shires, A., Chalkley, R., and Granner, D. K., *J. Biol. Chem. 250,* 4856 (1975).
105. Tobin, R. S., and Seligy, V. L., *J. Biol. Chem. 250,* 358 (1975).
106. Gorovsky, M. A., Pleger, G. L., Keevert, J. B., and Johmann, C. S., *J. Cell Biol. 57,* 773 (1973).
107. Easton, D., and Chalkley, R., *Exp. Cell Res. 72,* 502 (1972).
108. Adler, A. J., Fasman, G. D., Wangh, L. J., and Allfrey, V. G., *J. Biol. Chem. 249,* 2911 (1974).
109. Allfrey, V. G., Bautz, E. K. F., McCarty, B. J., Schimke, R. T., and Tissieres, A., eds. *The Organization and Expression of Chromosomes,* Berlin: Dahlem Konferenzen, 1976 (in press).
110. Baldwin, J. P., Boseley, P. G., Bradbury, E. M., and Ibel, K., *Nature 253,* 245 (1975).

Chapter 6

NUCLEAR NONHISTONE PROTEINS: CHEMISTRY AND FUNCTION

JEN-FU CHIU and LUBOMIR S. HNILICA

Department of Biochemistry
Vanderbilt University School of Medicine
Nashville, Tennessee 37232

I.   Introduction

Although all somatic cells contain the same complement
of genes, transcriptional specialization takes place during
cellular differentiation which selects the expression of
genetic information in specialized cells; e.g., only erythrocytes synthesize hemoglobin and only liver cells synthesize serum albumin. The variable gene activity concept may
explain why during embryogenesis and differentiation certain
genes function only at certain times and in particular tissues. In some way, the non-functional portion of the genome
must be prevented from expressing its genetic information.
Considerable evidence in the literature indicates that chromosomal proteins, i.e., both the histones and nonhistone
proteins participate in the regulation of gene activity.

Since the original proposal by Stedman and Stedman (1)
that histones might act as gene repressors, numerous experiments were performed to support this possibility. It was
shown by Allfrey *et al.* (2) that removal of histones from
chromatin will greatly increase its capacity to template for
in vitro RNA synthesis. Conversely, addition of histones to
free DNA will inhibit its transcription (3). However, only
minor differences have been found between histones from different cell types, or from the same tissue types of different
species. Thus, histones do not exhibit the heterogeneity or
tissue specificity which would be expected of protein involved
in regulating the expression of specific genes. Their relative constancy between diverse cell types and under different
physiological conditions has led to the general conclusion
that histones are not involved in regulating the activity of
specific genes (review, see Ref. 4).

On the other hand, nonhistone proteins from various
sources were found to be considerably heterogenous and variable, to bind selectively to homologous DNA and to stimulate

193

the transcription by RNA polymerase of isolated DNA or chromatin. Accumulated evidence, supported by chromatin reconstitution experiments, suggests an important role of the
chromosomal nonhistone proteins in selective regulation of
gene transcription (review, see Ref. 4-7). In this chapter,
we would like to review the biochemistry and possible functions of nuclear nonhistone proteins.

II. Isolation and Fractionation

Mirsky and Pollister (8) were first to report the presence of acidic nonhistone proteins in chromatin and their
solubilization with 1 M NaCl from isolated nuclei. The nonhistone proteins present in this extract could be partially
separated from DNA and histones by precipitation of the
deoxynucleohistone with 0.14 M NaCl. Early attempts to isolate and fractionate chromosomal nonhistone proteins were
described by Wang (9,10), who analyzed the nuclear residue
after extraction with 1.0 M NaCl. This residue contained
nonhistone proteins which could be solubilized in 2% deoxycholate and then further fractionated by various pH and
ammonium sulfate fractionations.

Bekhor, Kung and Bonner (11) introduced the use of 2.0 M
NaCl containing 5.0 M Urea for dissociation and reconstitution of chromatin. At pH 8.0, the dissociation of chromatin
was almost complete, and only a small amount of protein
remained with the DNA. Since then, a variety of methods for
chromatin protein isolation and analysis were developed.
After removal of histones from chromatin with acid, nonhistones can be separated from DNA by DNase digestion (12),
alkaline (13), salt (14), detergent (15,16) or phenol extraction (17,18,19). Chromatin can also be dissociated by sodium
dodecyl sulfate (20) or by a variety of salt solutions both
in the presence or absence of urea (21-31). The dissociated
proteins can then be further fractionated by chromatography
on QAE-sephadex A-50 (32,33), Bio-Rex 70 (23,34) or DEAE
cellulose (23).

If the biological activity is expected to remain with
the isolated chromosomal fractions, it is very important to
avoid the use of denaturing agents. Unfortunately, most of
the available methods for isolation and fractionation of
chromosomal proteins do not meet all these criteria. Thus,
harsh conditions involving acids, alkali, organic solvents,
or ionic detergents are suitable almost exclusively for analytical purposes only. Of the many modifications and vari-
ants in methodology for isolation and fractionation of
chromosomal nonhistone proteins only several can be discussed

here. A more detailed review was published in a book on "Acidic Proteins of the Nucleus", edited by Cameron and Jeter (35).

One of the techniques successfully employed by many investigators was developed by Teng et al. (18). It takes advantage of the solubility of many nonhistone proteins in buffered phenol. Isolated chromatin is first dehistonized by extraction with 0.25 M Tris HCl buffer, pH 8.4, containing 10 mM EDTA and 0.14 M 2-mercaptoethanol. The suspension is mixed with phenol saturated with the above buffer and extracted for several hours. Chromosomal proteins (mostly phosphoproteins) solubilized in the phenolic phase can be recovered by extensive dialysis. An additional fraction of highly phosphorylated nonhistone proteins, different from the phenol soluble fraction can be isolated from the phenol-buffer interface (36).

MacGillivray et al. (37,38) reported that histones, nonhistone proteins, and nucleic acids of chromatin dissociated in 2 M NaCl - 5 M urea - 1 mM sodium phosphate (pH 6.8) can be separated by column chromatography on hydroxylapatite. Differential elution of various components is accomplished by stepwise increase of phosphate buffer concentration in 2 M NaCl - 5 M urea elution solvent. In 1 mM phosphate starting buffer, histones do not adsorb to hydroxylapatite and are eluted in the breakthrough fraction. The nonhistone protein fractions are then eluted with 0.05, 0.1 and 0.2 M phosphate in elution buffer. All the applied DNA can be eluted with 0.5 M phosphate buffer. By polyacrylamide gel electrophoresis, all three nonhistone protein fractions are heterogeneous and differ significantly in their electrophoretic patterns. We have combined the methods of Chiu et al. (31) and MacGillivray et al. (37,38) to separate chromatin into four main fractions. The chromatin preparations are homogenized in 5.0 M urea-50 mM sodium phosphate buffer, pH 7.6, and stirred in an ice bath for 2 hrs. The homogenate is then centrifuged at 15,000 x g for one hour. The supernatant contains over 90% of the chromatin nonhistone proteins (designated as UP fraction). The sediment is then further fractionated on hydroxylapatite according to MacGillivray et al. (37,38). After elution of the unretained histone fraction (fraction HP) with 2 M NaCl-5 M urea-1 mM sodium phosphate, pH 6.8, a small nonhistone protein fraction (less than 5% of total chromatin protein) designated NP is eluted with 0.05 M sodium phosphate, pH 6.8, in 2 M NaCl-5 M urea at 2-4°C. Finally, DNA with a very small amount of associated proteins is eluted with 0.5 M sodium phosphate buffer, pH 6.8-2 M NaCl-5 M urea solution at 25°.

It is reasonable to expect that chromosomal proteins which have gene regulatory capacity or affect the DNA enzymatically or structurally, will exhibit affinity for this macromolecule. DNA affinity chromatography of nonhistone proteins was first developed by Alberts and associates (39) who were able to isolated specific DNA binding proteins from *E. coli* by passing DNA-free protein extracts through columns consisting of DNA absorbed onto cellulose. This technique was applied to eukaryotic systems in an attempt to fractionate chromosomal nonhistone proteins on cellulose, polyacrylamide, or agarose columns containing homologous or heterologous DNA. Using this technique, Kleinsmith *et al.* (40) isolated a small fraction (0.01%) of chromatin phosphoproteins which bound specifically to homologous DNA at physiological ionic strength. This technique also enabled Wakabayashi *et al.* (41) to isolated a fraction immunospecific nonhistone proteins from rat liver and various hepatomas which could only bind to homologous DNA.

Van den Broek (42) fractionated rat liver total chromosomal nonhistone proteins on heterologous and homologous DNA-cellulose columns. At first, proteins were applied to *E. coli* DNA-cellulose and then to rat liver DNA-cellulose columns. Only 3.9% of the total chromosomal nonhistone proteins could bind selectively to homologous DNA. This rat liver DNA-binding protein group was electrophoretically heterogenous and contained intermediate and low molecular weight proteins (below 66,000).

Recently, Allfrey *et al.* (43) fractionated calf thymus chromosomal nonhistone proteins on affinity columns containing DNA fractionated to different C_ot values. The calf thymus DNA fragments (200-400 base pairs) were covalently linked to aminoethyl-Sepharose 4B. Some proteins were found to bind selectively either to "unique" or to "repetitive" DNA sequences, as was indicated by their elution patterns from parallel columns containing DNA fractions with high, intermediate and low C_ot values. Differences were also noticed in the selective binding of some of these proteins to single-stranded or double-stranded DNA fractions of the same C_ot values. Since the elution of all the principal fractions was achieved by a stepwise increase of NaCl concentration, this affinity fractionation is based, at least to some extent, on electrostatic interactions between the DNA and nonhistone proteins.

Since chromatin nonhistone proteins are very heterogenous (perhaps hundreds of individual species), they may have many biological functions. Presently, there are no methods available which would offer a complete separation of all biologically active proteins (a standard procedure). Techniques taking advantage of particular biological properties of chromosomal nonhistone proteins (enzymatic activity, transcriptional regulation, immunological specificity, etc.) are most useful and probably indicate the future direction of this rapidly moving field.

III. Composition and Heterogeneity

Nuclear nonhistone proteins are also frequently called nuclear acidic proteins. This indicates that their amino acid composition, in contrast to the histones, shows an excess of dicarboxylic amino acid (glutamic acid and aspartic acid) over the basic amino acids (arginine, lysine and histidine). The ratio of acidic to basic amino acids of many chromosomal nonhistone proteins from various sources is over 1.0 (Table I, calculated without the consideration of amides). However, one group of nonhistone proteins recently isolated by Johns and his associates (44) from calf thymus chromatin contains proteins which have the ratio of acidic over basic amino acids about 0.8. Similarly, Chiu et al. (45) described a nonhistone protein fraction in rat liver chromatin, which ratio of acidic over basic amino acids was 0.82. Since some nonhistone proteins are slightly basic and since very little information is available on the contents of asparagine and glutamine in the individual "acidic" nonhistone protein fractions, the term "nonhistone proteins" appears more appropriate than "acidic proteins" and is exclusively used in this review.

Because chromosomal nonhistone proteins tend to aggregate and form complexes with histones and nucleic acids, their electrophoretic and chromatographic resolution is extremely difficult. The introduction of sodium dodecyl sulfate permitted the dissociation of nonhistone proteins and their electrophoretic resolution is sodium dodecyl sulfate containing polyacrylamide gels (14,16,17,29,38,46). As determined from their electrophoretic migration in sodium dodecyl sulfate containing polyacrylamide gels, the nonhistone proteins exhibit considerable heterogeneity (Fig. 1). They range in molecular weights from 5,000 to over 15,000 daltons. The exact number of individual nonhistone protein species is difficult to determine electrophoretically, but several lines of evidence suggest that there may be well over 100 nonhistone proteins in eukaryotic nucleus (47). The extent of chromosomal nonhistone protein heterogeneity was further emphasized by the introduction of two-dimensional

Table I. Amino acid composition of nonhistone proteins isolated from various sources.

Tissue	Rat Liver	Rat Liver	Rat Liver	Rat Liver	Mouse Liver	Calf Thymus	Ehrlich Ascites
Amino acid							
Lysine	6.0	6.3	7.0	7.8	7.7	18.5	6.7
Histidine	2.5	1.5	1.4	1.9	2.2	1.1	1.2
Arginine	5.9	5.6	5.5	8.0	8.3	4.1	7.6
Aspartic acid	10.3	9.5	9.8	7.5	8.0	7.7	9.1
Threonine	3.8	5.5	6.0	4.7	4.6	4.1	4.9
Serine	4.4	7.5	6.2	6.8	6.1	7.5	7.5
Glutamic acid	12.2	12.4	14.5	12.3	13.6	11.7	12.9
Proline	6.5	5.0	4.1	5.0	3.6	6.8	7.3
Glycine	15.3	7.5	7.0	10.7	9.9	7.1	11.6
Alanine	6.7	7.7	9.9	8.8	8.0	14.2	10.8
Half-cystine	-	1.3	1.1	-	1.9	0.6	-
Valine	6.1	5.6	6.2	5.6	5.8	4.4	7.7
Methionine	1.9	1.9	1.7	1.4	1.9	0.0	-
Isoleucine	4.1	4.3	5.4	4.2	4.5	1.9	2.8
Leucine	6.8	9.9	9.8	8.8	8.3	5.1	5.1
Tyrosine	3.3	2.6	1.8	3.7	2.6	1.6	1.3
Phenylalanine	4.0	3.9	2.7	2.8	3.2	2.7	3.5
Tryptophan	-	2.0	-	-	-	-	-
Acidic/basic	1.56	1.6	1.8	1.1	1.2	0.8	1.42
Reference	18	132	136	30	38	143	218

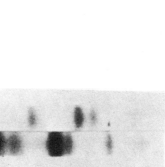

Figure 1. Nonhistone proteins of (1) calf thymus and (2) rat liver chromatins extracted with 5M urea–50 mM phosphate buffer, pH 7.6. Electrophoresis was performed in the presence of 0.1% SDS, Coomassie Brilliant Blue Staining. The origin of migration is at the top of the gels.

polyacrylamide gel electrophoresis. By this technique, Busch and associates separated chromosomal nonhistone proteins from rat liver and Novikoff hepatoma into at least 69 and 84 components, respectively (48).

The heterogenous pattern of chromosomal nonhistone proteins may to some extent depend on the concentration of detergent used for their extraction. For example, Marushige et al. (15) isolated a nonhistone protein fraction from rat liver chromatin and performed sedimentation equilibrium experiments to further characterize this fraction and to investigate the effects of sodium dodecyl sulfate on this protein. In 0.1% sodium dodecyl sulfate, its average molecular weight was 14,300 daltons; partial removal of sodium dodecyl sulfate caused this protein to aggregate and reach extreme molecular weight values in excess of 100,000 daltons. The number and type of nonhistone proteins can also vary with the method of their isolation and fractionation (49). Since the electrophoretic separation in sodium dodecyl sulfate-acrylamide gel systems is principally based on the molecular size of the analyzed proteins rather than their changes, polypeptides of similar electrophoretic mobility may contain any number of protein species of identical or very similar molecular weight. Conversely, since sodium dodecyl sulfate is a strong dissociating agent, several polypeptide bands on the gel may represent individual subunits of one native protein. Therefore, great caution must be exerted in interpretation of the heterogenous patterns obtained for nuclear nonhistones analyzed by gel electrophoresis or gel filtration in the presence of detergents, especially if the analysis was performed at rather low detergent concentrations.

The considerable heterogeneity of nonhistone proteins is at least partially contributed by numerous nuclear enzymes, such as DNA polymerases and ligases involved in DNA synthesis and repair (50-52), RNA polymerases active in RNA transcription (53), proteases involved in the processing and degradation of proteins (54-56) and enzymes involved in the modification of nucleic acids and proteins (methylases, acetylases, phosphokinases, demethylases, deacetylases, and phosphatases, etc.). In addition to these enzymatic components, the nonhistone protein fraction is also thought to contain proteins that have structural functions. Other nuclear proteins, probably present only in minute quantities, may function in gene regulatory capacity.

IV. Tissue and Species Specificity

Benjamin and Gellhorn (14) extracted nonhistone proteins

from rat and mouse liver nuclei with 4 M CsCl buffered to pH 11.6. They compared electrophoretic patterns of these proteins and found them similar. On the other hand, David and Burdman (57) working with rat brain and liver nuclei reported considerable electrophoretic tissue speicificity of chromosomal nonhistone proteins from these two tissues. More recently, Elgin and Bonner (16) investigated in detail the tissue and species specificity of nonhistone proteins in purified chromatin. Using sodium dodecyl sulfate polyacrylamide gel electrophoresis, they were able to resolve rat liver chromatin nonhistone proteins into thirteen major bands with molecular weights ranging approximately from 5,000 to 50,000 daltons. Corresponding peptide bands were found in electrophoretic patterns of chicken liver nonhistone proteins. However, there was an additional band found in preparations from chicken erythrocytes. Since the histones were removed by extraction with acid, it is unlikely that this additional protein band might have represented the erythrocyte specific H5 histone. Rat kidney differed from rat liver by lacking two and possessing one additional protein band. Recently, Chiu et al. (58) extracted about 90% of the total chromosomal nonhistone proteins (UP) from chicken reticulocytes and chicken brain chromatin with 5 M urea 50 mM phosphate buffer, pH 7.6. The remaining nonhistone proteins (NP) were separated from histones by hydroxylapatite chromatography. The nonhistone protein fractions (UP) were very similar in both chicken brain and reticulocyte chromatin by electrophoretic criteria. However, the proteins of the NP fraction were quite different in both tissues (Fig. 2). Wu et al. (59) also compared nonhistone proteins of rat liver, kidney, spleen, lung, thymus, thyroid and brain chromatins at a high level of resolution using two types of sodium dodecyl sulfate-polyacrylamide gels. Most of the nonhistone protein bands were present in all the tissues examined and only a few nonhistone proteins appeared to be tissue-specific. In particular, the brain nonhistone protein pattern differed most extensively from all others and included an exceptional number of high molecular weight protein bands. These results corroborated earlier experiments on mouse kidney, liver, spleen and brain by MacGillivray et al. (38). Considerable tissue specificity of chromosomal nonhistone proteins was also detected by two-dimensional gel electrophoresis (48).

The reason that so many similar chromosomal nonhistone proteins were found in various tissues may reflect their structural or enzymatic nature. Hence, it was of interest to know the extent to which the nonhistone chromosomal proteins are conserved among different species of higher animals. To

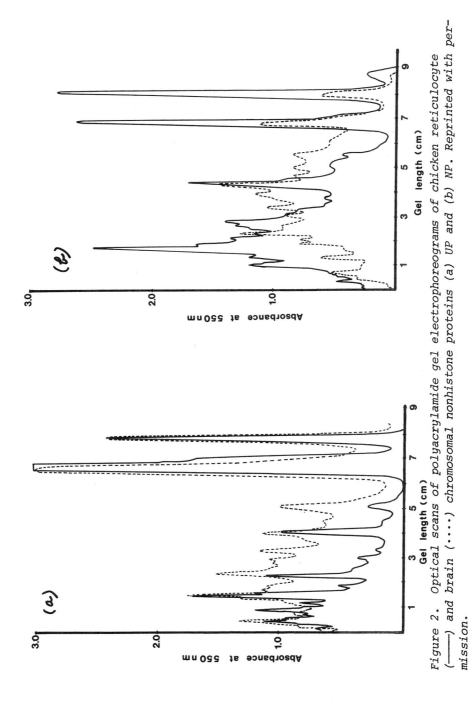

Figure 2. Optical scans of polyacrylamide gel electrophoreograms of chicken reticulocyte (——) and brain (····) chromosomal nonhistone proteins (a) UP and (b) NP. Reprinted with permission.

answer this question, Wu *et al.* (60) and Barret and Gould
(61) have compared the chromosomal nonhistone proteins of
the same tissue in different species and of different tissues
in the same species. Wu *et al.* (60) used sodium dodecyl sulfate polyacrylamide gel electrophoresis to compare the nonhistone proteins between liver and kidney in rat, cat, cow,
chicken, turtle and frog. These species represent examples
of all the major vertebrate classes except for fish. The gel
patterns indicated that the nonhistone proteins have changed
much more during evolution than have the histones. However,
a subset of bands appeared to be conserved in each tissue
investigated (Tables II and III). Barett and Gould (61) compared the nonhistone proteins from rat liver, chicken liver
and chicken reticulocytes using two-dimensional polyacrylamide gel electrophoresis. Great similarity in protein patterns was found between livers of the two species while the
chromosomal proteins in chicken liver and reticulocytes
differed considerably. Specific electrophoretic patterns
were also reported in bovine tissues by Platz *et al.* (62)
and in rat tissues by Teng *et al.* (18) and Gronow and
Thackrah (63).

The reports of tissue specific distribution of chromosomal nonhistone proteins detected by polyacrylamide gel
electrophoresis should be interpreted with caution. Some of
this specificity may simply reflect the aggregation, limited
proteolysis, or subunit dissociation of these proteins.
Perhaps the most elegant illustration of tissue specificity
of chromosomal nonhistone proteins was shown by Chytil and
Spelsberg (64). These authors used dehistonized chromatin to
elicit antibodies in rabbits and using quantitative microcomplement fixation (65), these authors found a considerable
specificity for the dehistonized or intact chromatin samples
from various tissues (Fig. 3). The immunological specificity
of chromosomal nonhistone proteins also was reported by Wakabayashi and Hnilica (66) who compared rat liver, calf thymus
and Novikoff hepatoma chromatins. Only the homologous
antigen-antibody systems were immunoreactive. In other words,
the Novikoff hepatoma chromatin fixed the completent extensively only in the presence of Novikoff hepatoma dehistonized
chromatin antiserum. Similar complexes from rat liver or
calf thymus chromatin were virtually inactive (Fig. 4).
These authors also found that the immunospecificity was due
to the complexes of nonhistone proteins with homologous DNA.
Either nonhistone proteins or DNA alone were not immunoreactive. Polyanions such as polyethylene sulfonate, polyglu-
tamic acid, dextran sulfate, yeast RNA and heterologous DNA
(i.e., sea urchin DNA) could not replace homologous DNA in

Table II. The Nonhistone proteins of liver from various animals.

Band	Rat	Cat	Cow	Chicken	Turtle	Frog	Common
π_1	(+)	+	+	+	0	+	0
π_2	++	0	+	+	++	++	0
	0	0	0	0	+	0	0
ν_1	++	++	+	+	+	+	*
ν_2	(+)	+	+	0	+	+	0
	0	+	0	++	0	0	0
λ_1	++	++	+	0	0	+	0
λ_2	0	+	0	0	+	0	0
	+	0	+	+	+	(+)	0
	(+)	+	0	0	(+)	(+)	0
κ_1	+	+	+	+	+	+	*
	0	0	0	0	+	0	0
κ_2	++	++	++	++	+	++	*
κ_3	0	?	?	0	+	0	0
κ_4	+	+	0	(+)	0	+	0
κ_5	++	+	+	+	(+)	+	*
κ_6	++	+	0	0	+	0	0
	0	0	0	+	0	0	0
κ_7	+	+	+	+	+	++	*
	0	0	0	0	++	0	0
θ_0	+	+	++	0	++	0	0
θ_1	++	++	++	++	+	++	*
θ_2	+	0	+	0	+	+	0
θ_3	+	+	+	+	+	0	0

Presence (+) or absence (0) of liver NHC proteins. ++ indicates a dominant band; (+) indicates a minor band; ? indicates no information; * indicates a common band. (From ref. 60).

Table III. The Nonhistone proteins of Kidney from various animals.

Band	Rat	Cat	Cow	Chicken	Turtle	Frog	Common
π_1	+	++	0	+	(+)	0	0
π_2	0	0	+	0	+	+	0
	0	0	0	+	0	0	0
ν_1	++	+	+	0	+	0	0
ν_2	+	+	0	+	0	+	0
λ_1	++	++	++	+	+	+	*
	0	0	0	0	(+)	0	0
λ_2	+	+	0	0	+	+	0
λ_3	+	+	+	+	+	+	*
	0	0	+	0	0	0	0
κ_1	0	+	0	+	+	+	0
κ_2	+	+	++	0	++	+	0
κ_3	0	0	0	?	?	0	0
κ_4	++	+	+	+	++	++	*
κ_5	+	0	+	0	(+)	+	0
κ_6	0	+	+	+	+	0	0
	0	0	0	0	+	0	0
κ_7	+	0	0	0	0	+	0
	0	0	0	+	0	0	0
θ_0	+	+	+	+	+	+	*
θ_1	++	++	++	++	++	++	*
θ_2	+	0	+	?	+	+	0
	0	+	0	+	0	0	0
θ_3	+	+	+	+	+	+	*

Presence (+) or absence (0) of kidney NHC proteins. ++ indicates a dominant band; (+) indicates a minor band; ? indicates no information * indicates a common band (60).

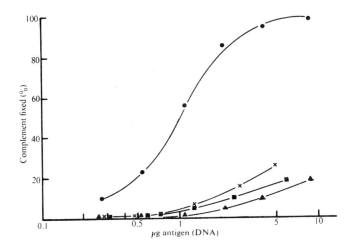

Figure 3. Complement fixation by varying quantities of non-histone protein-DNA complexes from chick oviduct and other organs in the presence of rabbit antiserum. Dehistonized chromatin from chick oviduct was used for immunizing rabbits. The rabbits were bled by cardiac puncture 7 days following the last injection and the resulting serum (diluted 1/400) was used for determination of microcomplement fixation by varying amounts of different antigens by the method of Wasserman and Levine (65). The anticomplementarity (binding of the complement in the absence of antiserum) was tested in the whole range of concentration and subtracted from the total complement fixation. Reprinted with permission (64).
●*, Oviduct;* ■*, liver;* ▲ *spleen;* ✕*, heart.*

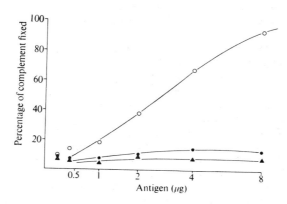

Figure 4. Complement fixation of NP-DNA complexes from Novikoff hepatoma, rat liver, and calf thymus in the presence of rabbit Novikoff hepatoma NP-DNA antiserum. The antigens were the nonhistone protein-DNA pellets (NP-DNA) from Novikoff hepatoma, rat liver and calf thymus chromatin. The immunospecificity of individual chromatin preparations was determined by the microcomplement fixation of Wasserman and Levine (65). The antisera were first purified on DEAE cellulose column and the antigen-antiserum incubation was carried out at 37°C for 90 min. All experimental points were corrected for anticomplementarity. Reprinted with permission (66).
  ○, Novikoff hepatoma; ●, calf thymus; ▲, rat liver.

the immunologically active complexes.

These findings were confirmed by other investigators who found that when chromatin or DNA-nonhistone protein complexes were used as antigens, antibodies could be obtained which were specific for the species (30,67-69), for the tissues (64,66,70), for normal or transferred cells (30,31,66,71-73), for differentiating tissues (69,74,75), and for normal or regenerating rat liver (75).

The most significant finding was the discovery that neoplasia changed the immunological tissue specificity of the nonhistone protein-DNA complexes from the type characteristic for the normal tissue to a new kind, typical for malignant growth. Wakabayashi and Hnilica (66) and Chiu et al. (72) reported that various experimental tumors did not differ much in the immunospecificity of their chromosomal nonhistone protein-DNA complexes (Table IV). The affinity of various rat hepatoma chromatin preparations for the Novikoff hepatoma NP-DNA antiserum approximately paralleled the degree of their differentiation as well as their individual growth rates. The fast growing and poorly differentiated 7777 and 3924A Morris hepatomas were more reactive than the slow growing and better differentiated 7800 and 7787 tumors. Hepatocarcinogenic diet rapidly changed the immunospecificity of normal rat liver chromatins to the type characteristic for malignant tumors (31). Using the *in vitro* reconstitution of chromatin protein fractions, it was shown that the immunological tissue specificity of chromatin depends on the presence of a protein fraction that binds only to homologous DNA. This specificity could be transferred from one chromatin preparation to another by reconstituting this protein fraction to the DNA and the remaining chromatin components (Fig. 5).

Using the antibodies against dehistonized chromatin and microcomplement fixation, Spelsberg et al. (74) demonstrated changes in antigenecity of chromatin during estrogen mediated growth and differentiation of chick oviduct. This observation may reflect the findings of Dierks-Ventling and Jost (76) who have shown that the stimulation by 17β-estradiol *in vitro* resulted within 3 hrs in the selective synthesis of two nonhistone nuclear proteins of molecular weights of 26,000 and 20,000. Shelton and Allfrey (17) also reported selective synthesis of a nuclear nonhistone protein (molecular weight 41,000) in liver cells after the injection of cortisol into rats. Chytil et al. (69) described changes in the antigenicity of chromosomal nonhistone protein-DNA complexes in rat liver during embryonic and postnatal development. These results demonstrated that alterations in the composition and probably structure of the nuclear nonhistone proteins are

Table IV. Complement fixation of chromatin preparations
from normal rat liver and various tumors.

The assays were performed in the presence of rat liver NP:
DNA antiserum or Novikoff hepatoma NP:DNA antiserum. The
reaction mixtures containing, in a total volume of 0.8 ml,
5 μg of DNA as chromatin, antiserum (0.1 ml of 200-times
diluted rabbit antiserum) and complement (0.2 ml of 50-times
diluted guinea pig serum) were incubated at 4° for 17 hours.
To each sample 0.2 ml of activated sheep erythrocytes was
added and the mixture was incubated at 37° for 20 minutes.
The extent of hemolysis was determined by reading the absor-
bance at 413 nm. All assays were corrected for anticomple-
mentarity. Reprinted with permission (31).

Chromatin preparations	% of complement fixation	
	Novikoff hepatoma NP:DNA antiserum	Rat liver NP:DNA antiserum
Normal rat liver	7	65
Novikoff hepatoma	50	9
As 30D hepatoma	38	5
Azo dye-produced hepatoma[a]	34	11
Ehrlich ascites	29	
Canine transmissible sarcoma	32	

[a]Liver chromatin prepared from rats fed 3'-MDAB for 111 days.

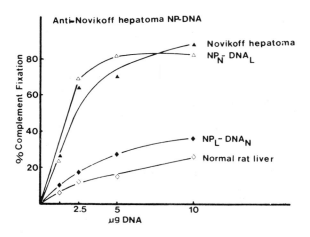

Figure 5. Complement fixation of normal and reconstituted NP-DNA complexes from rat liver and Novikoff hepatoma in the presence of antiserum against Novikoff hepatoma NP-DNA. All experimental points were corrected for anticomplementarity. (▲ —— ▲) Novikoff hepatoma chromatin (native); (△ —— △) reconstituted complex of Novikoff hepatoma NP and normal rat liver DNA (NP$_N$-DNA$_L$); (☐ —— ☐) normal rat liver chromatin (native); (● —— ●) reconstituted complex of rat liver NP and Novikoff hepatoma DNA (NP$_L$-DNA$_N$). Reprinted with permission (72).

closely associated with the process cell differentiation. Nuclear nonhistone proteins were also reported to change during the nitrosamine-induced carcinogenesis of rat liver (63) in virus transformed fibroblasts and in Morris hepatomas (77-80). Tuan et al. (81) described a tumor specific protein fraction in rat Walker carcinoma 256. This protein fraction was mitogenic to endothelial cells and when implanted into the rabbit cornea, it caused proliferation of vascular endothelium and formation of new blood vessels. This tumor angiogenesis factor was found to be associated with the nonhistone protein fraction of malignant cell nuclei but not with the corresponding fraction from normal cell.. Histone fractions had no activity in this bioassay.

Other examples of selective changes of nonhistone proteins were also reported in developing avian erythroid cells (82-85), in developing rat liver and brain (69,86), in developing Oncopeltus embryos (87), in resting and growing cells (24,88-90), during the cell cycle (91-93) and during rat spermatogenesis (94,95).

V. Nonhistone Proteins and the Cell Cycle

Similar to the histones, chromosomal nonhistone proteins are synthesized in the cytoplasm. Stein and Baserga (96) pulse labeled HeLa S_3 cells growing in suspension culture with ^3H-leucine and determined the radioactivity incorporated into proteins after chasing for various intervals. Their results indicated a decline in the cytoplasmic radioactivity with the simultaneous increase in the radioactivity of nuclear proteins. To minimize the effect of isotope reutilization, they also performed very short pulse label (30 sec) experiments which were immediately followed by cycloheximide inhibition of protein synthesis. A significant loss of cytoplasmic protein radioactivity and increased nuclear protein radioactivity was again noted. Fractionation of nuclear proteins showed increased radioactivity of both the histones and chromosomal nonhistone proteins.

Unlike the histones, whose synthesis is intimately coupled to that of DNA, the nonhistone proteins are synthesized during all phases of the cell cycle. Shapiro and Levina (97) studied chromosomal protein synthesis in phytohemagglutinin (PHA) stimulated lymphocytes which were partially synchronized with fluorodeoxyuridine. The nonhistone proteins incorporated ^3H-leucine throughout the entire cell cycle including mitosis and a 2-3 fold higher rate of synthesis was observed during the G_1 as compared with the S or G_2 periods. Subsequently, Shapiro and Polikarpova (98) using

[3]H-tryptophan, which is absent in histones, demonstrated chromosomal protein synthesis to take place in cultured Chinese hamster cells at all stages of the cell cycle. The incorporation was two fold higher in S and G_2 than in the G_1 period. Similarly, Sören (99) found by microinterferometry that both the nuclear and cytoplasmic mass increased considerably during blastogenesis in the PHA stimulated human leukocytes. A considerable part of the nuclear mass was found to accumulate before the onset of DNA synthesis, suggesting a preferential synthesis of nonhistone proteins prior to the DNA synthetic period of the cell cycle.

Using biochemical techniques, other laboratories have also come to the same conclusion. Stein and Baserga (100) have studied the synthesis of the nuclear nonhistone proteins during the S, G_2 and M phases of HeLa cells synchronized with a double thymidine block and found that these proteins (0.1 N NaOH extract of dehistonized chromatin) were synthesized throughout the entire cell cycle, including mitosis (Table V). The incorporation of labelled amino acids (using 60 min pulse) into chromatin of synchronized Chinese hamster DON-C cells was also found to occur during the entire cell cycle (101).

The biosynthesis of nonhistone proteins was not affected when DNA replication was blocked by inhibitors such as cytosine arabinoside and hydroxyurea (54). Sampetti-Bosselet *et al.* (103) reported that the inhibition of DNA synthesis in HeLa cells by a short hydroxyurea treatment (for 1 hour) resulted in about 50% reduction of the incorporation of ^{3}H-lysine into histones with essentially no effect on the biosynthesis of nonhistone proteins. Stein and Thrall (104) have found that in contact-inhibited human diploid WI-38 fibroblasts which are stimulated to proliferate by change of the medium, the synthesis of nonhistone proteins was not affected when DNA synthesis was blocked by cytosine arabinoside (Fig. 6). This was in contrast to the synthesis of histones, which was restricted to the S phase of the cell cycle and was completely inhibited when DNA synthesis was interrupted. It appears that a short exposure to the inhibitors is essential for the described effect. Malpoix (102) who used a longer exposure (4 hours) to hydroxyurea, reported a 74% decline of the ^{3}H-lysine incorporation into histones and a 60% reduction of incorporation into nonhistone proteins. He also showed that under these conditions, hydroxyurea inhibited DNA synthesis by 29%, total protein synthesis by 71% and RNA synthesis by only 9%.

Stimulation of cell populations to proliferate has long been a favorite method for studying variations in the biosyn-

Table V. Synthesis of Total Cellular and Chromatin Proteins During S, G_2 and Mitosis.

Protein Fraction	Phase of Cell Cycle		
	S	G_2	M
Total Cell Protein	7919	7258	2881
0.15M NaCl	61	50	21
0.35M NaCl	38	31	11
0.25N H_2SO_4	940	127	110
Residual Chromatin Fraction	1372	1316	1240

Protein synthesis was assayed by harvesting 300 ml of cells and resuspending them in 10 ml of leucine-free suspension medium supplemented with 2% fetal calf serum and 150 μCi of leucine-^3H. After incubation at 37°C for 15 minutes, the cells were washed 3 times in cold Spinner Salts Solution. Total cell protein synthesis was based on the 20% TCA precipitable leucine-^3H counts of the initial cellular homogenate. The other protein fractions were all obtained from chromatin preparations as described in Reference 100. The incorporation of leucine-^3H into the various protein fractions is expressed as cpm/μg DNA (from reference 100).

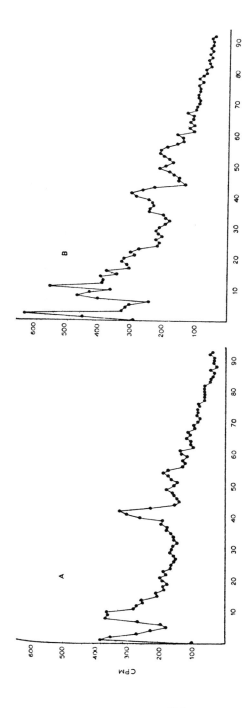

Figure 6. A) SDS polyacrylamide-gel electrophoretic profile of [³H] L-tryptophan-labeled G_1 total chromosomal proteins. B) Effect of cytosine arabinoside on the SDS polyacrylamide gel electrophoretic profile of [³H] L-tryptophan G_1 total chromosomal proteins.

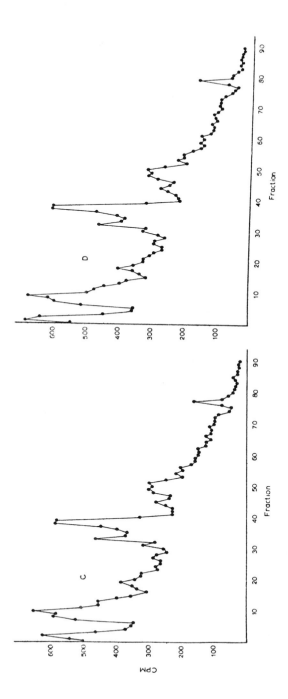

Figure 6. C) SDS polyacrylamide gel electrophoretic profile of [³H] L-tryptophan-labeled S-phase total chromosomal proteins. D) Effect of cytosine arabinoside on the SDS polyacrylamide gel electrophoretic profile of [³H] L-tryptophan-labeled S-phase total chromosomal proteins. Reprinted with permission (104).

215

thetic activity of nuclear nonhistone proteins during the G_1 and S phases. Teng and Hamilton (105) showed the early action of estrogen on the uterus of ovariectomized rats to include the stimulation of synthesis and the accumulation of chromosomal nonhistone proteins. Similarly, Chung and Coffey (106) reported that following castration and subsequent testosterone treatment, the maximal rate of nonhistone protein synthesis in ventral prostate occurred 24-48 hours prior to the peak of DNA synthesis. Using this system, Anderson et al. (107) also found an early increase in the synthesis of nonhistone proteins which was followed by a decline in synthetic activity at 12-24 hours after stimulation.

Spelsberg et al. (74) found the amounts of nonhistone proteins in chick oviducts to change during primary administration of diethylstibestrol but the newly synthesized proteins were not analyzed. According to Hemminki and Bloud (108), the nuclear proteins of estrogen primed chicks, as compared to progesterone primed chicks, were enriched in polypeptides of m.w. 50,000 and larger. A secondary stimulation with the steroid hormone increased the amino acid incorporation into histones and nonhistones two to six fold. Injection with estrogen resulted in preferential labeling of nonhistone proteins in the 50,000 to 60,000 m.w. range, regardless of the nature of primary stimulation. A secondary injection with progesterone increased the amount of highly labelled nonhistone proteins with a m.w. of over 70,000. These observed changes in the metabolism of nonhistone proteins in oviduct nuclei during secondary stimulation with steroid hormones may reflect either differential growth stimulation of the different cell types in the oviduct (109) or the expression of processes characteristic for a specific phase of the cell cycle (7).

The synthesis of chromosomal nonhistone chromatin proteins by rat testes and epididymis at various stages of development was studied by Kadohama and Turkington (110). The testicular nonhistone proteins were resolved reproducibly into 25 electrophoretic bands. At 5 days of age, the high molecular weight bands 1 through 5 were not detectable, but with the formation of spermatogonia and spermatocytes, these bands accumulated. There was also a marked increase in the biosynthetic rates of bands 1 through 9 with a simultaneous decrease in the phosphorylation of the majority of the nuclear proteins. Proteins of the epididymis at puberty were characterized by a marked decrease in bands 1, 2 and 20, and emergence of new bands 8 through 17 as the predominant components. This process was associated with a marked

increase in the biosynthetic rates and a reduced phosphorylation of the majority of chromosomal nonhistone proteins in
epididymis. The biosynthetic patterns of testicular nonhistone proteins were dependent on pituitary gonadotropic hormones, while those of the epididymal proteins were dependent
on testosterone.

Recently, Johnson *et al*. (111) reported a series of
alterations of chromatin proteins in cultures of purified
equine lymphoytes stimulated with concanavalin A (con A).
The rate of phosphorylation of nonhistone proteins increased
as much as 4 fold within 2 hours after exposure to con A,
reaching a maximum at 8 hours. Con A also stimulated the
synthesis of chromosomal nonhistone proteins. This is in
agreement with Levy *et al*. (24) who noted an immediate
increase in the synthesis of chromosomal nonhistone proteins
following the stimulation of lymphoid cells with PHA. The
synthetic activity of these proteins continued to increase
throughout the first 10 hours while the stimulation of DNA
synthesis did not occur until 24 hours after PHA.

Spivak (112) isolated chromatin from splenic tissue
during the early and late phases of erythropoiesis as well as
from non-anemic animals. The total protein content of chromatin from the early erythropoietic phase was greater than
that of chromatin from the erythroid phase or from non-anemic
controls. Electrophoretic analysis of chromosomal nonhistone
proteins in the pre-anemic, early and late erythroid spleen
revealed that certain proteins were common to all situations
with some quantitative variations. Of particular interest
was the m.w. 43,000 protein band which was most prominent in
the pre-anemic spleen and considerably diminished in the
erythropoietic spleen. In addition, there was also an
increase in the number of nonhistone protein bands in the
electrophoretic pattern of the erythroid spleen, particularly
between m.w. 47,000; 64,000; 78,000 and 94,000. Analysis of
the ^{3}H-valine incorporation into the nonhistone proteins
indicated that during early erythropiesis, there was a general increase in nonhistone protein synthesis. During the
last erythroid phase, the nonhistone protein synthesis diminished with the most marked decrease in the higher molecular
weight area.

Kostraba and Wang (113) have used regenerating liver as
an experimental model for studying the temporal relationships
between the chromosomal proteins and DNA synthesis. The nonhistone proteins showed an increased incorporation up to
approximately 12 hours after hepatectomy. This initial
increase was followed by a slight decrease in synthetic
activity ending at 18 hours. Following this, the synthesis

of all nonhistone proteins increased rapidly in conjunction with the increased DNA replication, peaking at about 25 hours. The biosynthetic activity remained high and declined only as the number of cells synthesizing DNA diminished. However, the electrophoretic patterns of nonhistone proteins analyzed during different time points after hepatectomy were strikingly similar (47). Using a more sensitive double isotope labeling, Garrard and Bonner (47) found that although all the polypeptides exhibited an increased rate of labeling, many appeared to be selectively stimulated, particularly proteins in the molecular weight range 55,000 to 45,000.

Synthesis of specific nonhistone chromosomal proteins in the uterus of ovariectomized rats was recently examined after treatment with 17β-estradiol (114). Biosynthetic stimulation of at least five nonhistone chromosomal proteins having molecular weights of 96,000, 70,500, 29,400, 20,700 and 16,400 was observed. The rate of synthesis of the chromosomal nonhistone protein having a molecular weight of 70,500 increased 1 hour after the hormone treatment. This was the first nonhistone protein to be induced by estrogen, and its induction was blocked by pretreatment with actinomycin D. The biosynthetic rates of nonhistone proteins with molecular weights of 96,000, 29,400, 20,700 and 16,400 were sequentially increased at 3.5; 24 and 24 hours, respectively, following the hormone treatment.

That a differential turnover of the nonhistone proteins may occur during the cell cycle was suggested by Gray et al. (115) who found that one of the two nonhistone bands in their gel electrophoretic system failed to incorporate ^{3}H-lysine or ^{3}H-arginine during labeling. Similarly, Shelton and Allfrey (17), who used a more sophisticated system for analysis of nonhistone proteins in rat liver chromatin, found the biosynthesis of a nonhistone protein fraction to be selectively stimulated *in vivo* by cortisol. The *in vivo* stimulation of immature chick livers by 17β-estradiol resulted in selective synthesis within 3 hours of two nonhistone nuclear proteins (m.w. 26,000 and 20,000)(76).

Weisenthal and Ruddon (116) reported the absence of higher molecular weight nonhistone proteins in nondividing chronic lymphocyte leukemia. However, when these cells were cultured and stimulated to divide with PHA, their nonhistone protein banding profile changed, emphasizing the middle and high molecular weight proteins. Stein and Baserga (117) observed an increase of nonhistone protein biosynthesis in mouse salivary glands after treatment with isoproterenol. The stimulation reached its peak in 12 hours. The increased synthetic activity of the ncnhistone proteins occurred primarily in a limited number of specific protein bands and there were quantitative changes in the specific activities of individual bands at different times following the stimulation (118). When RNA synthesis was inhibited by actinomycin D prior to isoproterenol stimulation, the early increase in the nonhistone protein synthesis was not affected. However, later nonhistone protein synthesis was inhibited together with the subsequent DNA replication. This suggests that non-histone proteins synthesized 1 hour after isoproterenol stimulation were necessary for the subsequent synthesis of nonhistone proteins and DNA. Activation of cultured WI-38 fibroblasts to proliferate also resulted in a biphasic increase of the nonhistone protein biosynthesis (119). The initial increase was evident within minutes after the stimu-lation and peaked approximately in 1 to 3 hours. Subse-quently, there was a decline in the nonhistone protein syn-thesis, followed by a second increase. This second peak of nonhistone protein synthesis occurred in conjunction with the peak of DNA synthesis. Similar results were also obtained with cultured fibroblasts (3T6) stimulated to proliferate (120). Electrophoresis of the nonhistone proteins from both 3T6 and WI-38 cells showed differences in specific activities of some of the individual protein bands in stimulated and unstimulated cells. Generally, there was a decrease in the synthesis of some higher molecular weight proteins with a concomitant increase in the synthetic activity of the middle and lower molecular weight proteins. These changes probably reflect progression of the cells through the cell cycle (121).

Bhorjee and Pederson (122) examined the qualitative and quantitative changes of the nonhistone proteins in relation to the cell cycle in synchronized HeLa cells. The amounts of some nonhistone proteins varied during the cell cycle by as much as 50%, while others remained essentially constant. A group of nonhistone proteins (band 11, m.w. 75,000) was greatly reduced just before the onset of DNA replication and returned to normal levels during the mid-S phase. One of the most conspicuous shifts occurred in band 15, which was reduced by about 50% during mid-S and G_2 periods. Relative duction in the heights of peaks 4 and 17 also occurred during G_2. Using cells synchronized by either the double thymidine block or by selective detachment, Stein and Borun (123) have also seen significant increases in specific activ-ities of the nonhistone proteins during the G_1 period of syn-chronized HeLa cells. During this time period, essentially all the high and middle molecular weight proteins were chased out by 120 minutes. During S phase, polypeptides of inter-mediate molecular weight were preferentially chased from the nucleus. Finally, during G_2, the higher and middle molecular weight protein fractions exhibited the highest turnover rate.

Using DNA cellulose chromatography to detect proteins with affinity for DNA, Salas and Green (124) reported the differential synthesis of ^{3}H-tryptophan-labeled proteins between resting (stationary) and exponentially growing (log) cells and during the cell cycle of mouse fibroblasts 3T3 and 3T6. Electrophoresis of the 0.15 M NaCl eluate of cell protein extracts resolved the DNA-binding proteins into eight distinct fractions (P1-P8). Three (P1, P2 and P6) of the eight peaks exhibited significant quantitative differences, depending on whether the initial cell extracts were obtained from resting or growing cells. In resting cells, the P1 and P2 proteins were actively synthesized whereas the P6 was practically absent. The converse relationship was found in growing cells, i.e., the rate of synthesis of P1 and P2 was greatly reduced while that of P6 was highly increased. Comparison of the electrophoretic profiles of these proteins from different phases of the cell cycle revealed that the proteins synthesized during S phase were similar to those obtained from growing cells with the exception of P6 which was present in larger quantities in synchronized S cells than in growing cells. The P1 decreased to about one half of its early G_1 value and disappeared altogether during S. From these results, Salas and Green suggested that not only was the synthesis of P6 coupled to DNA replication, but also that the P1, which was synthesized only in resting and serum starved cells and not during the S phase, may act as a repressor or inhibitor of DNA replication. Similar results were also obtained by Fox and Pardee (125) as well as Choe and Rose (126) who studied the DNA-binding proteins in CHO cells during the cell cycle. Stein (127) analyzed DNA binding proteins from normal human diploid cells WI-38 by DNA cellulose chromatography and polyacrylamide gel electrophoresis. When WI-38 cells in the replicative and stationary phases were compared, five proteins, P5b, P6a, P8, P9 and P10 (respective m.w. 8,700; 50,000; 33,000; 28,000; and 25,000) were extensively labeled in replicating cells and two proteins, P5C and P12 (m.w. 72,000 and 18,000) were extensively labeled in the stationary phase cells. In addition, several high molecular weight DNA binding proteins, tentatively identified as collagen and protocollagen, were preferentially labeled during the stationary phase cells. Stationary phase senescent WI-38 cells at or near the end of their *in vitro* lifespan contained much more of P8 protein as compared to stationary phase WI-38 cells at their early population doubling levels. Further characterization of the P8 protein showed that it binds preferentially to single-stranded DNA and represents more than 1% of the total soluble protein content in young cells during growth phase. Thus, the P8 protein of WI-38 cells appears to be comparable to the P8 protein in mouse 3T6 and human SB cells (128).

The turnover of nonhistone proteins and their phosphate groups was compared in normal and SV-40 transformed WI-38 human diploid fibroblasts (129). The relative amount of chromosomal nonhistone proteins present in the 30,000 to 51,000 molecular weight range was 1.5 times greater in the SV-40 transformed cells than in normal WI-38 cells. Conversely, the relative amount of nonhistone proteins present in the 142,000 to 200,000 molecular weight range was 1.6 fold greater in normal than in SV-40 transformed cells. The WI-38 cells were pulse labeled with ^{3}H-tryptophan and ^{32}P-orthophosphate for 30 minutes and the ^{3}H and ^{32}P activities of the various molecular weight classes of nonhistone proteins were determined during the first four hours following termination of the labeling. While a rapid turnover of high molecular weight nonhistone proteins (142,000 to 200,000) could be seen during the first hour in SV-40 transformed cells, specific activities of these proteins were not significantly altered in normal cells. In contrast, a rapid turnover of low molecular weight (30,000 to 51,000) nonhistone proteins occurred during the first hour in normal WI-38 cells with no corresponding decrease in the specific activities of these proteins in SV-40 transformed cells. There was no apparent turnover of nonhistone protein-bound phosphate in either normal or SV-40 transformed cells during the four hours following labeling.

Prominent responses of the biosynthesis and turnover of chromosomal nonhistone proteins to changes in cellular growth and proliferation are compatible with their postulated functions in the regulation of nuclear metabolism and genetic activity. However, as was already mentioned during the discussion of tissue and species specificity of chromosomal nonhistone proteins, the regulatory macromolecules of chromatin are most likely present in quantities too little to be detected by polyacrylamide gel electrophoresis of total nuclear protein extracts. More selective methods must be used for the accumulation and isolation of regulatory proteins in quantities sufficient for their analysis and characterization.

VI. Interaction of Nonhistone Proteins with DNA and Chromatin

During the process of differentiation, cells selectively

suppress their genetic information by complexing most of their DNA with chromosomal proteins to make it unavailable for transcription by RNA polymerases (2,3,130,131). The binding of nuclear proteins to DNA has, therefore, obvious functional consequences ranging from the repression to activation of RNA synthesis.

It is known from studies on prokaryotes that the interaction of proteins with DNA can be highly specific. The recognition of specific interaction sites on the DNA is achieved through the specificity of polypeptide sequences as well as through tertiary and quarternary protein conformation. In eukaryotes, the histones, which represent a major group of chromosomal proteins, are greatly limited in their heterogeneity and species and tissue specificity. There are five main histone fractions in mammalian chromatin and all are well characterized (4). Because of this lack of specificity, it is generally accepted that histones function as nonspecific repressors, perhaps through their structural effects on DNA conformation in chromatin. Chromosomal non-histone proteins, on the other hand, appear to possess all the heterogeneity and specificity expected from gene regulatory molecules. Because one can reasonably expect the regulatory molecule (protein) to exhibit affinity for the substrate which it regulates (DNA), the DNA-binding properties of various chromosomal nonhistone proteins are being investigated in several laboratories.

Kleinsmith et al. (40) isolated a very small fraction (<0.01%) of the nonhistone proteins from rat liver nuclei. Principally, the I^{125}-labeled proteins were chromatographed through a column of salmon sperm DNA-cellulose, the run-off peak was collected and its protein rechromatographed on a cellulose column containing rat liver DNA. The retained proteins were eluted stepwise with 0.14, 0.6, and 2.0M NaCl. The specific DNA binding nonhistone proteins were eluted at the 0.6M NaCl concentration. This fraction, associated strongly with rat DNA, but not with the DNA from the other sources (Fig. 7). In his more recent paper, Kleinsmith (132) showed that about 1% of the phosphoprotein fraction of rat liver nuclei binds selectively to homologous DNA. The binding ratio of about 1:100 (protein/DNA) suggested the formation of specific complexes. A lower, but still considerable binding to heterologous DNA preparations (mouse, calf, salmon, etc.) was also observed. According to the author, the molecular weights of the DNA binding proteins ranged from 30,000 to 70,000.

Teng et al. (18) described a fraction of nuclear phosphoproteins which were soluble in buffered phenol. These

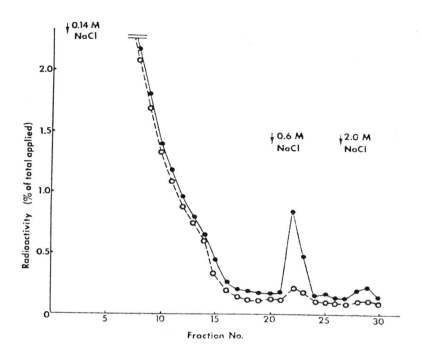

Figure 7. Chromatography of ^{32}P-labeled rat liver nuclear
phosphoprotins on columns of DNA-cellulose. Proteins which
bind to DNA are eluted by raising the ionic strength to
0.6M NaCl. Rat liver phosphoproteins bind much better to
columns made from rat DNA (●——●) than to columns made from
E. coli DNA (○--- ○). Kleinsmith, L. J. (132). Reprinted
with permission.

phosphorylated nonhistone proteins also associated preferably with homologous DNA, but their binding ratio was considerably higher (about 13%). The molecular weights of these proteins were similar to the DNA-binding proteins described by Klein- smith (132). Yet another group of nonhistone proteins with affinity for DNA was isolated by Patel and Thomas (133). According to their preliminary characterization, these pro- teins were phosphorylated and represented about 1.5-3.0% of the total nuclear proteins. In their electrophoretic heter- ogeneity, these DNA-binding nonhistone proteins resembled the total phosphoprotein fraction of chromatin. Similar to the finding of Kleinsmith (132), these proteins associated prefer- entially with homologous DNA but they also interacted with heterologous DNA (calf thymus and Flavobacterium).

Hnilica and his associates (30,31) have isolated a fraction of immunospecific nonhistone proteins (NP proteins) with affinity for native DNA. In an attempt to correlate the immunospecificity and the specificity of DNA binding, native DNA agarose-polyacrylamide gels were employed for the isola- tion of DNA-binding proteins NP. As documented by the elu- tion pattern shown in Fig. 8, a fraction (NP) of nonhistone proteins from rat liver chromatin was retained only on the column containing homologous (rat spleen) DNA. There was not detectable retention of the same protein on calf thymus DNA- containing columns. The relatively high (0.4M) KCl concen- tration necessary for the elution of these DNA binding non- histone proteins from their association with homologous DNA indicates a relatively strong interaction between these two macromolecular species. When the I^{125}-labeled NP peak eluted from rat spleen DNA columns was concentrated and subjected to polyacrylamide gel electrophoresis in the presence of sodium dodecyl sulfate, it resolved into three major polypeptide bands of relatively low molecular weight (12,000-15,000) and several minor higher molecular weight proteins. The inter- action of immunospecific NP fraction with DNA was further characterized by Chiu *et al.* (31,45) and Wang *et al.* (134). When the NP fraction which represents less than 5% of the total chromatin protein content was interacted with homolog- ous DNA in 10 mM NaCl, in 10 mM Tris-HCl buffer, pH 8.0, the DNA (rat spleen) binding sites available were saturated at the NP protein:DNA ratio of approximately 1.5:100 (W/W). The binding of NP proteins was species specific (Table VI). Both rat spleen or liver DNA bound the liver NP protein fraction equally well. Calf thymus DNA exhibited a small but signifi- cant binding while the affinity of rat liver NP proteins to chicken erythrocyte or *E. coli* DNA is negligible. The NP proteins associated preferentially with unique, double

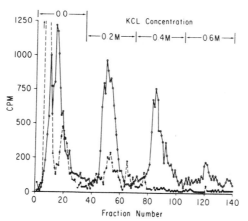

Figure 8. *Affinity chromatography of rat liver nonhistone protein and histone. Both protein mixtures were labeled in vitro with ^{125}I. Rat spleen DNA bound to agarose-polyacrylamide matrix was used for chromatography. Radioactivity of rat liver nonhistone protein (O – O – O), Radioactivity of histone (● – ● – ● – ●). Reprinted with permission (41).*

Table VI. Interactions of rat liver NP fraction with homologous and heterologous DNA.

The formation of DNA-protein complexes was assayed by sucrose density gradient centrifugation using 125I-labeled NP protein. The binding ratios represent weight percentages of protein retained by the DNA. (From reference 134).

Source of DNA	DNA	Protein Applied	Protein bound	Protein/DNA binding
		µg		ratio
Rat spleen	400	40	5.8	0.0145
Rat liver	400	40	5.6	0.0141
Calf thymus	400	40	1.0	0.0025
Chicken erythrocyte	400	40	0.3	0.0008
Escherichia coli	400	40	0.1	0.0003

stranded sequences of fractionated homologous DNA. The interactions were strong at low ionic strength (Km = 6.7 x 10^{-9})
and decreased with rising salt concentrations.

Bonner and his associates (135,136) also isolated a
fraction of chromosomal nonhistone proteins from rat liver
which interacted preferentially with rat DNA. This fraction
represented 3.9% of the total rat liver nonhistone protein
content and in its molecular weight, amino acid composition
and other properties resembled rat liver DNA-binding proteins
of NP fraction isolated by Wakabayashi *et al.* (30). Sevall
et al. (136) extended the studies on rat liver chromosomal
proteins to include binding specificity to various DNA preparations. At a relatively low ionic strength (110 mM), the
interactions were strong but nonspecific. At this ionic
strength, the authors found very little difference in binding
of the rat liver nonhistone protein subfraction to the DNA
from rat liver, calf thymus and *D. melanogaster*. However,
significant preference for homologous DNA and for the middle
repetitive fraction of sheared DNA (C$_{ot}$ between 0.02-200) was
observed when the salt concentration was increased to 260 mM.
Although the interaction forces decreased, the interaction
specificity improved significantly. These investigations
were further extended by Johnson *et al.* (137) who tried
to isolated DNA binding proteins from Novikoff hepatoma
nucleoplasmic and cytoplasmic extracts. They assumed that
cellular proteins which bind DNA reversibly must be in equilibrium between the soluble (sap) and particulate chromatin
components. If a regulatory protein acts in a manner analogous to the lac of lambda repressor and controls DNA transcription by binding to a unique site on the genome, only a
few copies per cell of such protein should be recovered from
isolated chromatin as a starting material. On the other hand,
many more copies may be present in the soluble phase (nuclear
or cytosol). These authors (137) obtained three groups of
DNA-binding proteins from soluble protein extracts of Novikoff
hepatoma cells. By first exposing the crude protein extract
to heterologous (*E. coli*) DNA, they could identify proteins
which bind nonspecifically. Subsequent exposure of the
unbound proteins to homologous (rat liver) DNA allowed identification of proteins which bound selectively. This second
class of proteins represented 0.5-1.0% of the total soluble
protein and exhibited a 7-17 fold preference for rat DNA
over *E. coli* DNA. A third class of DNA-binding proteins
(1-1.5% of the soluble proteins) associated with DNA so
strongly that their elution could not be effected with 4.0M
NaCl and these proteins had to be released by DNase I treatment. The DNA-binding protein fractions which eluted at 2.0M

NaCl was labeled with I^{125} or I^{131} and characterized by sodium dodecyl sulfate polyacrylamide gel electrophoresis and isoelectric focusing. These proteins represented a discrete subset of the total soluble protein complex. However, many electrophoretic similarities were noted between the major components of protein fractions which associated either with homologous or heterologous DNA.

Recently, Allfrey *et al.* (43) described a comprehensive technique for the fractionation of calf thymus nuclear non-histone proteins by affinity chromatography on aminoethyl Sepharose 4B columns charged with fragmented and fractionated DNA. They found that the nonhistone proteins differed in their affinities for sequences of DNA with low, intermediate and high $C_{o}t$ values. Additional differences were observed in protein binding to homologous or heterologous single-stranded and double-stranded DNA subfractions of the same $C_{o}t$ value. The selectivity of this method is illustreted in Figure 9. The authors concluded that various nonhistone proteins bind to the DNA with differing affinities and that the protein fractions eluted at different salt concentrations exhibit considerable complexity and size distribution. Some proteins differed in their affinities for high, intermediate and low $C_{o}t$ DNA sequences as well as for native or denatured DNA (Figure 10).

Based on Crick's model (138) of the specific interaction of protein regulators with the locally denatured DNA, Umansky *et al.* (139) suggested that during the genetic reprogramming of DNA in dividing cells, the complexing of protein regulators with denatured sites of DNA should most likely occur during the S-period. Because cyclic nucleotides mediate the effects of many hormones and other biologically active agents, it was of interest to know the effects of cyclic nucleotides upon binding of nuclear nonhistone proteins to the DNA. Johnson *et al.* (140) fractionated nuclear proteins on Bio-Rex 70 column and found six nuclear protein fractions to contain both cyclic GMP-binding and DNA-binding activities. Proteins in Bio-Rex peak II bound cyclic GMP with high affinity. At low concentrations (10^{-6}-10^{-9}) c-GMP stimulated binding of these proteins to DNA with a maximum (25-50% stimulation) increase at $10^{-8}M$-$10^{-9}M$ c-GMP. However, at high concentrations ($>10^{-5}M$) c-GMP inhibited the binding of these proteins to DNA. The authors suggested that the effect of c-GMP on DNA protein interaction may serve some regulatory purpose, since the c-GMP concentration in resting lymphocytes is about 10^{-8}-$10^{-9}M$. This suggestion is supported by the reports that cyclic GMP enhanced the activity of RNA polymerase I from human lymphocyte nuclei at $10^{-8}M$-$10^{-9}M$ and inhibi-

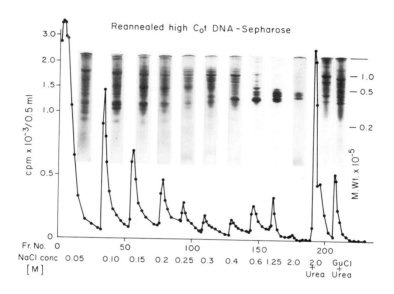

Figure 9. *Fractionation of nuclear nonhistone proteins from calf thymus on double-stranded high $C_{o}t$ (225-40,000) DNA covalently linked to aminoethyl-Sepharose 4B. The ^3H-labeled nuclear proteins were combined with the DNA under renaturing conditions. The proteins were eluted first with a discontinuous salt gradient and finally with urea and guanidine hydrochloride (GuCl), each fraction being monitored for radioactivity. The proteins eluted at each salt concentration were identified by electrophoresis in 10% polyacrylamide containing 0.1% sodium dodecyl sulfate. The patterns shown for each protein set at the appropriate peak of the elution diagram. The corresponding molecular weight scale is given at the right of the Figure. Reprinted with permission (43).*

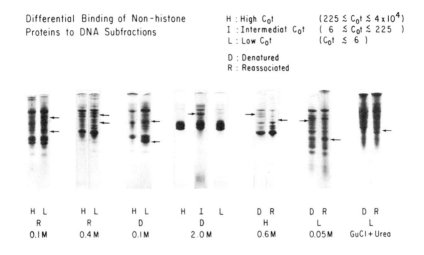

Differential Binding of Non-histone
Proteins to DNA Subfractions

H : High C_0t $\quad$ $(225 \le C_0t \le 4 \times 10^4)$
I : Intermediat C_0t $\quad$ $(6 \le C_0t \le 225)$
L : Low C_0t $\quad$ $(C_0t \le 6)$

D : Denatured
R : Reassociated

Figure 10. Comparison of electrophoretic banding patterns
of protein sets eluted at corresponding salt concentrations
from parallel aminoethyl-Sepharose 4B columns with DNA subfractions differing in C_0t value or in strandedness. Side by
side comparisons of proteins eluted at 0.1M NaCl and 0.4M
NaCl from high C_0t (H: 225-40,000) and low C_0t (L: <6) double
stranded DNA are shown in the first two panels. Examples of
differential binding of protein to high, intermediate (I: 6
<C_0t<225) and low C_0t single-stranded DNA subfractions are
shown in the third and fourth panels. The other panels
illustrate differences in protein elution patterns from
single-stranded and double-stranded DNA subfractions of the
same C_0t value: R, reassociated DNA strands; D, denatured DNA
strands. Major differences are indicated by the arrows;
smaller differences in relative concentrations are not indicated but many exist. Reprinted with permission (43).

ted its activity at higher c-GMP concentrations (141,142).

As was already mentioned, Johns and his associates (143) isolated a new group of DNA-binding proteins characterized by their high content of acidic and basic amino acids. In their recent papers (144,145), these authors reported the purification of a single protein designated as HMG1 from the 0.35M NaCl extract of chromatin. The molecular weight of the protein HMG1 was about 26,5000. The HMG1, unlike the histones, did not form an insoluble complex with DNA. However, sedimentation analysis of DNA-HMG1 complexes in the ultracentrifuge revealed that HMG1 binds strongly to calf thymus DNA and bacteriophage T7 DNA. The binding ratio of HMG1:DNA was about 4. The protein distributed evenly along the DNA chains and each binding site on the DNA was about 13-18 nucleotides long. The HMG1 binding to DNA was realized through a rapidly reversible equilibrium and depended on the ionic strength, suggesting interactions between the basic amino acids of HMG1 and phosphate groups of the DNA. This resembles the binding of histones to DNA and points to a possibility that HMG1 plays some role in the chromatin structure. According to their preliminary finding, the HMG1 protein will combine with histone H1, but the authors do not know yet whether this modifies the binding of H1 to the DNA.

Quite recently, Sheehan and Olins (146) studied the binding kinetics of rat liver nuclear nonhistone proteins to DNA using a membrane filter technique of Lin and Riggs (147). The binding of nonhistone proteins to homologous DNA was quite rapid (<3 min) and the complexes appeared to be stable for at least 50 minutes. The optimal salt concentration was 0.05M NaCl which is higher than that described for phenol-extracted phosphoprotein (18) and for the immunospecific non-histone protein fraction NP (134) and lower than that described by Patel and Thomas (133). The binding affinity increased when the pH was lowered. The authors suggested that the lower the pH may either change the efficiency of protein retention on filters or alter the protein-nucleic acid interactions.

It is obvious from the reviewed material that the interactions of chromosomal nonhistone proteins with DNA attract considerable attention and significant new information can be expected to result from this rapidly moving research.

VI. Phosphoproteins

Although the existence of proteins which rapidly incorporate radioactive phosphate has been known in a variety of tissues for many years, it was not until recently that serious

studies on nuclear phosphoproteins were initiated (148-152).
Phosphorylated cellular proteins are most concentrated in the
nucleus, where both histones and nonhistone proteins can be
phosphorylated (153,154). However, about 90% or more of the
nuclear-bound phosphorus is associated with nonhistone proteins. The major site of protein phosphorylation is the
hydroxyl group of serine residues, which accounts for about
90% of the total protein-bound phosphorus. The remaining
10% of the phosphorus in nuclear proteins is found as phospho-
threonine (153). The process of nuclear protein phosphorylation is independent from their biosynthesis. Phosphate
groups attached to the nonhistone proteins are not stable and
turn over rapidly. Current evidence suggests that separate
enzymes are involved in the phosphorylation and dephosphory-
lation of nonhistone proteins. In addition to ATP as a phosphate donor, a variety of other nucleoside and deoxynucleo-
side triphosphates were found to be capable of phosphorylating
the chromosomal phosphoproteins (155). However, the extent
of phosphorylation with nucleoside triphosphates other than
ATP is very low.

As the research on nuclear nonhistone protein phosphorylation proceeded over the years, it has become increasingly
evident that this process exhibits many of the characteristics involved in gene regulation. For example, nuclear phosphoproteins exhibit tissue and species specificity, changes
in nuclear phosphoproteins could be correlated with altered
genetic activity, specific binding to DNA, effects on RNA
synthesis *in vitro*, etc. The extent of nuclear protein
phosphorylation was found to increase after treatment with
biological stimulants such as hormones, carcinogens and
cyclic AMP. If the phosphorylation of nonhistone proteins is
related to the control of gene expression, then one would
expect to observe changes in phosphorylation which correlate
with changes in activity of the specific genes. Frenster
(156) fractionated calf thymus chromatin into extended (active
in RNA synthesis) and condensed (inactive in RNA synthesis)
chromatin fractions. He found that extended chromatin contains four times as much protein bound phosphate as condensed
chromatin. Kleinsmith and Allfrey (157) have also found that
liver nuclei contain three times as much phosphoprotein per
unit of DNA as calf thymus nuclei at the time when the thymus
is not active in RNA synthesis and undergoes involution.
During the maturation of avian erythrocytes, the chromatin
progresses from a diffuse configuration which is active in
RNA synthesis to dense and heterochromatin appearance which
is transcriptionally inactive. The levels of nuclear phosphoprotein kinases and protein bound phosphorus decrease several

fold during this process of terminal differentiation. Similar correlation between nuclear protein phosphorylation and cellular activity was also observed in differentiating cells of several species (158-161).

Another type of physiological situation where changes in nonhistone protein phosphorylation were observed to parallel the increase in gene activity is the mitogenic or hormonal stimulation. When lymphocytes are stimulated to proliferate by the exposure to phytohemagglutinin, their capacity for RNA synthesis increases dramatically during the first 24 hours. One of the earliest events which precedes the RNA synthesis is a substantial increase in the rate of nonhistone protein phosphorylation (162). Turkington and Riddle (163) reported an increase in the rate and extent of phosphorylation of nuclear nonhistone proteins in mammary glands treated with insulin, or a combination of hydrocortisone with prolactin under conditions stimulating RNA synthesis. Hormonal stimulation of RNA synthesis was closely coupled to an increased phosphorylation of nuclear nonhistone proteins. An increase in the phosphoprotein kinase activity of rat ventral prostate nuclei within 30 minutes after the injection of testosterone into orchiectomized rats was observed by Ahmed and Ishida (164). These authors also observed a similar increase in rat salivary glands stimulated by isoproterenol. In this system, the prereplication phase involved an early inhibition of RNA synthesis followed by a stimulation which correlated directly with a simultaneous increase in nonhistone protein phosphorylation. Increased nonhistone protein phosphorylation accompanying hormone stimulated cell differentiation and proliferation was reported in a variety of different systems, including ovaries stimulated by chorionic gonadotropin (165), cortisol and corticosterone stimulated liver (150,166), aldosterone stimulated kidney (167) and other systems (168-171).

The activity of phosphoprotein kinases in the cytoplasm and chromatin of rat liver increased several fold after the administration of an azo-dye carcinogen, N,N-dimethyl-p-(m-tolylazo) aniline (172). This increase was accompanied by the phosphorylation of several high molecular weight nonhistone proteins in liver chromatin of the treated rats (Fig. 11). At the same time, the ability of isolated liver chromatin to template for the *in vitro* RNA synthesis was found considerably higher than in the controls. These results are consistent with the hypothesis that chromosomal protein phosphorylation is closely coupled to the transcription of RNA and may be involved in the regulation of specific genes. A dramatic increase in phosphorylation of high molecular weight nonhistone proteins in SV 40 transformed WI-38 fibroblasts reported

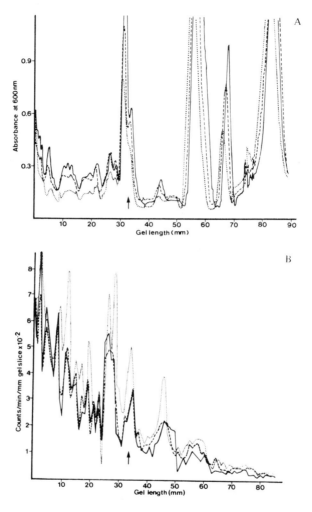

Figure 11. A) Optical scans of polyacrylamide gel electro-
phoreograms of liver chromatin proteins from normal 3'-MDAB
fed rats. B) Distribution of ^{32}P radioactivity in 1 mm
slices of polyacrylamide gels shown in Fig. 11A. (———) Con-
trol rats; (----) rats on 3'-MDAB diet for 15 days; (••••)
rats on 3'-MDAB diet for 28 days. The position of very lysine
rich KAP (F1 or I) histones indicated by an arrow. Reprinted
with permission (172).

by Pumo *et al.* (173) corroborates the observations of Chiu
et al. (172) and further supports the possibility that phosphorylation of selected chromosomal proteins is either the
prerequisite or a consequence of transcriptional changes
accompanying cellular proliferation. Increased phosphorylation of chromosomal nonhistone proteins was also seen in
α-1,2,3,4,5,6-hexachlorohexane stimulated rat liver (174)
after partial hepatectomy (175), and in folic acid induced
regeneration of kidney (176).

Direct evidence that phosphoproteins stimulated transcription of DNA and chromatin in cell free systems was
reported by several investigators (18,153,177-182). Initial
experiments showed that nuclear phosphoproteins can prevent
much of the histone mediated inhibition of *in vitro* RNA synthesis if interacted directly with the histones prior to
their association with DNA. Kamiyama and Dastugue (177)
separated the kinase activity from the phosphoprotein substrate and showed that the reversal by nonhistone proteins of
the histone mediated inhibition of RNA synthesis was not
directly related to the protein phosphokinase activity since
the protein kinase itself did not suppress this inhibition
(177). The ability of phosphoproteins to increase RNA synthesis in the presence of chromatin templates or DNA was
found to be directly related to the extent of their phosphorylation (183). Teng *et al.* (18) found that fractions of nuclear phosphoproteins which selectively associated with homologous DNA were particularly effective in stimulating the *in
vitro* RNA transcription. Shea and Kleinsmith (179) have further tested this correlation in a system employing rat DNA as
a template with purified rat liver RNA polymerase. They
found that the addition of rat liver phosphoproteins can more
than double the *in vitro* synthesis of RNA. This effect of
phosphoproteins was blocked by preincubating them with phos-
phatase. This suggests that the phosphate group of phosphoproteins was critical for the observed stimulation. The
effect of phosphoproteins on RNA synthesis was template
specific, since the synthesis of RNA from heterologous (calf
thymus) DNA source was not found to be stimulated by rat
liver phosphoproteins. Stimulatory effects of nuclear phosphoproteins on RNA synthesis supported by homologous DNA were
also observed by Wang and his associates (180-182,184). The
in vitro transcription of rat liver or Walker tumor chromatin
was stimulated by addition of nonhistone proteins of the same
tissue origin while the heterologous nonhistone proteins were
not stimulatory. DNA-RNA hybridization experiments with the
transcribed RNA showed that RNA synthesized from activated
chromatin was transcribed from additional DNA sequences.

Homologous nonhistone protein fractions activated the transcription of DNA sequences different from those activated by heterologous nonhistone protein fractions. Recently, Kostraba *et al.* (182) isolated a phosphoprotein fraction from Ehrlich ascites tumor cells which stimulated DNA-templated RNA synthesis *in vitro*. This stimulation was template-specific, i.e., effective only with DNA from Ehrlich ascites, but not from rat liver, calf thymus or chick erthrocytes. The stimulatory effect of these proteins appeared to be RNA polymerase-specific because they stimulated only Ehrlich ascites RNA polymerase and not RNA polymerase from *Micrococcus luteus*.

Since in prokaryotes expression of genetic information can be controlled by regulatory proteins which bind to specific regions of DNA, it can be expected that phosphoproteins which can regulate gene activity in eukaryotes would also bind to DNA in a selective fashion. The DNA binding properties of nuclear phosphoproteins were discussed in the previous section.

Relatively little is known about the enzymes which are responsible for the phosphorylation of nuclear phosphoproteins. Several investigators have attempted the fractionation of nuclear protein phosphokinases from rat liver and other tissues. Chromatography on phosphocellulose (185–188) or DEAE-cellulose chromatography (189,190) were used with variable success. Most references reported separation of nuclear protein phosphokinases into 2–4 fractions of similar general properties but differing in their substrate specificities. Kish and Kleinsmith (185) described the separation of calf liver nuclear protein phosphokinases into 12 fractions which differed in their substrate specificities and sensitivity to c-AMP.

In eukaryotic systems, c-AMP affects the transcription of individual genes, especially in hormone-dependent systems. The varying response of different nuclear protein phosphokinases to c-AMP (185,191,192) adds an additional dimension to the gene regulatory selectivity of the nonhistone protein-phosphokinase system. Kinases dependent on c-GMP were also reported in mammalian tissues (193). Proteins which bind c-AMP and c-GMP were obtained by chromatography of chromosomal nonhistone proteins on phosphocellulose (188) and on Bio-Rex 70 resin (140). These cyclic nucleotide receptor proteins inhibited much of the activity of cyclic nucleotide sensitive protein kinases and this inhibition could be diminished by the addition of cyclic nucleotides.

It was suggested by Allfrey and his associates (43) that the cyclic nucleotide effects on various cellular systems may be mediated through their binding to cyclic nucleotide receptor proteins which in turn affect the protein phosphokinases and their influence on DNA transcription. The cyclic nucleotide sensitive protein kinases (cytoplasmic) consist of regulatory (R) and catalytic (C) subunits. The cyclic nucleotide binds to the regulatory subunit which separates from the protein kinase holoenzyme. This leaves an activated catalytic subunit (194,195). In their recent work, Johnson *et al.* (140) identified the calf thymus R subunit of a c-AMP dependent protein kinase as a prominent c-AMP receptor protein and described conditions under which the protein kinase holoenzyme and the C subunits bind to the DNA. At least two protein kinases with c-AMP binding activity were also found in rat liver (196). These enzymes await, however, their more detailed characterization.

Very little work has been done on nuclear protein kinases in the nuclei of malignant cells although several investigators studied protein phosphokinases in the cytoplasm on cancerous tissues. Thomson *et al.* (186) fractionated and compared nuclear phosphoprotein kinases from normal and regenerating rat liver, Novikoff hepatoma, Ehrlich ascites and Walker tumor. Five kinase fractions were found in normal and regenerating rat liver. These enzymes differed in their ion, pH and substrate requirements. Neoplastic growth or liver regeneration changed profoundly the activity patterns of nuclear phosphoprotein kinases. An additional enzymatic activity which eluted at the later part of the 0.3M NaCl elution step was found in several tumors but not in normal, regenerating or embryonic liver (Fig. 12). The tumor associated phosphoprotein kinase was not detected in the cytoplasm of either normal or neoplastic tissues, implying that this enzyme fraction is a true nuclear protein kinase and not a cytoplasmic contaminant. The activity of tumor associated kinases represented 3-12% of the total nuclear kinase activity and was greatly stimulated (138-140% higher) when assayed in the presence of Mn^{2+} instead of Mg^{2+}. Conversely, there was only 30-60% of the original activity retained when Mg^{2+} was replaced in the other five phosphoprotein kinase fractions from normal or neoplastic tissues (Table VII). The phosphokinase specific for tumors selectively phosphoylated chromosomal nonhistone proteins present only in tumors and migrating in the high molecular weight portion of the gel (Fig. 13). There was virtually no phosphorylation of normal rat liver nuclear proteins by this enzyme. It appears that the nuclear phosphoproteins and their kinases are a highly dynamic system responding dramatically to any changes in the differentiated state of cells. Additional evidence that

237

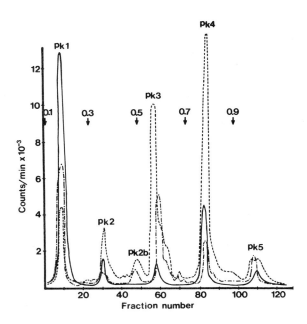

*Figure 12. Nuclear protein kinase activities eluted from
phosphocellulose column. Three mg of phosphoprotein fraction
from normal rat liver (———), Novikoff hepatoma (- - - -) or
Ehrlich ascites nuclei (·——·) were applied to a phosphocellulose column and eluted with 0.1M, 0.3M, 0.5M, 0.7M and 0.9M
NaCl as indicated by the arrows. The flow rate was 0.5 -
0.55 ml/min. One ml fractions were collected and their
protein kinase activity was assayed. Reprinted with permission (186).*

Table VII. Characterization of nuclear protein kinase (PK) fractions from phosphocellulose columns. The conditions for phosphoprotein kinase assays are described in Ref. 186, except that inactivated exogenous substrates (a) or different conc. of Mg^{2+} (b) or Mn^{2+} (c) were used. The values are averages of 2-4 experiments. NRL, normal rat liver; NH, Novikoff hepatoma; EA, Ehrlich ascites (from Ref. 186).

Enzyme peak	Tissue	(a) % of endogenous activity with addition of			(b) Optimal Mg^{2+} concn	(c) Optimal Mn^{2+} concn	Average ratio of activity at optimal Mn^{2+} and Mg^{2+} concn (Mn/Mg)
		Histones	Nonhistones	Casein			
PK_1	NRL	146	103	160	10	0.8	49
	NH	2145	308	-	10	1.0	56
	EA	1930	345	156	10	-	61
PK_2	NRL	102	90	170	20	1.6	66
	NH	373	122	-	10	1.2	113
	EA	298	138	499	25	-	115
PK_{2b}	NRL	-	-	-	-	-	-
	NH	275	950	-	20	1.4	210
	EA	359	1500	64	20	1.4	269
PK_3	NRL	246	294	156	25	0.8	53
	NH	750	181	-	10	0.7	111
	EA	543	475	253	20	-	99
PK_4	NRL	306	400	1161	25	1.6	42
	NH	360	301	-	25	1.0	42
	EA	415	210	469	20	-	25
PK_5	NRL	134	143	539	10	0.6	34
	NH	212	152	-	25	0.9	44
	EA	149	201	304	10	-	60

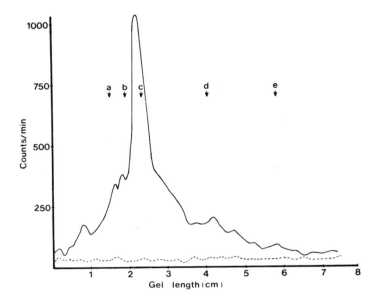

Figure 13. The radioactivity profiles in polyacrylamide gels of heated rat liver phosphoproteins (----) and of heated Ehrlich ascites phosphoproteins (——) which have been phosphorylated with the protein kinase fraction PK 2b isolated from Ehrlich ascites nuclei. 200 μl of the protein kinase PK 2b were incubated with 120 mg of heat inactivated phosphoprotein, 43 mM Tris, 1.4 mM MnCl, and 10 μl of [γ-32P] ATP solution (0.6 μM/ml, 0.5 Ci/mM in final volume 300 μl. After 10 min. at 30°C, the reaction was terminated by the addition of ultra-pure urea to the final concentration of 4M. The tubes were well mixed and the samples dialyzed against electrophoretic sample buffer (0.01M sodium phosphate buffer, pH 7.0, 8M urea, 1.2% sodium dodecyl sulfate, 10% β-mercaptoethanol). The samples were subjected to polyacrylamide gel electrophoresis. The molecular weight markers are indicated by the arrows (a-e). a, hemoglobin (160,000); b, bovine serum albumin (67,000); c, ovalbumin (45,000); d, chymotrypsinogen (25,000); e, cytochrome c (12,4000). Reprinted with permission (186).

phosphoproteins and their kinases indeed relate to cellular genetic activity was demonstrated by Keller *et al.* (197) who fractionated estrogen treated chick oviduct chromatin into template active and inactive components by sucrose gradient centrifugation and by ECTHAM-cellulose chromatography. The fractionated chromatin was assayed for protein kinase activity. The activity was accumulated in the transcriptionally active fraction of chromatin.

VII. Gene Regulation

Stimulated by findings that histone alone cannot restrict DNA in chromatin specifically, several investigators began an intensive search for macromolecules which, perhaps in association with histones, would restrict or activate specific segments of DNA. Evidence based on DNA-RNA hybridization experiments indicates that the DNA transcription in chromatin is tissue specific and that the specificity of this restriction is relatively unaffected by the isolation of chromatin (131,198,199). Further studies seeking the biochemical identity of macromolecules conferring tissue specificity to the DNA restriction by histones were considerably facilitated by the discovery that chromatin components can be dissociated with concentrated salt solutions and brought back together without a substantial loss of transcriptional specificity, providing the dissociation and reassociation take place in the presence of 5.0M urea (11,25,200,201). Dissociation experiments demonstrated that the removal of histones from chromatin increases both the rate of transcription as well as the number of RNA species transcribed (198,202,203). Reconstitution of purified DNA with histones produced transcriptionally inactive nucleohistones. However, if the reconstitution was performed with a mixture of histones, nonhistone proteins and DNA, the resulting artificial chromatin was capable of templating for RNA species similar to those transcribed by the native sample (200). It was concluded that although the histones are necessary to restrict the DNA quantitatively, the qualitative specificity of the restriction is determined by macromolecules present in the nonhistone protein fraction.

The essential role of nonhistone chromatin proteins in tissue specific DNA restriction was further supported by studies on "hybrid chromatin" composed of DNA and chromosomal nonhistone proteins from one tissue combined with the histones from another (25,204). The reconstitution of histones from the chromatin of one tissue to dehistonized chromatin from another tissue, as well as the substitution of histones from one species (cow) to another (rat), demonstrated that the transcriptional specificity of reconstituted chromatin was determined by the DNA-associated nonhistone protein fraction and not by the histones. The histone thus served only as general, quantitative repressors. A similar conclusion was reached by Gilmour and Paul (32) who used chromatography of QAE Sephadex A-50 to separate the chromatin proteins from histones. Again, the specificity of the RNA synthesis directed by reconstituted chromatins depended on the tissue origins of the nonhistone protein fraction with the histones contributing no specificity. However, the DNA-RNA hybridization conditions used in these experiments could produce only information concerning the repetitive fraction of DNA. More recent reports of Harrison *et al.* (205), Bishop *et al.* (206), and Bishop and Rosbash (207) showed that tissue specific globin mRNA is transcribed from unique DNA sequences and that there are only one, or at most a few, copies of the globin gene per genome. Therefore, if the distribution and genome content of most somatic protein genes in eukaryotes are similar to that of the globin gene, the *in vitro* transcripts of these genes could not be detected by low C_ot hybridization. Therefore, the properties of viral reverse transciptase were exploited to synthesize DNA complementary to isolated mRNA (208-210). These cDNA probes proved to be extremely useful for quantitation of specific genes and for the detection of specific mRNA's in eukaryotic cells and in the *in vitro* transcripts of their chromatins. This technique enabled Gilmour and Paul (211) to examine whether globin mRNA sequences can be detected in the RNA transcribed by *E. coli* RNA polymerase from the chromatin of mouse brain (nonhematopoietic) and mouse fetal liver (hematopoietic) by hybridization to globin cDNA. They found that the RNA transcribed from hematopoietic chromatin hybridized to 40% of the globin cDNA at an input ratio of 25,000 while the RNA from brain chromatin did not show significant hybridization above the background levels. Similar conclusions were reached by other investigators (58,212-215).

Using the technique of hybrid chromatins, Gilmour and Paul (211), Axel *et al.* (212), Steggles *et al.* (214), Barrett *et al.* (215) and Chiu *et al.* (58) were able to show that under the stringent hybridization conditions capable of detecting single gene transcripts, the transcriptional expression of the globin gene was directed by chromosomal nonhistone proteins specific for the tissues engaged in hematopoiesis or for nucleated avian reticulocytes.

Taking advantage of the fact that histone synthesis is confined to the S phase of the cell cycle, Stein *et al.* (216)

showed that only the S phase polysomes contain histone mRNA. The histone mRNA was isolated and polyadenylated *in vitro* to make it suitable for reverse transcription. The cDNA complementary to histone mRNA was then used to seek out RNA species similar to histone mRNA which were transcribed *in vitro* on chromatin templates from G_1 or S phase cells and their reconstituted hybrids (217). The S phase transcripts contained RNA species complementary to histone c-DNA. These RNA species were not present in the *in vitro* transcripts of G_1 chromatins (Fig. 14). However, reconstitution of S phase

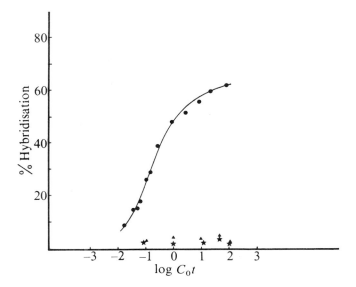

Figure 14. Kinetics of annealing of histone cDNA to in vitro *transcripts of chromatin from G_1 and S phase HeLa S_3 cells. 400 c.p.m. of ^3H-cDNA (27,000 d.p.m. ng^{-1}) were annealed at 52^oC to either 0.15 or 1.5 μg of RNA transcripts from G_1 (▲) or S phase (●) chromatin. 400 c.p.m. of cDNA were also annealed to 1.5 μg of E. coli RNA isolated in the presence of S phase chromatin (★). E. coli RNA was included in each reaction mixture so that the final amount of RNA was 3.75 μg. For details of the experiment, see Ref. 17. Reprinted with permission (17).*

nonhistone proteins to the G_1 histones and DNA produced chromatins capable of the *in vitro* transcription of histone-mRNA-like species. This work of Stein and his associates again emphasizes the important role of chromosomal nonhistone protein in the regulation of genes in eukaryotic organisms.

The search for specific gene regulatory proteins was further narrowed by Chiu *et al.* (58) who isolated a fraction of chromosomal nonhistone proteins from chicken erthrocytes (less than 10% of the total chromatin protein content). This fraction was qualitatively tissue specific and heterogeneous by electrophoretic criteria. Reconstitution of this reticulocyte fraction (NP proteins) to chicken brain DNA, histones and the bulk of nonhistone proteins (UP proteins) resulted in chromatin capable of the *in vitro* transcription of RNA species complementary to globin cDNA. On the other hand, reconstitution of chicken brain NP proteins to reticulocyte DNA, histones and the bulk of nonhistone proteins (UP) did not produce chromatins with transcriptionally active globin genes (Fig. 15). It is of interest that this chromosomal protein fraction NP also contains proteins which interact

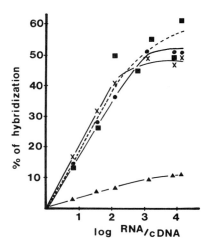

Figure 15. Hybridization of globin cDNA to RNAs transcribed from hybrid chromatins. Various chromatin fractions were exchanged between chicken reticulocyte and brain chromatins. UP; proteins extracted by 5M urea in 50 mM sodium phosphate buffer, pH 7.6 (over 90% of total nonhistone protein content of chromatin); HP, histones; NP, DNA-binding proteins (8 to 10% of total nonhistone protein content of chromatin); DNA, purified DNA. Hybrid chromatins: ●——●, chicken DNA and reticulocyte UP and reticulocyte HP and reticulocyte NP; ■——■, chicken DNA and reticulocyte UP and brain HP and reticulocyte NP; ✕——✕; chicken DNA and brain UP and brain HP and reticulocyte NP; ▲——▲, chicken DNA and brain UP and brain HP and brain NP. Reprinted with permission (58).

with homologous DNA to form immunologically tissue specific complexes. It was shown that the immunological specificity of these complexes changes with cell differentiation, development and carcinogenesis (30,31,66,69,72).

Not all the chromosomal nonhistone proteins act as gene activators. Kostraba and Wang (218) recently isolated a nonhistone protein present in the Ehrlich ascites chromatin. This protein had a minimum molecular weight of 10-11,000, was acidic (acidic/basic amino acids = 1.42), associated with DNA and inhibited its *in vitro* transcription by homologous RNA polymerase. This protein was highly phosphorylated (2.7% alkali-labile phosphorus) and it repressed the DNA transcription by inhibiting the initiation of RNA chains. The RNA elongation process *in vitro* was not affected by this protein. Low molecular weight nonhistone proteins with affinity for DNA were also found in condensed fractions of chromatin (219) and in the DNA-binding nonhistone protein fraction NP of Chiu *et al.* (31,45). Whether any of these low molecular weight proteins are similar to the inhibitory protein of Kostraba and Wang (218) remains to be determined.

ACKNOWLEDGEMENTS: Supported by NCI Contracts NO1 CB 53896, NO1 CP 65730 and USPHS Grant CA 18389.

REFERENCES

1. Stedman, E., and Stedman, E., *Nature 166,* 780 (1950).
2. Allfrey, V. G., Littau, V. C., and Mirsky, A. E., *Proc. Nat. Acad. Sci. U.S.A. 49,* 414 (1963).
3. Huang, R. C. C., and Bonner, J., *Proc. Nat. Acad. Sci. U.S.A. 48,* 1216 (1962).
4. Hnilica, L. S., *The Structure and Biological Functions of Histones,* Chemical Rubber Publishing Co., Cleveland, Ohio (1972).
5. Baserga, R., and Stein, G. S., *Fed. Proc. 30,* 1752 (1971).
6. Spelsberg, T. C., Wilhelm, J. A., and Hnilica, L. S., *Sub-Cell. Biochem. 1,* 107 (1972).
7. Stein, G. S., Spelsberg, T. C., and Kleinsmith, L. J., *Science 183,* 891 (1974).
8. Mirsky, A. E., and Pollister, A. W., *J. Gen. Physiol. 30,* 117 (1946).
9. Wang, T. Y., *J. Biol. Chem. 241,* 2913 (1966).
10. Wang, T. Y., *J. Biol. Chem. 242,* 1220 (1967).
11. Bekhor, I., Kung, G. M., and Bonner, J., *J. Mol. Biol. 39,* 351 (1969).
12. Wilson, E. M., and Spelsberg, T. C., *Biochim. Biophys. Acta 322,* 145 (1973).

13. Murphy, R. F., and Bonner, J., *Biochim. Biophys. Acta* *405*, 62 (1975).
14. Benjamin, W., and Gellhorn, A., *Proc. Nat. Acad. Sci. U.S.A.*,*59*, 262 (1968).
15. Marushige, K., and Bonner, J., *Biochemistry 7*, 3149 (1968).
16. Elgin, S. C. R., and Bonner, J., *Biochemistry 9*, 4440 (1970).
17. Shelton, K. R., and Allfrey, V. G., *Nature 228*, 132 (1970).
18. Teng, C. S., Teng, C. T., and Allfrey, V. G., *J. Biol. Chem. 246*, 3597 (1971).
19. LeStourgeon, W. M., and Rush, H. P., *Science 174*, 1233 (1971).
20. Shirely, T., and Huang, R. C. C., *Biochemistry 8*, 4138 (1969).
21. Graziano, S. L., and Huang, R. C. C., *Biochemistry 10*, 4770 (1971).
22. Hill, R. J., Poccia, D. L., and Doty, P., *J. Mol. Biol. 61*, 445 (1971).
23. Levy, S., Simpson, R. T., and Sober, H. A., *Biochemistry 11*, 1547 (1972).
24. Levy, R., Levy, S., Rosenberg, S. A., and Simpson, R. T., *Biochemistry 12, 224* (1973).
25. Spelsberg, T. C., and Hnilica, L. S., *Biochem. J. 120*, 435 (1970).
26. Richter, K. H., and Sekeris, C. E., *Arch. Biochem. Biophys. 148*, 44 (1972).
27. Arnold, E. A., and Young, K. E., *Biochim. Biophys. Acta 257*, 482 (1972).
28. Yoshida, M., and Shimura, K., *Biochim. Biophys. Acta 263*, 690 (1972).
29. Shaw, L. M. J., and Huang, R. C. C., *Biochemistry 9*, 4530 (1970).
30. Wakabayashi, K., Wang, S., and Hnilica, L. S., *Biochemistry 13*, 1027 (1974).
31. Chiu, J. F., Hunt, M., and Hnilica, L. S., *Cancer Res. 35*, 913 (1975).
32. Gilmour, R. S., and Paul, J., *FEBS Lett. 9*, 242 (1970).
33. Chaudhuri, S., *Biochim. Biophys. Acta*,*322*, 155 (1973).
34. Bonner, J., Chalkley, G. R., Dahmus, M., Fambrough, D., Fujimura, F., Huang, R. C. C., Huberman, J., Jensen, R., Marushige, K., Ohlenbush, H., Olivera, B. M., and Widholm, J., *Methods Enzymol. 12B*, 3 (1968).
35. Cameron, I. L., and Jeter, J. R., Jr., *Acidic Proteins of the Nucleus*, Academic Press, New York (1974).
36. Thomson, J. A., Brade, W. P., Chiu, J. F., Hnilica, L.S.,

and McGill, M., *Can. J. Biochem. 50,* 86 (1976).

37. MacGillivary, A. J., Carroll, D., and Paul, J., *FEBS Lett. 13,* 204 (1971).

38. MacGillivary, A. J., Cameron, A., Krauze, R. J., Rickwood, D., and Paul, J., *Biochem. Biophys. Acta 277,* 384 (1972).

39. Alberts, B. M., Amodio, F. J., Jenkins, M., Gutmann, E. D., and Ferris, F. L., *Cold Spring Harb. Symp. Quant. Biol. 33,* 289 (1968).

40. Kleinsmith, L. J., Heidema, J., and Carroll, A., *Nature 226,* 1025 (1970).

41. Wakabayashi, K., Wang, S., Hord, G., and Hnilica, L. S., *FEBS Lett. 32,* 46 (1973).

42. Van den Broek, H. W. J., Nooden, L. D., Sevall, J. S., and Bonner, J., *Biochemistry 12,* 229 (1973).

43. Allfrey, V. G., Inoue, A., Karn, J., Johnson, E. M., Good, R. A., and Hadden, J. W., In *The Structure and Function of Chromatin,* Ciba Foundation Symposium No. 28, p. 199 (1975).

44. Johns, E. W., Goodwin, G. H., Walker, J. M., and Saunders, C., In *The Structure and Function of Chromatin* Ciba Foundation Symposium No. 28, p. 95 (1975).

45. Chiu, J. F., Wang, S., Fujitani, H., and Hnilica, L. S., *Biochemistry 14,* 4552 (1975).

46. Shelton, K. R., and Neelin, J., *Biochemistry 10,* 2342 (1971).

47. Garrard, W., and Bonner, J., *J. Biol. Chem. 249,* 5570 (1974).

48. Yeoman, L. C., Taylor, C. W., Jordan, J. J., and Busch, H., *Biochem. Biophys. Res. Commun. 53,* 1067 (1973).

49. Dounce, A. L., and Ickowicz, R., *Arch. Biochem. Biophys. 131,* 359 (1969).

50. Howk, R., and Wang, T. Y., *Arch. Biochem. Biophys. 136,* 422 (1970).

51. Chiu, J. F., and Sung, S. C., *Biochem. Biophys. Res. Commun. 57,* 740 (1972).

52. Zimmerman, S. B., and Levin, C. J., *J. Biol. Chem. 250,* 149 (1975).

53. Miller, O. L., and Beatty, B. R., *J. Cell Physiol. 74, Suppl. 1,* 225 (1969).

54. Reid, B. R., and Cole, R. D., *Proc. Nat. Acad. Sci. U.S.A. 51,* 1044 (1964).

55. Chae, C. B., and Carter, D. B., *Biochem. Biophys. Res. Commun. 57,* 740 (1974).

56. Chong, M. T., Garrard, W., and Bonner, J., *Biochemistry 13,* 5128 (1974).

57. David, A. R., and Burdman, J. A., *J. Neurochem. 15,*

25 (1968).

58. Chiu, J. F., Tsai, Y. H., Sakuma, K., and Hnilica, L. S., *J. Biol. Chem. 250,* 9431 (1975).

59. Wu, F. C., Elgin, S. C. R., and Hood, L. E., *Biochemistry 12,* 2792 (1973).

60. Wu, F. C., Elgin, S. C. R., and Hood, L. E., *J. Mol. Evol. 5,* 87 (1975).

61. Barrett, T., and Gould, H. J., *Biochem. Biophys. Acta 294,* 165 (1973).

62. Platz, R. D., Kish, V. M., and Kleinsmith, L. J., *FEBS Lett. 12,* 38 (1970).

63. Gronow, M., and Thackrah, T., *Arch. Biochem. Biophys. 158,* 377 (1973).

64. Chytil, F., and Spelsberg, T. C., *Nature New Biol. 233,* 215 (1971).

65. Wasserman, E., and Levine, L., *J. Immunol. 87,* 290 (1961).

66. Wakabayashi, K., and Hnilica, L. S., *Nature New Biol. 242,* 73 (1973).

67. Zardi, L., Lin, J., Petersen, R. O., and Baserga, R., *Cold Spring Harbor Conferences on Cell Proliferation, Vol. 1,* pp. 729-741 (1974).

68. Okita, K., and Zardi, L., *Exp. Cell Res. 86,* 59 (1974).

69. Chytil, F., Glasser, S. R., and Spelsberg, T. C., *Develop. Biol. 37,* 295 (1974).

70. Spelsberg, T. C., Steggles, A. W., Chytil, F., and O'Malley, B. W., *J. Biol. Chem. 247,* 1368 (1972).

71. Zardi, L., Lin, J. C., and Baserga, R., *Nature New Biol. 245,* 211 (1973).

72. Chiu, J. F., Craddock, C., Morris, H. P., and Hnilica, L. S., *FEBS Lett. 42,* 94 (1974).

73. Zardi, L., *Eur. J. Biochem. 55,* 231 (1975).

74. Spelsberg, T. C., Mitchell, W. M., Chytil, F., Wilson, E. M., and O'Malley, B. W., *Biochim. Biophys. Acta 312,* 765 (1973).

75. Chiu, J. F., Wakabayashi, K., Craddock, C., Morris, H. P., and Hnilica, L. S. In *Cell Cycle Control* (eds. Padilla, G. M., Zimmerman, A. M. and Cameron, I. L.) Academic Press, New York, pp. 308-319 (1974).

76. Dierks-Ventling, C., and Jost, J., *Eur. J. Biochem. 50,* 33 (1974).

77. Stein, G., Criss, W. E., and Morris, H. P., *Life Sci. 14,* 95 (1974).

78. Lea, M. A., Koch, M. R., and Morris, H. P., *Cancer Res. 35,* 1693 (1975).

79. Wilson, B., Lea, M., Vidali, G., and Allfrey, V. G., *Cancer Res. 35,* 2954 (1975).

80. Chae, C. B., Smith, M. C. and Morris, H. P., *Biochem. Biophys. Res. Commun. 60,* 1468 (1974).

81. Tuan, D., Smith, S., Folkman, J., and Merler, E., *Biochemistry 12,* 3159 (1973).

82. Ruiz-Carrillo, A., Wangh, L. J., Littau, V. C., and Allfrey, V. G., *J. Biol. Chem. 249,* 7358 (1974).

83. Sanders, L. A., *Biochemistry 13,* 527 (1974).

84. Harlow, R., and Wells, J. R. E., *Biochemistry 14,* 2665 (1975).

85. Burchard, J., Mazen, A., and Champagne, M., *Biochim. Biophys. Acta 405,* 434 (1975).

86. Davies, R. H., Copenhaver, J. H., and Carver, M. J., *Int. J. Biochem. 6,* 399 (1975).

87. Teng, C. S., *Biochim. Biophys. Acta 366,* 385 (1974).

88. LeStourgeon, W., Nations, C., and Rusch, A., *Arch. Biochem. Biophys. 159,* 861 (1973).

89. Jeter, J. R., Jr., and Cameron, I. L., In *Acidic Proteins of the Nucleus* (eds. Cameron, I. L. and Jeter, J. R., Jr.) Academic Press, New York, pp. 213-245 (1974).

90. Yeoman, L., Taylor, C., Jordan, J., and Busch, H., *Cancer Res. 35,* 1249 (1975).

91. Bhorjee, J. S., and Pederson, T., *Proc. Nat. Acad. Sci. U.S.A. 69,* 3345 (1972).

92. Borun, T. W., and Stein, G. S., *J. Cell Biol. 52,* 308 (1972).

93. Nicolini, C., Ng, S., and Baserga, R., *Proc. Nat. Acad. Sci. U.S.A. 72,* 2361 (1975).

94. Platz, R. D., Grimes, S. R., Meistrich, M. L., and Hnilica, L. S., *J. Biol. Chem. 250,* 5791 (1975).

95. Kadahama, N., and Turkington, R. W., *J. Biol. Chem. 249,* 6225 (1974).

96. Stein, G., and Baserga, R., *Biochem. Biophys. Res. Commun. 44,* 218 (1971).

97. Shapiro, I. M., and Levina, L., *Exp. Cell Res. 47,* 75 (1967).

98. Shapiro, I. M., and Polykapova, S. T., *Chromosoma 28,* 188 (1969).

99. Sören, L., *Exp. Cell Res. 59,* 244 (1970).

100. Stein, G., and Baserga, R., *Biochem. Biophys. Res. Commun. 41,* 715 (1970).

101. McClure, M., and Hnilica, L. S., *Proc. Int. Cancer Cong. 10th,* 494 (1970).

102. Malpoix, P. J., *Exp. Cell Res. 65,* 393 (1971).

103. Zampetti-Basselet, F., Malpoix, P., and Fievez, M., *Eur. J. Biochem. 9,* 21 (1969).

104. Stein, G. S., and Thrall, C. L., *FEBS Lett.34,* 35 (1973).

105. Teng, C. S., and Hamilton, T. H., *Proc. Nat. Acad. Sci. U.S.A. 63*, 645 (1969).

106. Chung, L. W. K., and Coffey, D. S., *Biochim. Biophys. Acta 247*, 584 (1971).

107. Anderson, K. M., Slavik, M., Evans, A. K., and Couch, R. M., *Exp. Cell Res. 77*, 143 (1973).

108. Hemminki, K., and Bolund, L., *Cell. Diff. 3*, 347 (1975).

109. O'Malley, B. W., and Means, A. M., *Science 183*, 610 (1974).

110. Kadahama, N. K., and Turkington, R. W., *J. Biol. Chem. 249*, 6225 (1974).

111. Johnson, E. M., Karn, J., and Allfrey, V. G., *J. Biol. Chem. 249*, 4990 (1974).

112. Spivak, J. L., *Exp. Cell Res. 91*, 253 (1975).

113. Kostraba, N. C., and Wang, T. Y., *Int. J. Biochem. 1*, 327 (1970).

114. Cohen, M. E., and Hamilton, T. H., *Proc. Nat. Acad. Sci. U.S.A. 72*, 4346 (1975).

115. Gray, R., Herzog, R., and Steffensen, D., *Biochim. Biophys. Acta 157*, 344 (1968).

116. Weisenthal, L. M., and Ruddon, R. W., *Cancer Res. 32*, 1009 (1972).

117. Stein, G., and Baserga, R., *J. Biol. Chem. 245*, 6097 (1970).

118. Baserga, R., and Stein, G., *Fed. Proc. Fed. Amer. Soc. Exp. Biol. 30*, 1752 (1971).

119. Rovera, G., and Baserga, R., *J. Cell Physiol. 77*, 201 (1971).

120. Tsuboi, A., and Baserga, R., *J. Cell Physiol. 80*, 107 (1972).

121. Stein, G., and Matthews, D. E., *Science 181*, 71 (1973).

122. Bhorjee, J. S., and Pederson, T., *Proc. Nat. Acad. Sci. U.S.A. 69*, 3345 (1972).

123. Stein, G., and Borun, T. W., *J. Cell. Biol. 52*, 292 (1972).

124. Salas, J., and Green, H., *Nature New Biol. 229*, 165 (1971).

125. Fox, T. O., and Pardee, A. B., *J. Biol. Chem. 246*, 6159 (1971).

126. Choe, G. K., and Rose, N. R., *Exp. Cell Res. 83*, 271 (1973).

127. Stein, G. H., *Exp. Cell Res. 90*, 237 (1975).

128. Tsai, R. L., and Green, H., *J. Mol. Biol. 73*, 307 (1973).

129. Kraus, M. O., Kleinsmith, L. J., and Stein, G. S.,

Life Sci. 16, 1047 (1975).

130. Paul, J., and Gilmour, R. S., *J. Mol. Biol. 16,* 241 (1966).

131. Bonner, J., Dahmus, M. E., Fambrough, D., Huang, R. C., Marushige, K., and Tuan, Y. H., *Science 159,* 47 (1968).

132. Kleinsmith, L. J., *J. Biol. Chem. 248,* 5648 (1973).

133. Patel, G. L., and Thomas, T. L., *Proc. Nat. Acad. Sci. U.S.A. 70,* 2524 (1973).

134. Wang, S., Chiu, J. F., Klyszjeko-Stefanowicz, Fujitani, H., and Hnilica, L. S., *J. Biol. Chem.* in press.

135. Van den Broek, H. W. J., Nooden, L. D., Sevall, J. S., and Bonner, J., *Biochemistry 12,* 229 (1973).

136. Sevall, J. S., Cockburn, A., Savage, M., and Bonner, J. *Biochemistry 14,* 782 (1975).

137. Johnson, J. D., John, T. S., and Bonner, J., *Biochim. Biophys. Acta 378,* 424 (1975).

138. Crick, F., *Nature 234,* 25 (1971).

139. Umnasky, S. R., Kovalek, Y. I., and Tokarskaya, V. I., *Biochem. Biophys. Acta 383,* 242 (1975).

140. Johnson, E. M., Inoue, A., Crouse, L. J., Allfrey, V.G. and Hadden, J. W., *Biochem. Biophys. Res. Commun. 65,* 714 (1975).

141. Riggs, A. D., Reiness, G., and Zubay, A. G., *Proc. Nat. Acad. Sci. U.S.A. 68,* 1222 (1971).

142. Anderson, W., Schneider, A., Emmer, M., Perlman, R., and Pastan, I., *J. Biol. Chem. 246,* 5929 (1971).

143. Goodwin, G. H., Sanders, C., and Johns, E. W., *Eur. J. Biochem. 38,* 14 (1973).

144. Shooter, K. V., Goodwin, G. H., and Johns, E. W., *Eur. J. Biochem. 47,* 263 (1974).

145. Goodwin, G. H., Nicolas, R. H., and Johns, E. W., *Biochim. Biophys. Acta 405,* 280 (1975).

146. Sheehan, D. M., and Olins, D. E., *Biochim. Biophys. Acta 353,* 438 (1975).

147. Lin, S., and Riggs, A. D., *Nature 228,* 1184 (1971).

148. Kleinsmith, L. J., *J. Cell Physiol. 85,* 459 (1975).

149. Kleinsmith, L. J., in *Acidic Proteins of the Nucleus* (Cameron, I. L. and Jeter, J. R., Jr., eds.) Academic Press, New York, p. 103 (1974).

150. Allfrey, V. G., Johnson, E. M., Karn, J., and Vidali, G., in *Protein Phosphorylation in Control Mechanisms* Huijing, F. and Lee, E. Y. C., eds.) Academic Press, New York, p. 217 (1973).

151. Magun, B. E., in *Acidic Proteins of the Nucleus* (Cameron, I. L. and Jeter, J. R., Jr., eds.) Academic

Press, New York, p. 137 (1974).

152. LeStourgeon, W. M., in *Acidic Proteins of the Nucleus* (Cameron, I. L. and Jeter, J. R., Jr., eds.) Academic Press, New York, p. 60 (1974).

153. Langan, T. A., in *Regulation of Nucleic Acid and Protein Biosynthesis* (Koningsberger, V. V. and Bosch, L., eds.) Elsevin, Amsterdam, p. 233 (1967).

154. Kleinsmith, L. J., Allfrey, V. G., and Mirsky, A. E., *Proc. Nat. Acad. Sci. U.S.A. 55,* 1182 (1966).

155. Kleinsmith, L. J., and Allfrey, V. G., *Biochim. Biophys. Acta 175,* 136 (1969).

156. Frenster, J. H., *Nature 206,* 680 (1965).

157. Kleinsmith, L. J., and Allfrey, V. G., *Biochim. Biophys. Acta 175,* 123 (1969).

158. LeStourgeon, W. M., and Rush, H. P., *Science 174,* 1233 (1971).

159. Platz, R. D., Stein, G. S., and Kleinsmith, L. J., *Biochem. Biophys. Res. Commun. 51,* 735 (1973).

160. Richies, P. G., Harrap, K. R., Sellwood, D., Rickwood, D., and MacGillivary, A. J., *Biochem. Soc. Trans. 1,* 70 (1973).

161. Platz, R. D., and Hnilica, L. S., *Biochem. Biophys. Res. Commun. 54,* 222 (1973).

162. Kleinsmith, L. J., Allfrey, V. G., and Mirsky, A. E., *Science 154,* 780 (1966).

163. Turkington, R. W., and Riddle, M., *J. Biol. Chem. 244,* 6040 (1969).

164. Ahmed, K., and Ishida, H., *Mol. Pharmacol. 7,* 323 (1971).

165. Jungmann, R. A., and Schweppe, J. S., *J. Biol. Chem. 247,* 5535 (1972).

166. Bottoms, G. D., and Jungmann, R. A., *Proc. Soc. Exp. Biol. Med. 144,* 83 (1973).

167. Liew, C. C., Suria, and Gornall, A. G., *Endocrinol. 93,* 1025 (1973).

168. Palmer, W. K., Castagna, M., and Walsh, D. A., *Biochem. J. 143,* 469 (1974).

169. Bergink, E. W., Kloosterboer, H. J., Gruber, M., and Ab, G., *Biochim. Biophys. Acta 294,* 497 (1973).

170. Fugassa, E., Taningher, M., Gallo, G., and Orunesu, M., *Experientia 31,* 522 (1975).

171. Korenman, S. G., Bhalla, R. C., and Stevens, R. H., *Science 183,* 430 (1974).

172. Chiu, J. F., Craddock, C., Getz, S., and Hnilica, L. S., *FEBS Lett. 33,* 247 (1973).

173. Pumo, D. E., Stein, G., and Kleinsmith, L. J., *Biochim. Biophys. Acta 402,* 125 (1975).

174. Brade, W. P., Chiu, J. F., and Hnilica, L. S., *Mol. Pharmacol. 10*, 398 (1974).

175. Chiu, J. F., Brade, W. P., Thomson, J., Tsai, Y. H., and Hnilica, L. S., *Exp. Cell Res. 91*, 200 (1975).

176. Brade, W. P., Thomson, J. A., Chiu, J. F., and Hnilica, L. S., *Exp. Cell Res. 84*, 183 (1974).

177. Kamiyama, N., and Dastugue, B., *Biochem. Biophys. Res. Commun. 44*, 29 (1971).

178. Spelsberg, T. C., and Hnilica, L. S., *Biochim. Biophys. Acta 195*, 63 (1969).

179. Shea, M., and Kleinsmith, L. J., *Biochem. Biophys. Res. Commun. 50*, 473 (1973).

180. Kostraba, N. C., and Wang, T. Y., *Biochim. Biophys. Acta 262*, 169 (1972).

181. Kostraba, N. C., and Wang, T. Y., *Exp. Cell Res. 80*, 291 (1973).

182. Kostraba, N. C., and Wang, T. Y., *J. Biol. Chem. 250*, 1548 (1975).

183. Kamiyama, N., Dastugue, B., Defer, N., and Kruh, J., *Biochim. Biophys. Acta 277*, 576 (1972).

184. Kamiyama, N., and Wang, T. Y., *Biochim. Biophys. Acta 228*, 563 (1971).

185. Kish, V. M., and Kleinsmith, L. J., *J. Biol. Chem. 249*, 750 (1974).

186. Thomson, J. A., Chiu, J. F., and Hnilica, L. S., *Biochim. Biophys. Acta 407*, 114 (1975).

187. Ruddon, R. W., and Anderson, S. L., *Biochem. Biophys. Res. Commun. 46*, 1499 (1972).

188. Rikans, L. E., and Ruddon, R. W., *Biochem. Biophys. Res. Commun. 54*, 387 (1973).

189. Farron-Furstenthal, F., *Biochem. Biophys. Res. Commun. 67*, 307 (1975).

190. Gamo, S., and Lindell, T. J., *Life Sci. 15*, 2179 (1975).

191. Johnson, E. M., and Allfrey, V. G., *Arch. Biochem. Biophys. 152*, 786 (1972).

192. Beavo, J. A., Bechtel, P. J., and Krebs, E. G., *Proc. Nat. Acad. Sci. U.S.A. 71*, 3580 (1974).

193. Kuo, J. F., Sanes, J., and Greengard, P., *Fed. Proc. 29*, 601 (1970).

194. Gill, G. N., and Garren, L. D., *Biochem. Biophys. Res. Commun. 39*, 335 (1970).

195. Tao, M., Salas, M. L., and Lipmann, F., *Proc. Nat. Acad. Sci. U.S.A. 67*, 408 (1970).

196. Johnson, E., Hadden, J., Inoue, A., and Allfrey, V. G., *Biochemistry 14*, 3873 (1975).

197. Keller, R. K., Socher, S. H., Krall, J. F., Chandra, T. and O'Malley, B. W., *Biochem. Biophys. Res. Commun.*

66, 453 (1975).

198. Paul, J., and Gilmour, R. S., *J. Mol. Biol. 34*, 305 (1968).

199. Smith, K. D., Church, R. B., and McCarthy, B. J., *Biochemistry 8*, 4271 (1969).

200. Gilmour, R. S., and Paul, J., *J. Mol. Biol. 40*, 137 (1969).

201. Huang, R. C. C., and Huang, P. C., *J. Mol. Biol. 39*, 365 (1969).

202. Tau, C. H., and Miyagi, M., *J. Mol. Biol. 50*, 641 (1970).

203. Spelsberg, T. C., and Hnilica, L. S., *Biochim. Biophys. Acta 228*, 212 (1971).

204. Spelsberg, T. C., Hnilica, L. S., and Ansevin, A. T., *Biochim. Biophys. Acta 228*, 550 (1971).

205. Harrison, P. R., Hell, A., Birne, G. P., and Paul, J., *Nature 239*, 219 (1972).

206. Bishop, J. O., Pemberton, R., and Baglioni, C., *Nature New Biol. 235*, 231 (1972).

207. Bishop, J. O., and Rosbash, M., *Nature New Biol. 241*, 204 (1973).

208. Verma, I. M., Temple, G. F., Fan, H., and Baltimore, D., *Nature New Biol. 235*, 163 (1972).

209. Ross, J., Aviv, H., Scolnick, E., and Leder, P., *Proc. Nat. Acad. Sci. U.S.A. 69*, 264 (1972).

210. Kacian, D. L., Spiegelman, S., Bank, A., Terada, M., Metafora, S., Dow, L., and Marks, P. A., *Nature New Biol. 235*, 167 (1972).

211. Gilmour, R. S., and Paul, J., *Proc. Nat. Acad. Sci. U.S.A. 70*, 3440 (1973).

212. Axel, R., Cedar, H., and Felsenfeld, G., *Proc. Nat. Acad. Sci. U.S.A. 70*, 2029 (1973).

213. Young, B. D., Harrison, P. R., Gilmour, R. S., Birnie, G. D., Hell, A., Humphries, S. E., and Paul, J., *J. Mol. Biol. 84*, 555 (1974).

214. Steggles, A. W., Wilson, G. N., Kantor, J. A., Picciano, D. K., Flavely, A. K., and Anderson, W. F., *Proc. Nat. Acad. Sci. U.S.A. 71*, 1219 (1974).

215. Barrett, T., Maryanka, D., Hamlyn, P. H., and Gould, H. J., *Proc. Nat. Acad. Sci. U.S.A. 71*, 5057 (1974).

216. Stein, J. L., Thrall, C. L., Park, W. D., Mans, R. J., and Stein, G. S., *Science 189*, 557 (1975).

217. Stein, G., Park, W., Thrall, C., Mans, R., and Stein, J., *Nature 257*, 764 (1975).

218. Kostraba, N. C., and Wang, T. Y., *J. Biol. Chem. 250*, 8938 (1975).

219. Pederson, T., and Bhorjee, J. S., *Biochemistry 14*, 3238 (1975).

Chapter 7

REGULATION OF GENE EXPRESSION

IN CHICK OVIDUCT

MING-JER TSAI and B. W. O'MALLEY

Department of Cell Biology
Baylor College of Medicine
Houston, Texas 77030

I.   Introduction

In the past decade, a great deal of progress has been
made towards an understanding of the regulation of gene
expression in procaryotic systems (1). Much of this progress
has been achieved by studying the *in vitro* transcription of
well-defined bacterial and bacteriophage DNA templates with
purified bacterial RNA polymerase. Through the use of these
systems, the roles of many regulatory elements, such as
specific gene repressors, activators, and termination factors, have been elucidated. A complete understanding of the
positive and negative controls which function to regulate
the transcription of certain bacterial operons, notably the
lactose operon of *Escherichia coli,* seems close at hand.
Furthermore, the ability to faithfully reconstitute the cellular transcription of the lactose operon by manipulation of
the various purified components indicates the degree of
sophistication which has been attained. With the tools of
molecular biology and genetics currently available, studies
on the regulation of other genetic elements of bacterial and
bacteriophage-infected cells should continue to progress
rapidly.

Our current understanding of the regulation of gene
expression in eucaryotic organisms lags behind work in the
procaryotic field. Analogous studies on the *in vitro* transcription of eucaryotic genomes pose more difficult technical
problems. For instance, the genetic complexity of eucaryotic organisms is several orders of magnitude greater than
that of procaryotic organisms, making the detection of specific gene products a considerably more arduous task. While
the use of genetic mutations has proven to be an immeasurable

aid in defining the role of many regulatory elements in bacteria, few workable systems for obtaining genetic information exist for eucaryotes. Furthermore, the DNA of higher organisms is complexed with a large variety of nonhistone and histone proteins in a structure which remains largely undefined. Even the nature of the primary transcription product leading to the synthesis of messenger RNA is still in question. Such problems have inhibited progress towards understanding the regulation of gene expression in eucaryotes.

Our approach to study gene regulation has been to utilize chromatin as a template for *in vitro* transcriptional studies. Since the histone and nonhistone chromosomal proteins may play an integral role in gene expression, we feel the proper template for *in vitro* transcriptional studies is chromatin and not deproteinized DNA. Our initial efforts have been to characterize the interaction of purified RNA polymerase with unsheared chromatin. The process of initiation of RNA synthesis on chromatin has received special attention, as this step appears to provide the greatest potential for controlling gene expression. In this article, we will discuss our recent findings on the transcription of chromatin isolated from chick oviduct.

The chick oviduct provides an excellent model system for the study of the control of gene expression. Chronic administration of estrogenic compounds to the immature chick over a period of 10 to 12 days results in the growth and differentiation of the oviduct (2). Several specific proteins are produced by the oviduct after estrogen administration. These changes in the oviduct appear to be mediated mainly at the level of transcription (3). Dramatic changes in oviduct endogenous RNA polymerase activity (2,4), nuclear RNA synthesis (5,6,7) and chromatin template activity (2,4,8) occur following estrogen administration. Furthermore, the appearance of messenger RNA for a specific induced protein, ovalbumin, precedes the accumulation of that protein during estrogen-mediated growth (6,9,10). If the estrogen stimulated chick is withdrawn from steroid hormone, the mRNA for ovalbumin gradually disappears from the oviduct over a period of several days (7). During withdrawal, however, the oviduct appears to remain in a quasi-differentiated state, as readministration of estrogen results in the production of ovalbumin mRNA within 1 hour (7,11-13). Consequently, the chick oviduct provides an opportunity to study a specific eucaryotic tissue in which the expression of genetic information can be dramatically altered. Furthermore, a specific marker for following the change in genetic expression is available. We have, thus, undertaken a study of the *in vitro* transcription of chick oviduct chromatin by exogenously added RNA polymerase. We have been particularly interested in the transcription of the ovalbumin gene as assayed by hybridization to complementary DNA transcribed from ovalbumin messenger RNA, and the role of chromosomal proteins in regulation of the expression of this gene. From these studies, we hope to learn the mechanism by which gene expression is altered following hormone uptake by estrogen responsive cells. In so doing, it is likely that much will be learned about the basic process of transcription in eucaryotic organisms.

II. Kinetics of RNA Chain Initiation

The initiation of RNA synthesis can be divided into two basic processes. The first involves selection of the proper site on DNA where initiation is to occur and formation of a binary complex between RNA polymerase and DNA at this site. The second process is the actual initiation of an RNA chain by formation of the first phosphodiester bond between two nucleoside triphosphates. As a first step towards a study of the *in vitro* transcription of chromatin, the kinetics of these two processes on chick oviduct chromatin and on deproteinized DNA were examined utilizing *E. coli* RNA polymerase.

Chamberlin and co-workers (14) have extensively studied the interaction between *E. coli* RNA polymerase and bacteriophage T7 DNA leading to the formation of binary complexes capable of initiating RNA synthesis. Utilizing this system, the formation of binary complexes was shown to occur through several intermediate steps:

$$(1) \qquad\qquad (2) \qquad\qquad (3)$$

$$(\text{DNA-ENZ}) \rightleftarrows \text{DNA} + \text{ENZ} \rightleftarrows (\text{DNA-ENZ})_I \rightleftarrows (\text{DNA-ENZ})_{RS}$$

Nonspecific complex   I complex   RS complex

When RNA polymerase is incubated with DNA, the enzyme initially binds randomly and reversibly to DNA to form a series of nonspecific complexes (step 1). If such binding occurs at or near a true initiation site for RNA synthesis, an initial preinitiation complex (I complex) is formed (step 2). This complex must then undergo a transition, involving the local opening of the DNA duplex structure, to form a highly stable complex (RS complex) capable of rapidly initiating RNA synthesis (step 3).

To measure the formation of the highly stable RS complex, the bacterial RNA polymerase inhibitor, rifampicin can be utilized. Rifampicin is an inhibitor of RNA synthesis which

257

acts prior to the formation of the first phosphodiester bond, but which has no effect on RNA chain elongation (15,16). When RNA polymerase present in the RS complex is challenged with a mixture of the four ribonucleoside triphosphates and rifampicin, the enzyme is highly efficient at initiating RNA synthesis. The lack of inhibition of this complex has been shown to be due to the rapid rate of RNA chain initiation, rather than the RS complex having an absolute resistance to rifampicin (17,18). Enzyme molecules which are bound at nonspecific or I complexes or are free in solution, however, are largely incapable of initiating RNA synthesis in such a challenge experiment. Thus, the use of the rifampicin-ribonucleotide challenge experiment allows differentiation of RNA molecules present as an RS complex from other forms of the enzyme.

To follow the kinetics of formation of RS complexes, RNA synthesis is initiated by the simultaneous addition of rifampicin and ribonucleotides. The amount of RNA synthesis at each time point is directly proportional to the amount of RS complex formed. The results of such a time course of RS complex formation with either chick DNA or chick oviduct chromatin is shown in Figure 1. For this experiment, chromatin isolated from fully stimulated chicks (14 days of diethylstilbestrol [DES] treatment) was used. The formation of RS complexes on chick DNA reached a maximum level by about 10 minutes of preincubation. On chromatin, the maximum was not reached until much longer preincubation periods of about 40 minutes. As shown on the insets to Figure 1, semilogarithmic plots of the maximum level of RNA synthesis minus the level of RNA synthesis at each preincubation time versus the time of preincubation were linear for both templates. This result would only be expected if the formation of RS complexes occurred via a first order process. From these plots, the half time of formation of the RS complex ($t_{1/2}$) between *E. coli* RNA polymerase and chick DNA was found to be 1.3 minutes. On the other hand, the $t_{1/2}$ of formation of the RS complex on chromatin was 9.4 minutes or approximately seven times slower than on deproteinized DNA.

The slow rate of RS complex formation on chromatin might possibly be attributed to a gradual change in the chromatin structure during the preincubation period, leading to the exposure of new initiation sites. To test this possibility, chromatin was incubated in the absence of RNA polymerase for varying lengths of time. This pretreated chromatin was then preincubated with RNA polymerase for 15 minutes. The level of RNA polymerase activity (Figure 1B, open triangles) was constant for chromatin pretreated for periods up to 60 min-

258

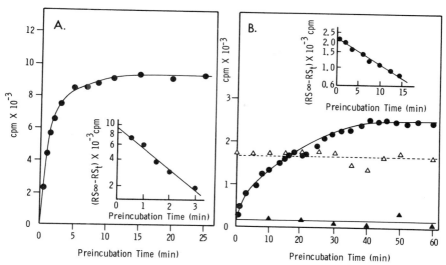

Figure 1. Time course of formation of RS complex. (A) E. coli RNA polymerase (1.4 µg) was incubated at 37°C with chick DNA (0.75 µg) in 0.1 ml of preincubation buffer containing 62.5 mM Tris-HCl, pH 7.9, 1.25 mM MnCl₂, 62.5 mM (NH₄)₂SO₄, 2.5 mM 2-mercaptoethanol, and 0.5 mg/ml bovine serum albumin. After the indicated time interval, 0.025 ml of ribonucleoside triphosphate mixture containing 0.75 mM each of ATP, CTP, GTP, [³H] UTP (0.2 mC/ml), and 0.2 mg/ml rifampicin was added. RNA synthesis was carried out at 37°C for 1.5 minutes and reactions terminated by the addition of 5 ml of cold 5% tri-chloroacetic acid containing 0.01 M sodium pyrophosphate. Samples were analyzed for incorporation of [³H] UTP into acid-insoluble material as described elsewhere (19). (B) E. coli RNA polymerase (7.0 µg) was incubated at 37°C with chick ovi-duct chromatin (5.0 µg) in 0.2 ml of preincubation buffer for the indicated time intervals (●). Controls were run in which 5.0 µg of chromatin was incubated without any added enzyme (▲) to test for endogenous RNA polymerase activity and in which 5.0 µg of chromatin was preincubated for the indicated time interval without any added enzyme, followed by the addi-tion of 7.0 g of RNA polymerase for 15 minutes (△) to test for changes in chromatin during incubation. RNA synthesis was initiated by the addition of 0.05 ml of ribonucleoside triphosphate mixture. Chromatin was isolated by the method of Tsai et al. (22) from the oviducts of chicks which had re-ceived daily injections of ethylstilbestrol (2.5 mg) for 14 days (14 day DES stimulated). All other details were as des-cribed above. Reprinted with permission.

utes and closely corresponded to the activity of chromatin which had received no pretreatment. Thus, it is unlikely that any artifactual modification of the chromatin structure occurs during 60 minutes of preincubation. The level of endogenous RNA polymerase activity in the chromatin was also checked by measuring RNA synthesis in the absence of added *E. coli* RNA polymerase. The endogenous RNA polymerase activity (closed triangles) did not contribute significantly to the total level of RNA synthesis and thus the $t_{1/2}$ measured actually represents the formation of RS complexes between *E. coli* RNA polymerase and chromatin. Since this $t_{1/2}$ is dramatically slower than that for deproteinized chick DNA, chromosomal proteins must somehow interact with RNA polymerase to retard the rate of RS complex formation. The probable nature of this interaction will be discussed in a later section.

The actual initiation of RNA chain synthesis, involving the formation of a 5' to 3' phosphodiester linkage between two nucleoside triphosphates, occurs directly from the RS complex. By the use of the ribonucleotide-rifampicin challenge technique, the rate constant of RNA chain initiation can be measured. As demonstrated by Mangel and Chamberlin (18), the formation of the first phosphodiester bond in the RNA chain and the attack of rifampicin are competing reactions for any available RS complexes. The rate constant of RNA chain initiation from preformed RS complexes can be determined by measuring RNA polymerase activity in the presence of varying concentrations of rifampicin in such a challenge experiment. The theoretical basis for this determination is expressed in the equation:

$$\frac{C_O}{C^*} = \frac{k_2 [R]}{k^*} + 1 \qquad (1)$$

where C_O represents the total concentration of RS complex present at the time of addition of ribonucleotide and rifampicin; C^* equals the concentration of RS complexes which are able to initiate RNA synthesis at a given concentration of rifampicin, R; K_2 represents the second order rate constant of rifampicin attack on RS complexes; and k^* is the apparent first order rate constant for RNA chain initiation at a fixed concentration of ribonucleotide (18). In the absence of any secondary initiations from a single site, the ratio of C_O to C^* should be proportional to the ratio of RNA polymerase activity in the absence of rifampicin (V_O) to that in the presence of rifampicin (V^*). Thus, from the slope of the plot of V_o/V^* versus the concentration of rifampicin, the value of k_2/k^* can be obtained. The second order rate constants of rifampicin attack on RS complexes between *E. coli* RNA polymerase and either chick DNA or chick oviduct chromatin have been measured (19). The k_2 values for chick DNA and chromatin are 2.3×10^3 sec^{-1} M^{-1} and 0.8×10^3 sec^{-1} M^{-1}, respectively. Therefore, it should be possible to obtain the rate constant for RNA chain initiation from RS complexes for these two templates from a plot of V_o/V^* versus the concentration of rifampicin.

The results of such a determination utilizing either chick DNA or chick oviduct chromatin are shown in Figure 2. For this experiment, RNA polymerase and template were first preincubated to allow maximum formation of RS complexes. RNA synthesis was initiated by the addition of 0.15 mM each of the four ribonucleoside triphosphates and varying concentrations of rifampicin, as indicated. The incubation period for RNA synthesis was limited to 90 seconds to minimize any secondary initiation events in the absence of rifampicin. As shown on the inset, a plot of V_o/V^* versus rifampicin concentration was linear on both DNA and chromatin, as predicted by equation (1). The intercepts on the ordinates for these plots, however, which should theoretically be 1.0, were 1.4 for DNA and 4.0 for chromatin. The most likely explanation for this difference is that in the absence of rifampicin secondary initiation events from a single site occur even within 90 seconds. Such reinitiations could not occur in the presence of rifampicin. Therefore, the value of V_o if artifactually high and must be corrected by a factor equal to the reciprocal of the intercept. When this correction is made, the apparent k^* values obtained were 0.98 sec^{-1} for chick DNA and 0.57 sec^{-1} for chromatin. These values correspond to a $t_{1/2}$ for RNA chain initiation of 0.7 seconds for DNA and 1.2 seconds for chromatin. Thus, the rate of RNA chain initiation from RS complexes was much faster than the rate of formation of such complexes. Furthermore, the k^* values obtained for chick DNA and chick oviduct chromatin were not remarkably different. The rate of formation of RS complexes on chromatin, however, was seven times slower than that for DNA.

As discussed earlier, the sequence leading to the formation of the RS complex involves the following steps: (1) dissociation of enzyme from nonspecific binding sites; (2) binding of the enzyme at an initiation site form the I complex; (3) conversion of the I complex to the highly stable RS complex through a local denaturation of DNA. Any of these steps can be rate-limiting in the formation of the RS

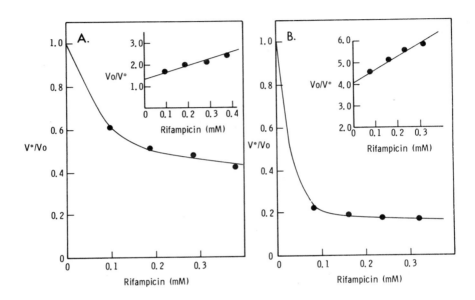

Figure 2. Determination of the rate constants of RNA chain initiation.(A) E. coli RNA polymerase (2.0 µg) and chick DNA (1.5 µg) were incubated for 15 minutes at 37°C in 0.2 ml of preincubation buffer to allow formation of RS complex. RNA synthesis was initiated by the addition of 0.05 ml of ribonucleoside triphosphate mixture in which the concentration of rifampicin was varied to yield the final concentrations indicated. V^*, the level of $[^3H]UTP$ incorporation in the absence of rifampicin, was equal to 22,500 CPM. Other conditions were as described in the legend to Figure 1. (B) RS complex was formed by preincubating 7.0 µg of E. coli RNA polymerase with 5.0 µg of chick oviduct chromatin (14 day DES stimulated) in 0.2 ml of preincubation buffer for 40 minutes. V^* was equal to 2700 CPM of $[^3H]UTP$ incorporation. Other conditions were the same as above.

complex. For chromatin, the $t_{1/2}$ of RS complex formation
was seven times longer than that for DNA. To better understand this difference between chromatin and DNA, we have
attempted to determine which step in the formation of RS complexes is rate limiting to the overall process.

The effect of the temperature of preincubation on RS
complex formation can be informative in attempting to determine the rate-limiting step. The transition from I to RS
complex (step 3) involves local openings of the DNA strands
and is thus strongly dependent on temperature. Although we
have no direct data on the temperature dependencies of
steps (1) and (2), we assume from the nature of the interactions involved that these steps would not be highly
dependent on temperature. Mangel and Chamberlin (18) have
demonstrated that the formation of I complex between $E.\ coli$
RNA polymerase and T_7 DNA is not markedly influenced by
temperature. Therefore, if step (3) is rate limiting,
decreasing the temperature of preincubation should increase
the $t_{1/2}$ of RS complex formation. On chick DNA this was the
case (Figure 3A). The $t_{1/2}$ of RS complex formation increased
from 1.3 to 8.3 minutes as the temperature of preincubation
decreased from 37°C to 9°C. Concomitantly, the maximum
level of RNA synthesis observed (RS_{max}) decreased in a reciprocal fashion as the temperature of preincubation was lowered.
The strong temperature dependency of RS_{max} is consistent with
the requirement for opening of the DNA strands for RS complex
formation. The temperature dependency of $t_{1/2}$ suggests that
at lower temperatures, the conversion from I to RS complex
is the rate-limiting step in formation of RS complexes on
DNA. At 37°, the rate-limiting step on DNA has been shown
to be step (1), the dissociation of enzyme from nonspecific
binding sites (19).

When chick oviduct chromatin was used as template, the
effects of temperature on RS complex formation were dramatically different (Figure 3B). The dependency of RS_{max} on the
temperature of preincubation was relatively poor compared to
DNA (note scale change). Thus, the local opening of the DNA
strands during the conversion of I to RS complex is not as
highly dependent on temperature as that seen for DNA. Furthermore, the $t_{1/2}$ of RS complex formation on chromatin was
completely independent of preincubation temperature. Therefore, the conversion from I to RS complex (step 3) is not
the rate limiting step in RS complex formation on chromatin
throughout the temperature range tested.

The formation of I complex (step 2) can also be eliminated as a possible rate-limiting step in RS complex formation
on chromatin. If step 2 were rate-limiting, the formation of

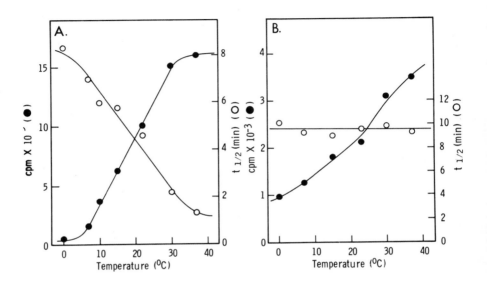

Figure 3. Temperature dependency of RS complex formation.
(A) E. coli RNA polymerase (1.4 µg) was incubated with chick
DNA (0.75 µg) in 0.1 ml of preincubation buffer at the
indicated temperature. At various time periods, RNA synthesis was initiated by the addition of 0.025 ml of ribonucleoside triphosphate mixture and carried out for 15 minutes at
37°C. The maximum level of [³H]UTP incorporation (●) and
t₁/₂ (o) were determined as shown in Figure 1. (B) E. coli
RNA polymerase (7.0 µg) was incubated with 5.0 µg of chick
oviduct chromatin (14 day DES stimulated) at the indicated
temperature. After various time periods, RNA synthesis was
initiated by the addition of 0.05 ml of ribonuceloside triphosphate mixture. Other conditions were as described above.

Table I. Effects of varying conditions of preincubation on RS complex formation.*

	Temperature	$(NH_4)_2SO_4$ (62.5 mM)	ANS (0.5 mM)	Chick Oviduct Chromatin	Chick DNA (Native)	Chick DNA (Denatured)
$t_{1/2}$	37°C	+	-	9.4	1.3	1.5
$t_{1/2}$	37°C	-	-	10.9	2.7	-
$\dfrac{\text{RS Max (+AmSO}_4)}{\text{RS Max (−AmSO}_4)}$				1.1	1.1	-
$t_{1/2}$	37°C	+	+	5.5	1.3	-
$\dfrac{\text{RS max (+ANS)}}{\text{RS max (−ANS)}}$				1.0	1.0	-
$t_{1/2}$	0°C	+	-	10.1	8.3	1.9
$\dfrac{\text{RS max (37°)}}{\text{RS max (0°)}}$				3.5	36	2.2

*RS_{max} and $t_{1/2}$ were determined as described in Figure 1.

RS complexes should occur via a second order process. As indicated in Figure 1, however, RS complex formation occurs via a first order process. Furthermore, varying the concentration of RNA polymerase and chromatin (as DNA) at a fixed ratio had no effect on the $t_{1/2}$ of RS complex formation. Therefore, neither step 2 nor step 3 is rate limiting in the formation of RS complex on chromatin.

Consequently, the rate-limiting step in the formation of RS complexes on chromatin appears to be the dissociation of enzyme from nonspecific interactions (step 1). These interactions could be with the DNA itself or with chromosomal proteins. The nature of these nonspecific interactions was more closely examined by testing the effect of various agents on the $t_{1/2}$ of RS complex formation. Changing the ammonium sulfate concentration in the preincubation mixture from 62.5 mM to 0 mM had little effect on the $t_{1/2}$ of RS complex formation on chromatin (Table I). This would not be expected if ionic interactions between RNA polymerase and chromatin were determining the rate-limiting step. Thus, the interaction between enzyme and chromatin may be hydrophobic in nature. To test this possibility, the effect of 1-anilino-8-naphthalene sulfonate (ANS) on the $t_{1/2}$ of RS complex formation was tested (Table I). ANS is shown to bind to hydrophobic regions of proteins and thus may interfere with hydrophobic interactions between proteins. The $t_{1/2}$ was measured at concentrations of ANS which had no effect on the total RNA polymerase activity. In the presence of 0.5 mM ANS, the $t_{1/2}$ of RS complex formation on chromatin decreased by 3.9 minutes. The same concentration of ANS did not alter the $t_{1/2}$ of RS complex formation on DNA. Thus, it is possible that hydrophobic interactions with chromosomal proteins may play a major role in the rate at which RNA polymerase can locate initiation sites in the chromatin. These nonspecific interactions cause the rate of formation of RS complexes on chromatin to be significantly slower than RS complex formation on deproteinized DNA.

III. Determination of the Number of Initiation Sites on Chromatin

The use of the ribonucleotide-rifampicin challenge technique provides a method for estimating the number of available initiation sites for RNA polymerase in a given template (20, 21,22). The method involves first preincubating RNA polymerase and template for a period of time sufficient to allow maximum formation of RS complex. RNA synthesis is then initiated by the addition of the four ribonucleoside triphosphates with 40 µg/ml rifampicin and allowed to proceed to completion. At this concentration of rifampicin, initiation from the RS complex is highly efficient, while secondary initiations are strongly inhibited (18,19,22). Thus, the number of RNA chains synthesized, which can be determined from the amount, length and base composition of the RNA produced, will be equivalent to the number of initiation sites at which RNA polymerase is bound. By adding increasing amounts of RNA polymerase to a fixed amount of template, the total number of available initiation sites can be determined from the saturation level of RNA synthesis obtained when all high affinity binding sites are occupied.

The results of such an analysis for chick DNA are shown in Figure 4. Increasing amounts of *E. coli* RNA polymerase holoenzyme or core enzyme (without sigma factor) were preincubated for 15 minutes at either $0°$ or $37°C$. Following preincubation, RNA synthesis was carried out for 15 minutes at $37°C$. The number of initiation sites available for holoenzyme after preincubation at $37°$ (curve A) was calculated to be about 1.3×10^6 per picogram of DNA (based on a number average chain length of 450 nucleotides and a base composition of 40.6% UMP). This corresponds to an average of one initiation site per 700 nucleotide base pairs of DNA. After preincubation at $0°$, neither holoenzyme nor core enzyme (curve C, D) is very efficient at formation of RS complexes. We attribute this to the high temperature dependency of the conversion from I to RS complex. When core enzyme is preincubated at $37°$ (curve B), the number of RNA chains initiated is still only 30% of that seen for holoenzyme preincubated at $37°$. Thus, both sigma factor and preincubation at $37°$ are required for the most efficient formation of RS complexes between *E. coli* RNA polymerase and chick DNA.

The number of initiation sites in chromatin isolated from a fully stimulated chick oviduct was analyzed in an analogous manner (Figure 5). In this experiment, chromatin (5 µg) was preincubated for 40 minutes with RNA polymerase. RNA synthesis was carried out in the presence of rifampicin and heparin. Heparin is a polyanion added to inhibit any RNase activity which, if present, would decrease the average length of the RNA produced. As the amount of RNA polymerase was increased, a transition point was observed which corresponded to the number of available high affinity binding sites in chromatin. The level of RNA synthesis at this point was used to calculate the number of sites available for the initiation of transcription. The number of initiation sites available for holoenzyme after preincubation at $37°C$ was calculated to be about 22,500 per picogram of DNA (based on a number average chain length of 750 nucleotides and a base

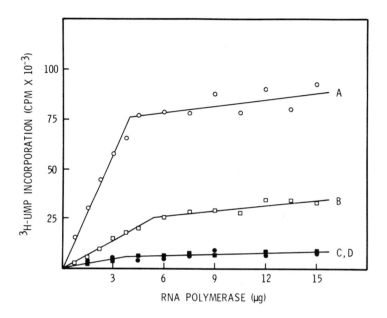

Figure 4. RNA initiation sites on chick oviduct DNA.
E. coli *RNA polymerase holoenzyme (o,●) or core enzyme (□ , ■)*
was incubated with 1.5 μg of chick oviduct DNA in 0.2 ml
preincubation buffer. After incubation for 15 minutes at
either 37°C (o, □) or 0°C (● , ■), RNA synthesis was initi-
ated by the addition of 0.05 ml ribonucleoside triphosphate
mixture containing 0.2 mg/ml of rifampicin. RNA synthesis
was carried out at 37°C for 15 minutes. Core enzyme was pre-
pared by phosphocellulose chromatography of purified holo-
enzyme as described by Burgess and Travers (41). Reprinted
with permission.

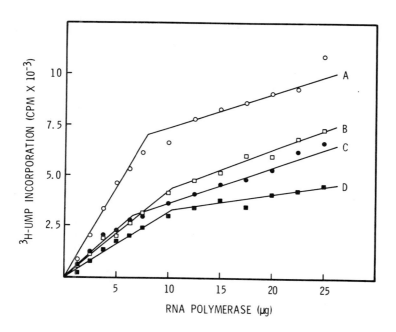

Figure 5. RNA initiation sites on chick oviduct chromatin.
E. coli RNA polymerase holoenzyme (o,●) or core enzyme (□,■)
was incubated with 5.0 µg of chick oviduct chromatin (14
days DES stimulated) in 0.2 ml of preincubation buffer.
After incubation for 40 minutes at either 37°C (O, □) or
0°C (●, ■), RNA synthesis was initiated by the addition of
0.05 ml ribonucleoside triphosphate mixture containing
0.2 mg/ml rifampicin and 4.0 mg/ml heparin. Other condtiions
were as described in the legend to Figure 4. Reprinted with
permission.

composition of 37.9% UMP). This corresponds to an average of one initiation site for every 40,000 base pairs of DNA. Thus, chromatin from the estrogen-stimulated chick oviduct contained only about 2% of the initiation sites found on deproteinized DNA. Furthermore, the formation of RS complexes on chromatin was much less dependent on either the presence of sigma factor or the temperature of preincubation than that for DNA. Therefore, the initiation sites available for RNA polymerase in chromatin may be in a state in which the conversion from I to RS complex is greatly facilitated. One possible explanation is that certain chromosomal proteins are present at initiation sites which help to destabilize the duplex structure of the DNA.

The capacity of oviduct chromatin to serve as a template for *E. coli* RNA polymerase has previously been shown to increase following estrogen administration of the immature chick (4,8). Such measurements, however, do not distinguish between the many steps of the RNA synthetic process which could influence determinations of total template capacity. We have, therefore, separately measured the number of high affinity initiation sites, the rate of RNA chain propagation and the chain length of RNA synthesized from chromatin isolated from various stages during the primary stimulation of chick oviduct (23). The number of initiation sites available for *E. coli* RNA polymerase was shown to increase by 50% as early as 8 hours after estrogen administration (Table II). The increase in initiation sites reached a maximum by about 4 days of estrogen treatment and then declined to a plateau level approximately two times higher than untreated animals by 10 days. By this time, estrogen-mediated differentiation is nearing completion in the chick oviduct. Over the same period, the average rate of RNA chain elongation for the first minute of RNA synthesis and the average chain length of the RNA product did not vary significantly. Thus, the estrogen-induced increase in total chromatin template capacity is mainly due to an increased availability of sites at which RNA polymerase can bind and form RS complexes.

During primary estrogen stimulation of the chick oviduct, differentiation of three distinct epithelial cell types occurs from a homogeneous population of primitive mucosal cells (2). Thus, changes in the number of initiation sites in oviduct chromatin during primary stimulation might simply reflect alterations in the tissue population of specific cell types. To overcome this difficulty, the number of available initiation sites for RNA polymerase was measured during estrogen withdrawal and acute (secondary) estrogen stimulation (24). During secondary stimulation, estrogen-mediated

Table II. Effect of primary estrogen stimulation on initiation sites for RNA synthesis.

Days of Hormone Treatment	Size of DNA Product	Rate of Elongation (Nucleotides/sec)	Initiation Sites Per pg of DNA
0	660	6.0	10,600
0.3	700	6.0	16,800
1	810	6.0	33,450
4	640	6.0	72,000
8	725	-	42,200
12	680	-	21,200
18	700	5.8	28,600

changes in RNA and protein synthesis are not dependent on cell proliferation (25). As shown in Table III, the level of available initiation sites for *E. coli* RNA polymerase decreases after withdrawal of hormone from stimulated chicks to a level similar to that observed for unstimulated chicks. As early as 30 minutes after injection of a single dose of diethylstilbestrol, the number of initiation sites increased from 8,100 to 17,150. By 1 to 2 hours, a plateau level had been reached which was about 3 times higher than the withdrawal level. Measurements of the average chain length of the RNA product and the rate of RNA chain elongation were similar for all preparations tested. Thus, the change in chromatin template activity during secondary stimulation actually reflects an increase in the availability of initiation sites for RNA polymerase.

During secondary stimulation with estrogen, simultaneous measurements were also made of the level of endogenous nuclear receptor and the amount of intracellular ovalbumin mRNA. The level of nuclear bound receptor, measured by a [^{3}H] estradiol exchange assay (26), increased to a transient maximum shortly after 20 minutes of readministration and then declined to a plateau level at 1 to 2 hours (Figure 6). The increase in the concentration of nuclear receptor molecules immediately preceded the increase in available initiation sites on chromatin prepared from the same nuclei. The measurement of the intracellular concentration of ovalbumin mRNA was made using a [^{3}H]-labeled complementary DNA probe (7). A detectable increase in ovalbumin mRNA occurred at 30 to 60 minutes, followed by a linear accumulation beginning at 1 hour. The increase in the intracellular level of specific induced mRNA, thus, occurs shortly after the increase in the number of available chromatin initiation sites for RNA synthesis. This increase of ovalbumin messenger RNA sequences was not sensitive to the protein synthesis inhibitor cycloheximide (unpublished observation). This suggests that *de novo* protein synthesis is not necessary to initiate the events of secondary stimulation which we have measured. The interesting temporal relationship to the three measured parameters makes it tempting to speculate a casual relationship between the events. Therefore, it is possible that after translocation to the nucleus, hormone-receptor complexes may act directly on chromatin such that previously repressed DNA sequences on chromatin become accessible for transcription by RNA polymerase.

Table III. Effect of estrogen withdrawal and secondary estrogen stimulation of initiation sites for RNA synthesis.

Hormone Treatment	Size of RNA Product	Rate of Elongation (Nucleotide/sec)	Initiation Sites Per PG of DNA
18 day DES stimulated	700	6.0	26,700
18 day DES stimulated + 12 day withdrawn	873	8.5	8,100
18 day DES stimulated + 12 day withdrawn + Secondary DES stimulation for:			
20 min	810	7.0	9,800
30 min	750	7.7	17,100
60 min	621	8.0	24,800
2 hr	777	8.0	26,900
4 hr	803	7.0	25,700

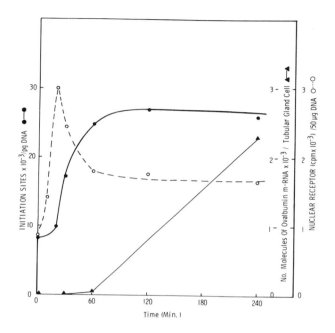

Figure 6. Effect of secondary estrogen stimulation on the stimulation of the initiation of RNA synthesis and the level of nuclear receptors for estrogen. Chicks which had received daily injections of 2.5 mg of DES for 18 days were withdrawn from hormone treatment for 12 days. Chicks were then given a single 2.5 mg dose of DES and at the indicated times, oviducts were removed and frozen. Chromatin was prepared from a portion of each oviduct sample for the measurement of initiation sites as described in the text and Table II. On the other portion of oviduct, the number of molecules of ovalbumin mRNA per tubular gland cell and the level of nuclear estrogen receptor were measured as described in the text.

IV. *In vitro* Transcription of Ovalbumin Gene

Previous progress in the study of *in vitro* transcription in higher organisms has been hindered by the high complexity of eucaryotic genome and the lack of suitable techniques for assaying specific transcription products. Thus, the pioneering work was mainly done using the method of filter hybridization which only allows "highly abundant sequences" in the RNA transcripts or RNA sequences complementary to repetitive DNA sequences to hybridize with DNA on the filter. Recently, studies on the *in vitro* synthesis of specific mRNA's have been facilitated by the discovery of RNA-dependent DNA polymerase (reverse transcriptase) from RNA tumor viruses. This enzyme allows the preparation of complementary DNA (cDNA) copies to specific mRNA molecules. With such cDNA probes, it became possible to detect and to estimate the concentration of a specific mRNA sequence present in a minute quantity in a total RNA population. Using cDNA to globin mRNA as the hybridization probe, Gilmour and Paul (27), Axel *et al.* (28) and Steggles *et al.* (29) have demonstrated the presence of globin sequences in RNA transcripts of chromatin isolated from erythropoietic tissues but not in the transcripts from chromatin of brain or other nonerythropoietic tissues.

In the oviduct system, we have used a radioactively labeled complementary DNA to ovalbumin mRNA ($cDNA_{ov}$) as a probe. The results of hybridization of this probe with either pure $mRNA_{ov}$ or RNA transcribed *in vitro* from estrogen stimulated chick oviduct chromatin is shown in Figure 7. The annealing reactions were run for a sufficient length of time to insure complete hybridization of any complementary $mRNA_{ov}$ sequences. Varying amounts of either $mRNA_{ov}$ or RNA transcribed by *E. coli* RNA polymerase from oviduct chromatin were incubated with a constant amount of cDNA for approximately 100 hours and the hybridization data are plotted as percent hybrid vs. the $RNA/cDNA_{ov}$ ratio. For pure ovalbumin mRNA, the percent of hybrid formation vs. $RNA/cDNA_{ov}$ ratio is linear at a RNA/cDNA ratio below 2.0 (less than 40% of cDNA hybridized) and has a slope of 0.23. The initial slope of this type of hybridization curve is proportional to the concentration of ovalbumin mRNA in the RNA preparation. Therefore, the concentration of $mRNA_{ov}$ sequences in any RNA preparation can be determined by comparing the initial slope of their respective titration curves relative to that observed with pure $mRNA_{ov}$. For example, if a RNA preparation contains 1% $mRNA_{ov}$, the initial slope for its hybridization curve should be 1% of that of pure mRNA or 0.0023. *In vitro* RNA transcribed from chromatin isolated from hormone stimulated

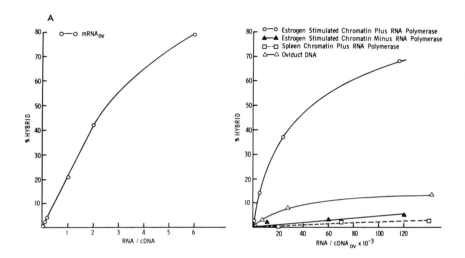

Figure 7. Titration of $cDNA_{OV}$ with ovalbumin mRNA or RNA transcripts from chromatin or DNA. (A) Different concentrations of ovalbumin $mRNA_{OV}$ were hybridized for 96 hours with 1.5 ng of $cDNA_{OV}$ and the amount of S_1 resistant-material was determined as described by Harris et al. (24). (B) Titration of RNA transcribed from chick DNA or chromatin with E. coli RNA polymerase; -o-o-, 16 day estrogen stimulated oviduct chromatin; -- ◘-- ◘--, spleen chromatin; -▲-▲-,16 day estrogen-stimulated chromatin minus RNA polymerase in the reaction mixture, tRNA was added to the incubation in an amount equal to that normally synthesized with added RNA polymerase; Δ-Δ-Δ, oviduct DNA.

chick oviduct had an initial slope of hybridization of 0.32 x 10^{-4}. Thus, 0.01% of the *in vitro* synthesized RNA was ovalbumin mRNA.

To establish that we have a valid system for measuring $mRNA_{ov}$ sequences synthesized *in vitro*, it was necessary to meet several criteria. First, it was imperative to demonstrate that the ovalbumin mRNA sequences detected in the RNA synthesized *in vitro* are a result of *de novo* synthesis by the added RNA polymerase and not from contaminating RNA sequences which associate with the isolated chromatin. To measure the endogenous contaminating sequences in the chromatin transcripts, a RNA sample from oviduct chromatin was prepared under normal conditions except that the RNA polymerase was omitted from the reaction (Figure 7b). Yeast RNA equivalent to the amount of RNA normally synthesized was added to the mixture and the sample was then processed and analyzed identically to the sample containing RNA polymerase. By comparing the initial slopes of the hybridization curves, it was shown that the endogenous $mRNA_{ov}$ sequences could account for only about 10% of the ovalbumin mRNA sequences found in the chromatin transcripts. Therefore, the majority of the RNA which hybridized with the cDNA probe must have been synthesized *in vivo* by the *E. coli* RNA polymerase. Also included in Figure 7b is the hybridization data obtained from an RNA sample synthesized from spleen chromatin (non-target tissue) using *E. coli* RNA polymerase. A second criterion for demonstrating the validity of the *in vitro* chromatin transcription system for ovalbumin mRNA sequences is tissue specificity. Ovalbumin mRNA transcription should be restricted to the target tissue of oviduct. Only background amounts of reaction with $cDNA_{ov}$ were detectable with spleen chromatin transcripts and thus, the *in vitro* RNA synthesis system was tissue specific with respect to the source of the chromatin. A third parameter for testing the *in vitro* transcription system is that of restriction of the chromatin template. The RNA synthesized from chick oviduct chromatin should represent a subfraction of the sequences present in DNA which are specific for that tissue and should include RNA transcribed from the ovalbumin gene. The RNA synthesized *in vitro* from a chick oviduct DNA template from which all chromosomal proteins have been removed should also contain RNA sequences transcribed from the ovalbumin gene. However, in this case, the specific $mRNA_{ov}$ sequences will be diluted by RNA transcribed from all the other sequences uncovered during removal of the chromosomal proteins. In Figure 7b, the results of the titration curves for RNA synthesized *in vitro* from estrogen-stimulated chromatin are compared with RNA synthesized *in vitro* from deproteinized chick DNA. By comparing the initial slope of the titration curves, it is evident that the concentration of $mRNA_{ov}$ sequences in the RNA transcribed from deproteinized DNA is 11% of that produced from chick oviduct chromatin. This level is consistent with the fact that a subset of sequences was synthesized using oviduct chromatin. These results are interpreted to indicate that there is a specific restriction of sites for *in vitro* synthesis of RNA in chromatin, and that the ovalbumin gene in the estrogen-stimulated oviduct chromatin resides in the unrestricted or "open" region of the oviduct chromatin.

A final test of the validity of the system was the demonstration that the activation of the ovalbumin gene is estrogen dependent. RNA was synthesized *in vitro* from chromatins isolated from unstimulated, estrogen-stimulated and hormone-withdrawn chick oviduct tissues. These RNA samples were then hybridized with the $cDNA_{ov}$ probe and the concentration of $mRNA_{ov}$ sequences in the RNA preparations was determined by the standard titration method (Figure 8). Although unstimulated oviduct chromatin is capable of supporting substantial levels of RNA synthesis using *E. coli* RNA polymerase, there are little or no RNA sequences complementary to the $cDNA_{ov}$ in the RNA synthesized *in vitro* using unstimulated chick oviduct chromatin. This is consistent with the hypothesis that the ovalbumin gene is indeed repressed or "turned off" in the unstimulated chick oviduct. In contrast, RNA synthesized *in vitro* using estrogen-stimulated chromatin hybridized quite efficiently with the $cDNA_{ov}$ probe. *In vitro* RNA made from oviduct chromatin isolated from chicks withdrawn from hormone for 12 days hybridized only slightly to $cDNA_{ov}$. Comparing the initial slope of the titration curve with that of pure mRNA, only 0.0015% of *in vitro* RNA transcribed from this withdrawn chromatin is represented by ovalbumin mRNA sequences. Therefore, the concentration of ovalbumin mRNA sequences in the *in vitro* RNA transcripts from withdrawn chromatin was 10% of that of estrogen-stimulated chromatin. Since withdrawn chromatin could support only 1/2 to 1/3 of the rate of total synthesis as compared to the stimulated chromatin, the absolute rate of $mRNA_{ov}$ synthesis would be only 3-5% of that synthesized from estrogen-stimulated chromatin.

From these hybridization studies, it can be concluded that isolated chick oviduct chromatin maintains the tissue specificity with regard to ovalbumin mRNA transcription which is observed *in vivo*. Furthermore, the control of the expression of the ovalbumin gene resides primarily at the transcriptional level. This conclusion militates against post-

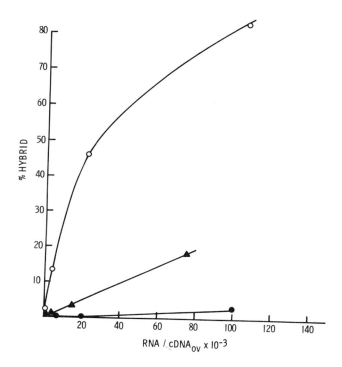

Figure 8. Titration of RNA transcribed from chick oviduct chromatin isolated from chick during different stages of estrogen treatment. Experiments were carried out as described in Figure 7: -o-o-, 14 day estrogen-stimulated oviduct chromatin; -▲-▲-, chromatin isolated from chick receiving 14 days estrogen stimulation then withdrawn from hormone for 12 days; -●-●-, unstimulated chick oviduct chromatin.

transcriptional control as the primary mechanism for steroid hormone regulation of specific mRNA synthesis in the chick oviduct system.

V. Role of Chromatin Proteins in Regulating Transcription of the Ovalbumin Gene

The chick oviduct provides .a good system for studying the role of chromatin proteins in the regulation of gene expression. As shown earlier, isolated chromatin still retains the specificity of transcription *in vitro* which is inherent to the tissue from which it was prepared. Furthermore, transcripts of deproteinized DNA contain 10 to 20 times more different sequences than chromatin transcripts. Thus, chromatin proteins play a major role in the determination of tissue specificity of transcription. Since there is substantial difference in the concentration of ovalbumin messenger RNA sequence in the *in vitro* transcripts of hormone-stimulated and withdrawn chick oviduct chromatins, these two chromatins can be used to study which chromosomal proteins are involved in the regulation of the expression of the ovalbumin gene.

Chromatin proteins were fractionated into 3 major fractions - histone (H), nonhistone (NH) and DNA-tightly bound nonhistone protein (DNA-TBP) complex essentially as described by Chiu *et al.* (30). The tightly bound protein constituted about 20% of the total nonhistone protein in the original chromatin. The extractable nonhistone proteins and histone proteins from estrogen-stimulated chromatin or hormone-withdrawn chromatin were reconstituted to their homologous tightly bound nonhistone protein-DNA complexes by gradient dialysis as described by Tsai *et al.* (31). Chromatin reconstituted in this manner from completely homologous constituents is essentially identical to native chromatin in terms of the number of initiation sites as measured by the rifampicin-nucleotide challenge assay described earlier (22). Therefore, this method of reconstitution seems to give rise to a proteins-DNA complex similar to the original chromatin.

However, in order to establish that this reconstitution method can be used to study the control of chromatin proteins in gene expression, functional studies on these chromatins as templates should be carried out. Using *E. coli* RNA polymerase, RNA was synthesized from reconstituted and native chromatin isolated from estrogen-stimulated chicks and from withdrawn chicks. These *in vitro* transcripts were then hybridized to complementary DNA transcribed from pure ovalbumin messenger RNA ($cDNA_{ov}$) as discussed in the previous section. As shown in Figure 9a, on native chromatin, 0.012% of the RNA transcripts from stimulated chromatin and 0.0015%

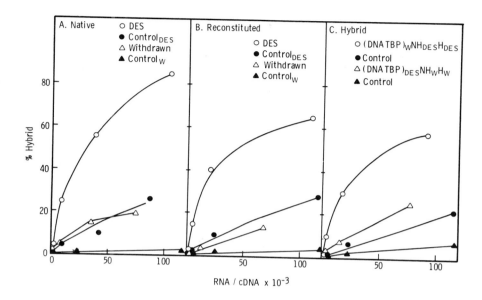

Figure 9. *Titration of ovalbumin messenger RNA in the in vitro RNA transcripts of native and reconstituted chromatin. Varying concentrations of in vitro RNA transcribed from chick oviduct chromatin using E. coli RNA polymerase were hybridized for 96 hrs. with 1.5 μg [³H] cDNA and the amount of S_1 nuclease resistant material was determined as described in Figure 7. In cases without E. coli RNA polymerase, tRNA was added as carrier during extraction of RNA in an amount equal to the RNA normally synthesized with added RNA polymerase. Loosely bound nonhistone proteins (NH), histones (H), and tightly-bound nonhistone protein-DNA complex (TBP-DNA) were fractionated from stimulated (DES) and withdrawn (W) chromatins as described by Chiu et al. (30). Extractable histone proteins and histones from estrogen-stimulated or hormone-withdrawn chromatin were reconstituted to their homologous (B) or heterologous (C) tightly-bound nonhistone protein-DNA complexes by gradient dialysis of Spelsberg et al. (43).*

of the RNA transcripts from withdrawn chromatin were homologous to ovalbumin mRNA. The reconstituted chromatin transcripts contained 0.011% and 0.0015% ovalbumin mRNA sequences for hormone-stimulated and withdrawn chromatin, respectively (Figure 9b) in close agreement with the values obtained for native chromatin. This excellent correlation is not an artifact resulting from endogenous contaminating RNA sequences. The endogenous contaminating RNA isolated from the incubation mixture without added RNA polymerase could account for less than 14% of the hybridizable mRNA sequences (Figure 9b). These findings suggest that the specificity of both gene restriction and gene transcription with respect to mRNA$_{OV}$ was not impaired during reconstitution.

The role of chromatin proteins in the regulation of *in vitro* RNA synthesis was then studied by reconstitution of hybrid chromatins using protein components of heterologous origin. The RNAs were synthesized from the hybrid chromatins and hybridized to cDNA$_{OV}$. The amount of ovalbumin mRNA sequences present in RNA synthesized *in vitro* from hybrid chromatin containing extractable nonhistone and histone proteins from estrogen-stimulated chromatin and tightly bound nonhistone protein-DNA complex from withdrawn chromatin (DNA-TBP$_W$-NH$_{DES}$-H$_{DES}$) was similar to that synthesized from either the homologous reconstituted or the native estrogen-stimulated chromatins (Figure 9c). The concentration of ovalbumin mRNA sequences as calculated from the initial slope was about 0.011% of the *in vitro* transcripts. Endogenous RNA sequences in the chromatin can only account for 12% for the putative *in vitro* synthesized mRNA sequences. Thus, the presence of extractable nonhistone and/or histone proteins of estrogen-stimulated chromatin was required for the transcription of the ovalbumin gene.

When hybrid chromatin was reconstituted from extractable nonhistone protein-DNA complex isolated from stimulated chromatin (DNA-TBP$_{DES}$-NH$_W$-H$_W$), the *in vitro* RNA transcripts behaved like the reconstituted withdrawn chromatin with respect to transcription of ovalbumin mRNA sequences. These results again indicate that the tightly bound protein-DNA complex does not play a determining role in the expression of the ovalbumin gene.

We next attempted to determine whether the histones or the extractable nonhistone proteins were the major determinants in regulating expression of the ovalbumin gene. Histones were isolated from estrogen-stimulated and withdrawn chick oviduct chromatins and then reconstituted to total nonhistone protein DNA complexes from withdrawn or stimulated chick oviducts. Switching of the histones from stimulated chromatin to withdrawn chromatin or vice versa did not affect the concentration of ovalbumin mRNA sequences in the transcripts of these hybrid chromatins as compared to that of the original chromatin (Figure 10). The concentration of ovalbumin mRNA sequences was 0.012% for hybrid chromatin with

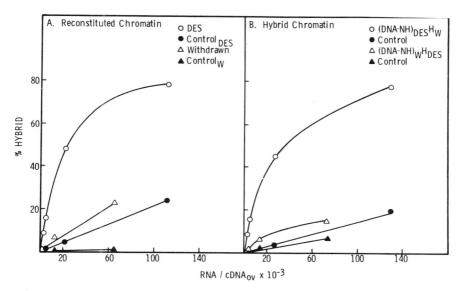

Figure 10. Titration of ovalbumin mRNA sequences in the in vitro transcripts of reconstituted chromatin in which histone fractions interchanged. Fractionation of nonhistone proteins and histone proteins from tightly-bound nonhistone protein-DNA complex and reconstitution of these components were carried out as described in Figure 3. Reprinted with permission.

nonhistone proteins from stimulated chromatin and histone proteins from withdrawn chromatin (H_W-NH_{DES}-DNA-TBP_{DES}). On the transcripts from hybrid chromatin in which the source of proteins was reversed (H_{DES}-NH_W-DNA-TBP_W), the ovalbumin mRNA concentration was calculated to be 0.0014%. The endogenous contaminating RNA sequences accounted for less than 7% of the *in vitro* transcripts. These data strongly implicate that the extractable nonhistone proteins contain the essential factors for regulating the transcription of the ovalbumin gene. These regulators could either be activators present in the nonhistone protein fraction of estrogen-stimulated chick oviduct chromatin and/or repressors present in the fraction isolated from hormone withdrawn chromatin.

Similar results have been obtained in other laboratories with respect to control of expression of the globin gene and histone genes (30,32-34). Recently, we have separated the nonhistone proteins of chromatin into several distinct fractions. These fractions have differential binding specificities with regard to DNA sequences. This procedure should now enable us to explore in more detail the involvement of chromosomal proteins in the regulation of gene expression.

VI. Comparison of Bacterial RNA Polymerase with Homologous RNA Polymerase

As described in an earlier section (Section II), the initiation sites on chromatins isolated from chick oviduct changed dramatically during hormonal stimulation and withdrawal. One major question raised in these studies was whether the number of initiation sites for other RNA polymerases, especially homologous RNA polymerase, also followed the same pattern of changes. To answer this question, enzyme saturation curves were carried out to measure the number of initiation sites available to *E. coli* RNA polymerase core enzyme and hen oviduct RNA polymerase II on chromatin isolated from hormone-stimulated and withdrawn chicks. When core enzyme was used, the enzyme saturation curve on chromatin showed two phases (Figure 11) similar to the saturation curves of *E. coli* holoenzyme. With stimulated chromatin (Panel A), *E. coli* core polymerase incorporated 44 pmole of UMP at the transition point. This level decreased to 16 pmole on chromatin isolated from chick oviduct after hormone withdrawal for 12 days (Panel B). A similar 3-fold decrease of RNA synthesis was previously observed when initiation sites for *E. coli* holoenzyme were measured (Section II).

For studies with oviduct RNA polymerase II, the rifamycin derivative, AF/013, was used instead of rifampicin, since rifampicin does not inhibit animal RNA polymerases. AF/013 inhibits initiation of RNA synthesis prior to the formation of the first phosphodiester bond (35,36). This compound has been previously used to inhibit the reinitiation of RNA synthesis in order to allow a comparison of *E. coli* and calf thymus RNA polymerases in selection of initiation sites on calf thymus or SV40 DNA (37,38). The enzyme saturation curve with hen oviduct RNA polymerase II on hormone-stimulated and withdrawn chromatin also had two phases (Fig. 12). With stimulated chromatin (Panel A) oviduct RNA polymerase II incorporated 100 pmole of UMP at the transition point. This level decreased to 24 pmole on withdrawn chromatin. A 3- to 4-fold decrease in RNA synthesis was also observed with

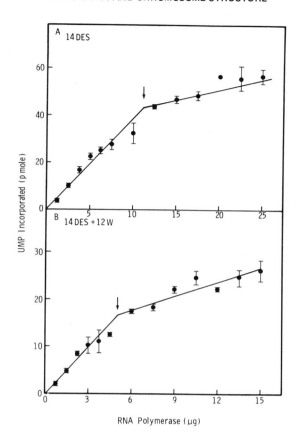

Figure 11. *Enzyme saturation curves on chick oviduct chromatins isolated from oviduct tissues of stimulated and withdrawn chicks by* Escherichia coli *RNA polymerase core enzyme.* E. coli *RNA polymerase core enzyme (0 to 25 μg) was preincubated with 5 μg of chromatin in 0.2 ml of preincubation mixture as described in the legend of Fig. 4 at 37º. At the end of 40 min, 50 μl of nucleotides and rifampicin mixture were added and further incubated at 37º for 15 min. RNA synthesized was precipitated with 5% trichloroacetic acid and counted in toluene base scintillation fluid. A, chromatin isolated from 14 days diethylstilbestrol-stimulated chicks (14 DES). B, chromatin isolated from 12 days withdrawn chicks after 14 days diethylstilbestrol (14 DES + 12 W).*

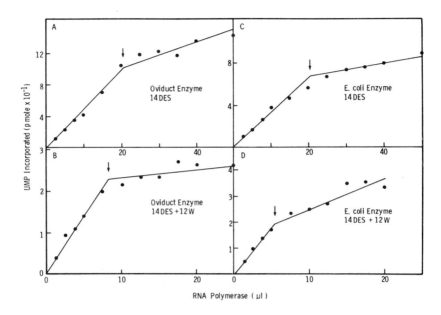

*Figure 12. Enzyme saturation curves on 14 days diethylstil-
bestrol stimulated and withdrawn chromatin by hen oviduct
RNA polymerase II and* Escherichia coli *RNA polymerase holoenzyme. Hen oviduct RNA polymerase II (0 to 50 µl with a
concentration of 400 µg/ml) was preincubated with 5 µg of
chromatin at 37° for 15 min. RNA synthesis was started with
150 nmole of UTP (575cpm/pmol) and 50 µg of AF/013. At the
end of 15 min, RNA synthesized was measured according to the
method described in the legend of Fig. 11. A, oviduct RNA
polymerase II, 14 day diethylstilbestrol-stimulated (14 DES)
chromatin; B, oviduct RNA polymerase II, withdrawn chromatin
(14 DES + 12 W)); C, E. coli RNA polymerase holoenzyme 14
days diethylstilbestrol stimulated chromatin; D, E. coli RNA
polymerase holoenzyme, withdrawn chromatin.*

E. coli holoenzyme when AF/013 was used (Panel C and D). At the present time it is not possible to assess the absolute number of initiation sites for either oviduct RNA polymerase II or core enzyme. Such an analysis would involve more detailed kinetic studies as well as size determinations of the chromatin transcripts. However, since these parameters have not varied between stimulated and withdrawn chromatins in previous transcription studies, the ratio of RNA synthesis at the transition point on two chromatins will be equal to the ratio of initiation sites available. Therefore, regardless of the polymerase used, the relative changes in the level of initiation sites during hormone stimulation and withdrawal followed a similar pattern.

It was of interest to determine whether oviduct RNA polymerase II and bacterial RNA polymerase utilize the same or different sites on chick DNA and chromatin. This problem can be approached by adding increasing amounts of the two polymerases in the rifampicin challenge assay. In this type of competition assay, a preinitiation complex is formed between RNA polymerase and DNA at the initiation site prior to the onset of RNA synthesis. Therefore, if both enzymes utilize the same sites for initiation, competition should occur between the enzymes for available initiation sites under conditions where RNA polymerase is in excess. In these conditions, the level of RNA synthesis for the mixture of two enzymes should be equal to the average value for the level of RNA synthesis for either enzyme alone. On the other hand, if both enzymes utilize different sites for initiation, no competition for the formation of preinitiation complexes should occur. In this case, the level of RNA synthesis for the mixture should be equal to the sum of the levels of RNA synthesis of either enzyme alone. The theoretical activity of RNA synthesis for the two extreme cases of complete competition or no competition can be calculated according to the following equations. If both enzymes use different sites:

$$V_{X_cY_{c'}} = V_{X_c} + V_{Y_{c'}} \qquad (1)$$

If both enzymes use the same sites:

$$V_{X_cY_{c'}} + 1/2 \ (V_{X_{2c}} + V_{Y_{2c'}}) \qquad (2)$$

Where $V_{X_cY_{c'}}$ is the enzyme activity when RNA polymerase X and Y are mixed together at concentrations c and c', respectively; V_{X_c} and $V_{X_{2c}}$ are the enzyme activities of RNA polymerase X at concentrations c and 2c; and $V_{Y_{c'}}$ and $V_{Y_{2c'}}$ are the enzyme activities of RNA polymerase Y at concentrations of c' and

2c'.

The initiation sites utilized by oviduct RNA polymerase II and *E. coli* holoenzyme on purified chick DNA were first examined. Enzyme saturation curves of 1.5 µg of DNA for either polymerase alone or a mixture of the two enzymes are shown in Figure 13. The stock concentrations of enzyme used in this experiment were 130 units/ml for oviduct enzyme and 400 µg/ml for *E. coli* enzyme. For oviduct RNA polymerase alone, the incorporation of [^{3}H] UMP reached a saturation level of approximately 43 pmole (Curve A), while with *E. coli* RNA polymerase alone, a plateau level corresponding to 500 pmole was obtained (Curve B). This substantial difference in the saturation levels of UMP incorporation indicates that hen oviduct RNA polymerase II is only capable of utilizing a subclass of the initiation sites on DNA which *E. coli* RNA polymerase may utilize. This is consistent with the charac- teristics of RNA polymerase II from higher organisms, which are known to be relatively inefficient at transcribing intact native DNA templates. Based on the enzyme saturation curves of either polymerase alone and the equations (1) and (2), theoretical curves were calculated representing the expected levels of RNA synthesis if both polymerases use different sites (C) or the same sites (D) on DNA. The experimental points for the mixture of the two enzymes (solid circles) followed the theoretical curve C quite closely. These data suggested that *E. coli* RNA polymerase II utilize different initiation sites on chick DNA. It should be noted that at the highest enzyme concentrations used, the experimental points for the mixture of polymerases became intermediate between curves C and D. If enzyme concentrations were increased further, the competition between the enzymes be- came even more prominent (data not shown). These results are explicable since at very high enzyme to DNA ratios RNA polym- erase will stack up one by one along the DNA strands and may thus sterically hinder the transcription process.

The initiation sites utilized by *E. coli* RNA polymerase holoenzyme, core enzyme and oviduct RNA polymerase II on chick oviduct chromatin were next examined. Chromatin pro- teins undoubtably play a major role in the regulation of gene expression. It is thus possible that proteins in the chroma- tin might be involved in the selection of initiation sites for RNA polymerase. If this is true, it would be expected that these different RNA polymerases would utilize the same sites for initiation in chromatin, despite having unique specificities on DNA. The results of a competition experi- ment between *E. coli* holoenzyme and core enzyme on oviduct chromatin from estrogen-stimulated chicks are shown in Figure

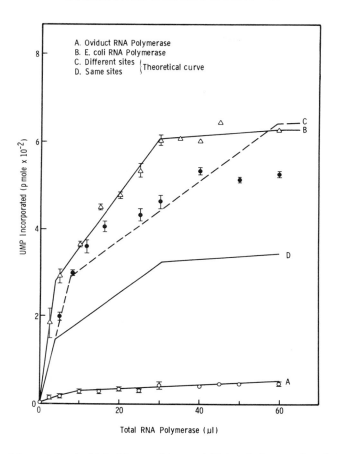

Figure 13. RNA initiation sites utilized by Escherichia
coli *RNA polymerase holoenzyme and oviduct RNA polymerase II
on chick oviduct DNA. Oviduct RNA polymerase (o, Curve A)
and E. coli RNA polymerase (Δ, Curve B) or a mixture of the
two enzymes in equal volumes (●) was incubated at 37°C with
1.5 µg of chick oviduct DNA in 0.2 ml of preincubation buffer
for 14 min. Stock concentrations used were 130 units/ml for
hen RNA polymerase II and 400 µg/ml for holoenzyme. After
15 min. RNA synthesis was initiated by the addition of AF/013
and nucleotides according to the procedure described in Fig.
12. Theoretical curves were constructed from experimental
Curves A and B assuming either no competition (Curve C) or a
complete competition (Curve D) of the two enzymes for RNA
initiation sites.*

289

14. In this particular experiment, RNA synthesis was carried out for 1.5 minutes. Both holoenzyme and core enzyme transcribe chromatin with similar efficiencies in the rifampicin challenge assay. With holoenzyme, incorporation of [^{3}H] UMP into RNA reached saturation at 29 pmole, while on core enzyme, the plateau level was 20 pmole. When holoenzyme and core enzyme were mixed together at equal concentrations, the experimental points followed very closely the theoretical curve D which was constructed assuming both enzymes compete for the same initiation sites. Therefore, with or without σ factor *E. coli* RNA polymerase utilized the same initiation regions on chromatin.

The results of an analogous competition experiment utilizing *E. coli* holoenzyme and oviduct RNA polymerase II are shown in Figure 15. Again, the experimental points for the mixture of the two polymerases followed very closely the curve predicted by assuming complete competition between the two enzymes for initiation sites. Thus, both homologous and bacterial RNA polymerases utilized the same regions on chromatin for initiation of RNA synthesis.

The levels of RNA synthesis observed for the three RNA polymerases tested were not remarkably different on chromatin. The differences in the levels of RNA synthesis which were observed could be accounted for by slightly different efficiencies of the enzymes in escaping rifamycin inhibition or small variations in the size of the RNA product. Little competition should occur by the technique during the actual process of RNA synthesis even if overlapping sequences are transcribed since only one round of transcription from any site can occur and ample time is allowed for this to take place. It should be noted, however, that if two polymerases initiate at different sites within a limited region (50 to 80 nucleotides) of chromatin, they would appear to compete with each other by this technique. Thus we interpret our results to show that different RNA polymerases utilize in general the same regions and possibly identical sites for RNA chain initiation on chromatin. Since these enzymes did not compete with each other on native chick DNA, the chromosomal proteins must be a major determinant of the specificity with which the enzymes initiate RNA synthesis.

The conclusion that different RNA polymerases utilize the same initiation regions on chromatin is in contradiction with some previous studies in the literature. Butterworth *et al.* (39) have reported that rat liver RNA polymerase II and RNA polymerase from *Micrococcus luteus* bind to and transcribe from different sites on rat liver chromatin. The methods utilized for this study also invoked mixing experi-

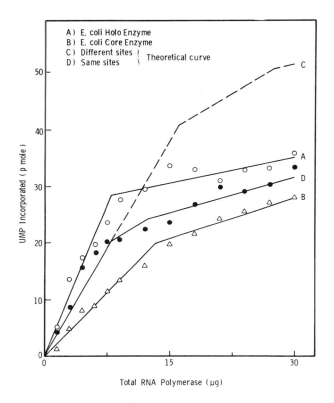

Figure 14. RNA initiation sites utilized by Escherichia coli
*RNA polymerase holoenzyme and core enzyme on chick oviduct
chromatin.* E. coli *holoenzyme (0, Curve A), core enzyme
(Δ, Curve B), or a mixture of the two enzymes in equal concentrations (●) was incubated at 37⁰ with 5.0 µg of chick
oviduct chromatin (14 days diethylstilbestrol-stimulated) in
0.2 ml of preincubation buffer for 40 min. RNA synthesis was
initiated by the addition of 0.05 ml of ribonucleoside triphosphate mixture containing 0.2 mg/ml of rifampicin and
4.0 mg/ml of heparin, and carried out for 1.5 min. at 37⁰.
Theoretical curves were constructed from the experimental
Curves A and B assuming either no competition (Curve C) or
complete competition (Curve D) between the two polymerases
for RNA initiation sites.*

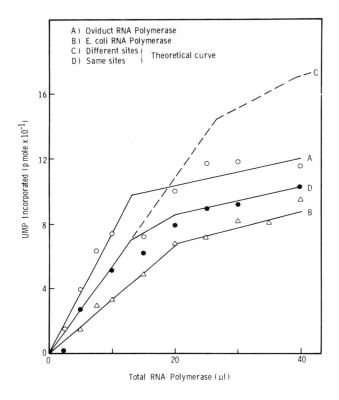

Figure 15. RNA initiation sites utilized by E. coli holo-enzyme and hen oviduct RNA polymerase II on chick oviduct chromatin. Hen oviduct RNA polymerase II (0, Curve A). E. coli holoenzyme (Δ, Curve B) or a mixture of the two enzymes in equal volumes (•) was incubated at 37°C with 5.0 μg of chick oviduct chromatin (14 days DES stimulated) in 0.2 ml of preincubation buffer for 15 min. Stock concentrations used were 130 units/ml for hen RNA polymerase and 400 μg/ml for holoenzyme. After 15 minutes, RNA synthesis was initiated by the addition of 0.05 ml of a mixture containing 3.0 mM each of ATP, CTP, GTP, 0.3 mM [³H] UTP (0.4 mC/ml) and 1 mg/ml rifamycin AF/013. RNA synthesis was carried out for 15 minutes at 37°C. Theoretical curves were constructed from experimental Curves A and B assuming either no competition (Curve C) or complete competition (Curve D) of the two polymerases for RNA initiation sites. Hen oviduct RNA polymerase II was purified by a modified procedure of Kedinger et al. (44,45).

ments between the two enzymes; however, the experiments were performed in the absence of any inhibitor of reinitiation. Thus, it is possible that during the course of the ten minute incubation a given initiation site was utilized sequentially and repeatedly by the two different RNA polymerases without any significant competition occurring in terms of total RNA synthesis. Recently, Cedar (40) has reported that calf thymus chromatin can support 30 times fewer specific sites for the homologous RNA polymerase than for the bacterial enzyme. The method utilized by Cedar involved the preincubation of enzyme and template with three of the four ribonucleotides, followed by simultaneous addition of the fourth ribonucleotide and a high concentration of ammonium sulfate to inhibit any further initiation. The explanation for the discrepancy between the results of Cedar and the results reported here is presently unclear. One possibility is that the presence of a high concentration of ammonium sulfate removed a portion of the calf thymus RNA polymerase II bound to chromatin. The effects of ammonium sulfate on transcription by eucaryotic RNA polymerases have not been well characterized but we have noted that the animal polymerases are more easily removed from chromatin by heparin than *E. coli* enzyme. However, this is only speculation at the present time and further characterization of the two methods utilized will be necessary to fully understand the differences observed.

VII. Summary

The chick oviduct can be induced by the administration of steroid hormones to undergo dramatic changes which appear to be mediated at the level of transcription. For this reason, it is an excellent system for studying the regulation of gene expression in a eucaryotic organism. Our approach has been to utilize chromatin isolated from chick oviduct as a template for *in vitro* transcriptional studies. To understand the mechanism of estrogen-mediated alteration of gene expression, we first felt it necessary to more fully characterize the interaction between chromatin and purified RNA polymerase.

The formation of binary complexes capable of rapidly initiating RNA synthesis (RS complexes) between *E. coli* RNA polymerase and chromatin differs in several respects from the formation of such complexes on deproteinized DNA. The half time of RNA complex formation on chromatin was approximately seven times slower than that for DNA. The rate limiting step in the formation of RS complexes on chromatin appears to be the rate of dissociation of enzyme from nonspecific binding

sites in the chromatin. Studies with ANS suggested that
these interactions may be largely hydrophobic in nature.
Thus, the presence of chromosomal proteins tends to retard
the rate by which RNA polymerase locates an initiation site
on chromatin. Such random, reversible interaction, however,
could conceivably serve a biological function *in vivo* to
trap the RNA polymerase molecules in the proximity of chromosomal DNA.

As opposed to the formation of RS complexes, the rate of
RNA chain initiation from preformed RS complexes is a fairly
rapid event. Furthermore, the rate constants of RNA chain
initiation for chromatin and DNA are not remarkably different. Thus, once the RS complex is formed, the actual event
of RNA chain initiation may be quite similar for the two
templates.

The total number of initiation sites available in chromatin for RNA polymerase increased after primary or secondary
estrogenic stimulation of the chick oviduct. Simultaneous
measurements of the rate of RNA chain elongation and RNA
chain length were fairly constant throughout the period of
estrogen administration. Thus, increases in the "template
capacity" of chromatin measured previously reflected an
increase in the available DNA sequences for RNA polymerase
transcription. During primary stimulation, a maximum level
of initiation sites, seven-fold higher than the unstimulated
chick, was reached at about four days of stimulation. This
level then decreased to a plateau level about two-times
higher than the unstimulated animal by about ten days of
stimulation. The high level of initiation sites at four days
corresponds with the period in which maximal growth and differentiation of cell types are occurring. Thus, it appears
that cells in the actual process of differentiation may be
transcribing a much larger number of genes than either the
undifferentiated cell types. Following completion of the
differentiation process, the need for this increased expression of genetic information appears to subside, since the
number of initiation sites for RNA synthesis decreases.

During the secondary stimulation of chick oviduct with
estrogen, the level of nuclear receptors for estrogen
increased immediately prior to an increase in the number of
initiation sites, which in turn preceded an increase in the
level of intracellular ovalbumin mRNA. This temporal relationship suggested a possible direct interaction between
hormone-receptor complex and chromatin such that new DNA
sequences may be accessible for transcription.

Using complementary DNA to ovalbumin mRNA, we could
estimate the concentration of ovalbumin mRNA sequences in

chromatin transcripts synthesized *in vitro*. Although chromatin from unstimulated chick oviduct was capable of supporting a substantial amount of RNA synthesis, no detectable ovalbumin mRNA sequences could be found in the transcript. In contrast, chromatin from estrogen-stimulated chick oviducts was capable of supporting synthesis of ovalbumin mRNA. After hormone was withdrawn, the chromatin isolated from these chicks could only support very low levels of synthesis of ovalbumin mRNA sequences. These data strongly suggest that primary gene derepression is the most likely mechanism for estrogen induction of ovalbumin synthesis.

The role of RNA polymerase in transcribing chromatin was examined. From the rifamycin-nucleotide challenge assay, we observed that both bacterial and homologous RNA polymerases competed for the same initiation region on chromatin. These results suggested that chromatin proteins are involved in determining the selection of initiation sites on chromatin by RNA polymerase and thus, may dictate the genes available for transcription. The chromatin proteins involved in regulation of gene expression were studied in more detail by reconstitution experiments. Reconstitution of extractable nonhistone protein-DNA complexes of hormone-withdrawn chromatins resulted in a hybrid chromatin which was capable of synthesizing a substantial amount of ovalbumin mRNA sequences. Therefore, it is the easily extractable nonhistone chromatin protein fraction rather than the DNA, the histone proteins or the tightly bound nonhistone protein fractions which is involved in the control of expression of ovalbumin gene.

Our current working hypothesis concerning the mechanisms of estrogen action of gene expression can be summarized as follows: After the hormone-receptor complexes enter the nuclei during secondary stimulation, they bind to the chromatin and somehow cause a physicochemical change in the chromatin by direct interaction with either (or both) the chromosomal proteins or the DNA backbone. This change in the conformation of chromatin leads to the expression of certain hormone-dependent genes. Rearrangement of chromatin proteins, histones and/or nonhistones, also could be involved in this process. This hypothesis is consistent with the rapid increase in the number of hormone-induced initiation sites for RNA synthesis. Following these events, accumulation of new specific mRNAs begins. These rapid fluctuations in gene expression during secondary stimulation by estrogen appear not to require the synthesis of protein intermediates. However, the gradual increase and decrease of initiation sites during the primary hormone-stimulation and withdrawal suggest that protein and/or DNA synthesis could be necessary for sequential expression of other sets of hormone-dependent genes. We are currently in the process of purifying the ovalbumin gene from total cellular DNA. With the purified gene for ovalbumin, we should be able to clearly define the molecular mechanisms of control of a specific gene *in vitro*. We will then attempt to reconstitute the *in vivo* regulation process using purified components of the transcription apparatus.

REFERENCES

1. Lewin, B., in *Gene Expression, Vol. 1, Bacterial Genomes* John Wiley and Sons, London (1974).
2. O'Malley, B. W., McGuire, W. L., Kohler, P. O., and Korenman, S. G., *Rec. Prog. Horm. Res. 25,* 105 (1969).
3. O'Malley, B. W., and Means, A. R, *Science 183,* 610 (1974).
4. Cox, R., Haines, M., and Carey, N., *Eur. J. Biochem. 32,* 513 (1973).
5. O'Malley, B. W., and McGuire, W. L., *Proc. Nat. Acad. Sci.U.S.A., 60,* 1527 (1968).
6. Means, A. R., Comstock, J. P., Rosenfeld, G. C., and O'Malley, B. W., *Proc. Nat. Acad. Sci.U.S.A., 69,* 1146 (1972).
7. Harris, S. E., Rosen, J. M., Means, A. R., and O'Malley, B. W., *Biochemistry 14,* 2072 (1975).
8. Spelsberg, T. C., Mitchell, W. M., Chytil, F., Wilson, E. M., and O'Malley, B. W., *Biochim. Biophys. Acta 312,* 765 (1973).
9. Comstock, J. P., Rosenfeld, G. C., O'Malley, B. W., and Means, A. R., *Proc. Nat. Acad. Sci.U.S.A. 69,* 2377 (1972).
10. Rhoads, R. E., McKnight, G. S., and Schimke, R. T., *J. Biol. Chem. 248,* 2031 (1973).
11. Chan, L., Means, A. R., and O'Malley, B. W., *Proc. Nat. Acad. Sci.U.S.A., 70,* 1870 (1973).
12. Cox, R., Haines, M., and Emtage, J. S., *Eur. J. Biochem. 49,* 225 (1974).
13. Palmiter, R. D., *J. Biol. Chem. 248,* 8260 (1974).
14. Chamberlin, M. J., *Ann. Rev. Biochem. 43,* 721 (1974).
15. Sippel, A., and Hartman, G., *Biochim. Biophys. Acta 157,* 218 (1968).
16. Umezawa, H., Mizuno, S., Yamazaki, H., and Nitta, K., *J. Antibiotics 21,* 234 (1968).
17. Hinkle, D. C., Mangel, W. F., and Chamberlin, M. J., *J. Mol. Biol. 70,* 209 (1972).
18. Mangel, W. F., and Chamberlin, M. J., *J. Biol. Chem. 249,* 2995 (1974).

19. Hirose, M., Tsai, M. J., and O'Malley, B. W., *J. Biol. Chem. 251*, 1137 (1976).
20. Bautz, E. K. F., and Bautz, F., *Nature 226*, 1219 (1970).
21. Chamberlin, J. J., and Ring, J., *J. Mol. Biol. 70*, 221 (1972).
22. Tsai, M. J., Schwartz, R. J., Tsai, S. Y., and O'Malley, B. W., *J. Biol. Chem. 250*, 5165 (1975).
23. Schwartz, R. J., Tsai, M. J., Tsai, S. Y., and O'Malley, B. W., *J. Biol. Chem. 250*, 5175 (1975).
24. Tsai, S. Y., Tsai, M. J., Schwartz, R. J., Kalimi, M., Clark, J., and O'Malley, B. W., *Proc. Nat. Acad. Sci. U.S.A. 72*, 4228 (1975).
25. Palmiter, R. D., Christensen, A. K., and Schimke, R. T., *J. Biol. Chem. 245*, 833 (1970).
26. Anderson, J., Clark, J. H., and Peck, E. J., Jr., *Biochem. J. 126*, 561 (1972).
27. Gilmour, R. S., and Paul, J., *Proc. Nat. Acad. Sci.U.S.A. 70*, 3440 (1973).
28. Axel, R., Cedar, H., and Felsenfeld, G., *Proc. Nat. Acad. Sci.U.S.A. 70*, 2029 (1973).
29. Steggle, A. W., Wilson, G. N., Kantor, J. A., Picciano, D. J., Falvey, A. K., and Anderson, W. F., *Proc. Nat. Acad. Sci.U.S.A. 71*, 1219 (1974).
30. Chiu, J. F., Tsai, Y. H., Sakuma, D., and Hnilica, L. S., *J. Biol. Chem. 250*, 9431 (1975).
31. Tsai, S. Y., Harris, S. E., Tsai, M. J., and O'Malley, B. W., *J. Biol. Chem.* (In press).
32. Paul, J., Gilmour, R. S., Affara, N., Birnie, G., Harrison, P., Hell, A., Humphries, S., Windass, J., and Young, B., *Cold Spring Harbor Symp. Quant. Biol. 38*, 885 (1973).
33. Barrett, T., Maryanka, D., Hamlyn, P. H., and Gould, H. J., *Proc. Nat. Acad. Sci. U.S.A. 71*, 5057 (1974).
34. Stein, G., Park, W., Thrall, C., Mans, R., and Stein, J., *Nature 257*, 764 (1974).
35. Meilhac, M., Typser, Z., and Chambon, P., *Eur. J. Biochem. 28*, 291 (1973).
36. Tsai, M. J., and Saunders, G. F., *Proc. Nat. Acad. Sci. U.S.A. 70*, 2072 (1973).
37. Meilhac, M., and Chambon, P., *Eur. J. Biochem. 35*, 454 (1973).
38. Mandell, J. L., and Chambon, P., *Eur. J. Biochem. 41*, 379 (1974).
39. Butterworth, P. H. W., Cox, R. F., and Chesterton, C.J., *Eur. J. Biochem. 23*, 229 (1971).
40. Cedar, H., *J. Mol. Biol. 95*, 257 (1975).

41. Burgess, R. R., and Travers, A. A., In *Methods in Enzymology Vol. XXI(D)*, Grossman, L., and Moldave, K. (eds.), Academic Press, New York, pp. 500-506 (1971).

42. Harris, S. E., Schwartz, R. J., Tsai, M. J., O'Malley, B. W., and Roy, A., *J. Biol. Chem. 251*, 1960 (1976).

43. Spelsberg, T. C., Steggles, A. W., Chytil, F., and O'Malley, B. W., *J. Biol. Chem. 247*, 1368 (1972).

44. Kedinger, C., Gissinger, F., Gniazdowski, M., Mandel, J., and Chambon, P., *Eur. J. Biochem. 28*, 283 (1972).

45. Kedinger, C., and Chambon, P., *Eur. J. Biochem. 28*, 283 (1972).

Chapter 8

LOW MOLECULAR WEIGHT NUCLEAR RNA:

SIZE, STRUCTURE AND POSSIBLE FUNCTION

RU CHIH C. HUANG

Department of Biology
Johns Hopkins University
Baltimore, Maryland 21218

NOMENCLATURE AND STRUCTURAL FEATURES OF SnRNA

Low molecular weight nuclear RNA (SnRNA) has been found in a wide variety of eukaryotic cells (1). Among this class of RNA, ranging in size between 80 to 320 nucleotides long, several species have recently been isolated in sufficient quantity and purity to allow structural analysis. These are termed with different nomenclature by different investigators. For example, Busch and his colleagues (1-4) named SnRNA of rat Novikoff hepatoma U_3 (215 nucleotides), U_2 (196 nucleotides), U_1 (171 nucleotides), 5S (120 nucleotides), 4.5S (96 nucleotides) and 4S (80 nucleotides). Zapisek *et al.* (5) named them as VII, VI, IV, III and II. Huang and her colleagues used C_1, C_2, C_3, C_4, C_5, C_6 and C_7 for mouse myeloma SnRNAs (Fig. 1) (5). Zieve and Penman termed these species of SnRNA of HeLa cells SnA, SnB, SnC, SnD, SnG and SnH (7,8). The molecules assigned are identical or nearly identical in size, such as SnRNAs U_3, C_1 and SnA; U_2, VII, C_2 and SnC; U_1, VI, C_3 and SnD; 5S, C_4, IV and SnG; 4.5S, C_6, III and SnH; or among 4S, C_7 and II.

Busch and his colleagues initiated the structural studies of these defined SnRNA species. Most significantly, they have found, for instance, that the 5' termini of U_1, U_2 and U_3 all have modified nucleotide bases. The proposed 5'-terminal sequences for U_1, U_2 and U_3 are $m_3G^{5'}pp^{5'}$Amp Ump Ap----, $m_3G^{5'}pp^{5'}$ Amp Ump Cp--- and $m_3G^{5'}pp^{5'}$ Amp Ap---, where both Am and Um are 2'-0-methyl nucleotides and m_3G is the unusual nucleoside $N^2,^7$-trimethyl guanosine (9). Very recently, Cory and Adams, working with SnRNA of mouse myeloma have confirmed the existence of a 5'-5' phosphate linkage at

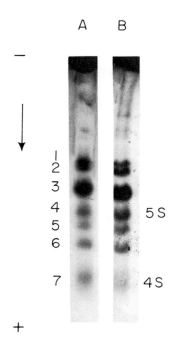

Figure 1. Gel electrophoretic patterns of low molecular
weight RNAs. $^{32}PO_4$-labeled myeloma chromatin RNA (extracted
with 0.35 M NaCl followed by phenol) and 4-8S RNA from the
saline-EDTA wash of the myeloma nuclei were fractionated by
gel electrophoresis on 10% polyacrylamide gels as described
(6). The gels were sliced longitudinally and autoradio-
graphed. Exposure time was 1 hr; (A) RNA from chromatin;
(B) RNA from saline-EDTA wash. Eight bands (0-7) were observed. Marzluff, W. F., White, E. L., Benjamin, R., and Huang,
R. C. C. (1975) Biochemistry 14, 3715. Reprinted with permission of the American Chemical Society.

the 5'-termini of U_1, U_2 and U_3 RNA; but rather than diphosphate links, they found that the 5'-termini of U_1, U_2 and U_3 consist of 5',5' triphosphate linkages (10). This finding seems to be extremely significant in view of their possible function. Similar 5' termini structures have been discovered in both viral and cellular messenger RNAs (mRNAs) (11-19). It is possible that U_1, U_2 and U_3 are synthesized and/or processed in the same manner as mRNA. One other rather surprising feature of U_2 RNA is its high content of pseudouridylic acid residues (11 nucleotides and its 2'-O-methyl ribose residues (10 nucleotides) at its 5' end (20).

The primary sequence of another species of SnRNA of Novikoff hepatoma, the 4.5S, was also established by Busch and his colleagues (4). It initiates with guanosine triphosphate. The whole sequence consists of a purine-rich 5' end and a pyrimidine-rich 3' end. This structure must contain a high internal complementarity. One other SnRNA was identified as 5S RNA which is identical to 5S rRNA as judged by its size, by its hybridization properties, and by T_1 fingerprint analysis (6).

SUBCELLULAR LOCALIZATION OF SnRNA

The subcellular localization of these SnRNAs has been investigated in several systems. Based upon their content in several subcellular organelles, Ro-Choi and Busch (1) have suggested that U_1, U_2, U_3 and 4.5S SnRNAs are nuclear in origin. 65-75% of the total SnRNA is associated with the chromatin fraction of rat liver. In the mouse myeloma system, Marzluff et al. (6) have made a gross fractionation of nucleoplasm and chromatin by extracting the myeloma nuclei with 0.075 M NaCl, and 0.024 M EDTA, pH 7.9. Under this condition, at least 50% of the SnRNA is still associated with chromatin and perhaps nucleoli. SnRNA C_1 (U_3), C_3 (U_1), C_5 (4.5S) are present in chromatin and nucleoli in a much higher proportion than the rest of the SnRNA species (Fig. 1). Zieve and Penman (7), working with HeLa cells, have concluded that at least seven species of SnRNA (SnA, SnB, SnC, SnF, SnG, SnH) are chromatin-bound and SnA (U_3) is localized within the nucleolus. They have found that, as the ionic strength of the lysis buffer is increased, greater amounts of the species are found in the cytoplasmic fraction. This release could either be an artifact or a true aspect of their metabolism. Goldstein and Ko (21), in a series of elegant experiments, established that several SnRNA species of Ameba can shuttle between nucleus and cytoplasm and that the shuttling behavior reflects a normal activity in Ameba.

COMPARISON OF EXTRACTION METHODS FOR SnRNA

One of the difficulties in obtaining large quantities of defined species of SnRNA for both the structural and functional studies is the lack of a quantitative procedure for RNA isolation. We have examined several methods in isolating SnRNA from chick embryo chromatin (Table I) (6). These methods have been designed to allow recovery of RNA under relatively mild conditions at neutral pH. We found that extraction with 0.35 M NaCl or 0.15 M deoxycholate, as originally reported by Mayfield and Bonner (22), followed by phenol deproteinization, gave the best yields of chromatin RNA. In contrast, other methods which involved dissociation of chromatin components in solutions of high ionic strength followed by high speed centrifugation, while giving excellent recovery of chromosomal proteins, resulted in an extremely low recovery of SnRNA.

STABILITY OF SnRNA

SnRNAs of myeloma chromatin were very stable. We have examined the radiospecific activity of these SnRNAs either after 20 hrs of labeling of tissue culture cells, or after 20 hrs of labeling followed by 8 hrs of chase. It was found that, during a chase period of 8 hrs, there was essentially no turnover of these SnRNA species. The relative amount of labeling of these RNA species also remained the same, indicating that there was no preferential turnover of any single species of RNA. In HeLa cells, the SnRNA is also very stable. Zieve and Penman (7) isolated a rather high yield of SnRNA after 5-6 hrs labeling followed by a 10 hr period of chase. Rubin and Hogness (23) have examined the synthesis and stability of SnRNA in *Drosophila* after a heat shock treatment. They found that, when cultures were raised from $25^{O}C$ to $37^{O}C$, most of the nuclear SnRNA and 5.8S rRNA were not formed. This block may result from inhibition of SnRNA synthesis, or equally possible, from interruption of a regular processing of high molecular weight precursors. These authors also observed synthesis of a 5S rRNA precursor. An additional fifteen nucleotides are linked to the 5S rRNA at the 3' end. During the heat shock from $25^{O}C$ to $37^{O}C$, the formation of 5S rRNA was greatly reduced while its precursor was accumulated.

GENES CODED FOR THE SnRNAs

We have further examined the sequence complexity and have titrated the gene copies of the SnRNAs in the mouse

Table I. Extraction of RNA from Chick Embryo Chromatin*

Method	mg of RNA Recovered/mg of DNA	% RNA Recovered
3 M NaCl-DEAE	0.004	5
3 M NaCl-7 M urea	0.005	6
4 M Gdn-HCl	0.005	6
0.2 M NaCl-phenol	0.025	30
0.35 M NaCl-phenol	0.045	60
0.5 M NaCl-phenol	0.030	40
0.015 M Na deoxycholate-phenol	0.045	60
0.015 M Na deoxycholate-DEAE	0.030	40
0.35 M NaCl-DEAE	0.030	40

*
Data obtained from Marzluff et al. (6).

genome. Individual species of SnRNAs were isolated and eluted from the gel after separating them electrophoretically. Each species was allowed to hybridize with mouse DNA. From the hybridization kinetics, the complexity of each species was calculated. For each species of SnRNA, the results were consistent with a single sequence, at least not more than 2 or 3.

Gene copies of SnRNA, however, vary between 100 copies for C_1 (U_3) and C_2 (U_2) RNA to about 2000 copies for C_6 RNA (Fig. 2, Table II) (6). There are large numbers of these RNAs in each myeloma nucleus, estimated at about 200,000 to 1,000,000 molecules per nucleus. These six species represent 80% of the total SnRNA in mouse myeloma. The remaining 20% are made up of RNAs in lower quantity but higher complexity. Since it is difficult to isolate them in substantial amounts, structural features of these RNAs are not established at the present time.

POSSIBLE FUNCTIONS OF SnRNA

It seems likely that the SnRNA is complexed with non-histone protein in chromatin. So far, we have examined chromatins isolated from sources with different nuclear synthetic activities: calf thymus, chick embryo, mouse myeloma, pig cerebellum and chick erythrocytes (24). These chromatins also contain different amounts of SnRNAs (Table III) (24). We have examined these chromatins using circular dichroism (CD). We have confirmed some of the earlier studies (25) and found that the CD of chromatin above 250 nm is different from that of free DNA in solution. In addition, we also found that SnRNA and/or aromatic chromopheres in non-histone protein contribute significantly to the spectra in this region (Fig. 3, Table IV) (24). They represent the major difference found in the CD spectra of these five chromatins. Low salt (0.2 M - 0.35 M) extracts substantial amounts of SnRNAs and non-histone proteins. Low salt extractions also causes a considerable attenuation of the 273 nm peak in all chromatins. This difference, however, reflects only the abundance of SnRNA and non-histone proteins in these chromatins, but not the structural difference in the chromatin DNA. We have compared the circular dichroism of native chromatin from chick embryos with the mathematically contructed CD of chromatin, combining contributions to the CD of low salt extracts and of low salt extracted chromatin from the same system. The identical CD profiles were observed (Fig. 4). This agreement indicates that the ellipticity of the chromosomal DNA is the same while the difference of CD profile before and after extraction

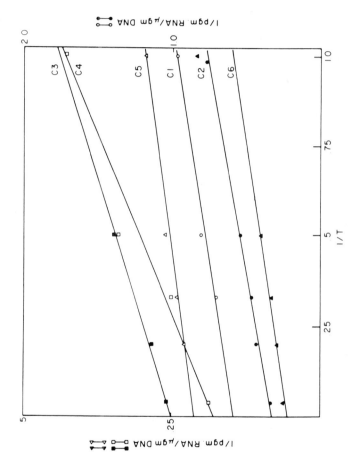

Figure 2. Hybridization of chromatin RNAs. Purified myeloma RNA fractions were hybridized to mouse DNA on nitrocellulose filters in 6 X SSC–50% formamide at 58°. 1/T is expressed in hours. Specific activity was 650,000 cpm/µg. (o) C1, 0.040 µg/ml; (●) C2, 0.045 µg/ml; (■) C3, 0.02 µg/ml; (□) C4, 0.016 µg/ml; (Δ) C5, 0.042 µg/ml; (▲) C6, 0.007 µg/ml. Marzluff, W. F., White, E. L., Benjamin, R., and Huang, R. C. C. (1975) Biochemistry 14, 3715. Reprinted with permission of the American Chemical Society.

305

Table II. Gene Copies for Myeloma Chromatin RNAs as Determined by DNA-RNA Hybridization[a].

Bands	Expt 1	Expt 2	Av	$C_T t_{1/2} \times 10^3$
C_1	110	100	105	0.37
C_2	125	165	145	0.30
C_3	295	270	282	0.22
C_4	550	550	550	0.32
C_5	560	475	517	0.28
C_6	3370	1820	2600	0.17

[a] $^{32}PO_4$ chromatin RNAs were purified and hybridized to mouse DNA as described (6). The copies of band C_4 (5 S) were defined as 550 in each experiment and the specific activity of the RNAs calculated assuming this value from the saturation value (cpm/μg of DNA) of C_4 in each experiment. Saturation values were determined from the intercepts in Figure 2 and at $t_{1/2}$ from plots similar to those of Figure 2. At least 80% saturation was reached in all determinations. C_T is the nucleotide concentration in moles/liter and $t_{1/2}$ is the time for half-saturation to be reached in seconds. Data obtained from Marzluff *et al.* (6).

Table III. RNA to DNA Mass Ratio of Salt-Extracted Chromatin[a]

Source	CHR[b]	Sample NP LS[c]		PLS[d]
Chick embryo brain	0.102 ± 0.020	0.064 ± 0.016	0.043	0.006
Pig cerebellum	0.122 ± 0.020	0.084 ± 0.034		
Myeloma	0.120 ± 0.040	0.080		
Calf thymus	0.044 ± 0.008	0.030		
Chicken erythrocyte	0.030			

[a]See Hjelm and Huang (24) for detailed description.

[b]Chromatin.

[c]Chromatin after low salt (0.2–0.35 M NaCl) extraction.

[d]Low salt extract from c.

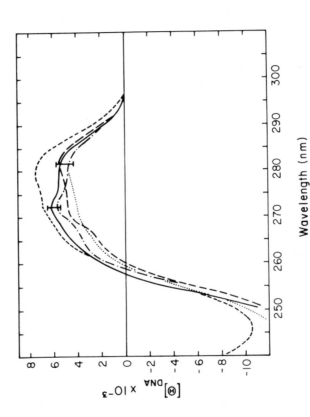

Figure 3. Circular dichroic spectra of various chromatins. (——) Chick embryo brain; (-·-·-) pig cerebellum and myeloma; (— —) chick erythrocyte; (·····) calf thymus; (– – –) calf thymus DNA. The spectra are averaged from several determinations but retain the characteristics of the individual spectrum. Molar ellipticities are expressed as (deg cm²)/dmol of DNA. Hjelm, R. P., Jr., and Huang, R. C. C. (1975) Biochemistry 14, 1682 (24). Reprinted with permission of the American Chemical Society.

Table IV. Ellipticities at 273 nm and 283 nm of Chromatin and Nucleoproteins and their Spectral Ratios.

Source	Chromatin		NP Low Salt[a]	
	273 nm	283 nm	273 nm	283 nm
Chick embryo brain	6220 ± 620 1.19 ± 0.09[b] (18)[c]	5100 ± 480	4400 ± 560 0.96 ± 0.05[b] (18)	4550 ± 490
Pig cerebellum	5800 ± 470 1.16 ± 0.08[b] (17)	4670 ± 480	4740 ± 600 1.08 ± 0.08[b] (17)	4270 ± 480
Myeloma	5750 ± 630 1.15 ± 0.03[b] (9)	4620 ± 660	3898 0.87[b] (1)	4455
Calf thymus	4000 ± 480 0.93 ± 0.08[b] (7)	4650 ± 610	3850 0.90[b] (1)	4275
Chicken erythrocyte	4790 ± 290 0.86 ± 0.08[b] (6)	5310 ± 570		

[a]Nucleoprotein low salt: Chromatin after low salt (0.2-0.35 NaCl) extraction.

[b]Ratio of ellipticities at 273 and 283 nm

[c]Number of determinations given in parentheses.
Data obtained from Hjelm and Huang (24).

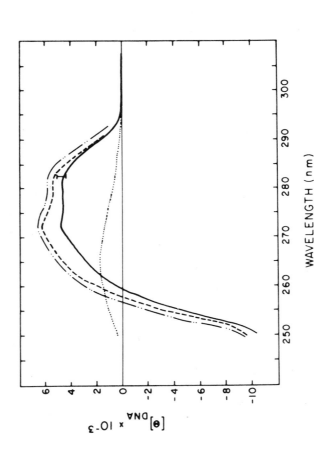

Figure 4. Circular dichroism of nucleoprotein low salt and mathematically constructed chromatin and native chromatin from chick embryo brain (——); NPLS (••••); PLS X 0.041; (– – –) calculated chromatin CD: (–•–) native chromatin. The error bar is plus and minus one standard deviation and applies approximately for both the NPLS and chromatin spectra. See Table III for the descriptions of NPLS and PLS. Hjelm, R. P., Jr., and Huang, R. C. C. (1975) Biochemistry 14, 1682. Reprinted with permission of the American Chemical Society.

represents the contributions from SnRNAs and non-histone proteins. Using the ratio of ellipticities at 273 nm and 283 nm
one may be able to estimate the SnRNA content of a given
chromatin (24).

Functions of these SnRNAs are totally unknown at the
moment. An attractive hypothesis is that these SnRNAs in
some way effect gene transcription. SnRNAs U_1, U_2 and U_3
share the same structural feature of mRNAs because they are
capped at 5' termini. Logically, they can be "mini RNAs" by
themselves or may be portions of mRNA precursors. In this
case, an SnRNA represents the partial sequences of an mRNA
precursor at its 5' end. Products of translation of SnRNA,
if indeed translatable, would be small polypeptides much
smaller than the regular proteins. They may function as control elements in nuclear events. Alternatively, an SnRNA may
function as a signal for starting the transcription on chromatin. Two pieces of evidence support this hypothesis: (a) the
amount of SnRNA correlates well with the chromatin activity
of the tissue; for instance, chick embryo chromatin, an
active system, has a high content of SnRNA while chick erythrocyte chromatin, a repressed system, has nearly none (24);
(b) based upon studies of reconstituted chromatin, SnRNAs and
non-histone proteins are shown to be essential for maintaining an active transcriptional unit (25). Data are still
missing, however, in respect to the effect of SnRNA on transcription of specific genes. Very recently, several specific
gene products have been examined. Transcription of three
specific genes, 5S, tRNA and immunoglobulin Kappa light chains
of mouse myeloma have been extensively studied (26-29). We
know a great deal about the sequence of 5S RNA. Therefore,
precise initiation, DNA strand selection and proper chain
termination in transcription can be examined. Low salt
extracted myeloma chromatin has been shown to have very
little RNA synthetic activity even after the addition of
homologous RNA polymerase (28 and unpublished data). It will
be extremely interesting to see if proper insertion of SnRNA
back to the extracted chromatin could cause the restoration
of transcriptional activity.

Recently, Thomas et al. (30) have found conditions under
which RNA could be hybridized to double-strength DNA and
formed R-loops.   R-loop structure can also exist
in the native chromatin. SnRNA may be a natural candidate
for R-loop involving partial regulatory DNA sequence of
chromatin. Experiments to test this possibility are now in
progress. With several assay systems recently established
(26-29) and with the knowledge gained on the structure of
SnRNA, it is hoped that we will be able to assign functional

roles to SnRNA in the near future.

REFERENCES

1. Ro-Choi, T. S., and Busch, H., *Low-Molecular Weight nuclear RNA's* in *The Cell Nucleus, Vol. III,* Academic Press, New York, San Francisco and London, 151 (1974).
2. Shibata, H., Ro-Choi, T. S., Reddy, R., Choi, Y. C., Henning, D., and Busch, H., *J. Biol. Chem. 250,* 3909 (1975).
3. Reddy, R., Ro-Choi, T. S., Henning, D., and Busch, H., *J. Biol. Chem. 249,* 6486 (]974).
4. Ro-Choi, T. S., Reddy, R., Henning, D., Takano, T., Tayler, C. W., and Busch, H., *J. Biol. Chem. 247,* 3205 (1972).
5. Zapisek, W. F., Saponara, A. G., and Enger, M., *Biochemistry 8,* 1170 (1969).
6. Marzloff, W. F., White, E. L., Benjamin, R., and Huang, R. C., *Biochemistry 14,* 3715 (1975).
7. Zieve, G., and Penman, S., *Cell 8,* 19 (1976).
8. Weinberg, R. A., and Penman, S., *J. Mol. Biol. 38,* 289 (1968).
9. Ro-Choi, T. S., Choi, Y. C., Henning, D., McCloskey, J., and Busch, H., *J. Biol. Chem. 250,* 3921 (1975).
10. Cory, S., and Adams, J. M., *Mol. Biol. Reports 2,* 287 (1975).
11. Furnichi, Y., and Miura, K., *Nature 253,* 374 (1975).
12. Wei, C. M., and Moss, B., *Proc. Nat. Acad. Sci. U.S.A. 72,* 318 (1975).
13. Urushibara, T., Furnichi, Y., Nishiwura, C., and Miura, K., *FEBS Lett. 49,* 385 (1975).
14. Furnichi, Y., Morgan, M., Muthukrishnan, S., and Shatkin, S. J., *Proc. Nat. Acad. Sci. U.S.A. 72,* 362 (1975).
15. Abraham, G., Rhodes, D. P., and Banerjee, A. K., *Cell 5,* 51 (1975).
16. Adams, J. M., and Cory, S., *Nature 255,* 28 (1975).
17. Furnichi, Y., Morgan, M., Shatkin, A. J., Jelinek, W., Salditt-Georgieff, M., and Darnell, J. E., *Proc. Nat. Acad. Sci. U.S.A. 72,* 1904 (1975).
18. Penny, R. P., Kelley, D. E., Friderici, K, and Rottman, F., *Cell 4,* 387 (1975).
19. Muthukrishnan, W., Both, G. W., Furnichi, Y., and Shatkin, A. J., *Nature 255,* 33 (1975).
20. Reddy, R., Ro-Choi, T. S., Henning, D., Shibata, H., Choi, Y. C., and Busch, H., *J. Biol. Chem. 247,* 7245 (1972).

21. Goldstein, L., and Ko, C., *Cell 2,* 259 (1974).
22. Mayfield, J. E., and Bonner, J., *Proc. Nat. Acad. Sci. U.S.A. 68,* 2652 (1971).
23. Rubin, G. M., and Hogness, D. S., *Cell 6,* 207 (1975).
24. Hjelm, R. P., Jr., and Huang, R. C. C., *Biochemistry 14,* 1682 (1975).
25. Huang, R. C. C., and Huang, P. C., *J. Mol. Biol. 39,* 365 (1969).
26. Marzluff, W. F., Murphy, E., and Huang, R. C. C., *Biochemistry 12,* 3440 (1973).
27. Marzluff, W. F., Murphy, E., and Huang, R. C.C., *Biochemistry 13,* 3689 (1974).
28. Marzluff, W. F., and Huang, R. C. C., *Proc. Nat. Acad. Sci. U.S.A. 72,* 1082 (1975).
29. Smith, M. M., and Huang, R. C. C., *Proc. Nat. Acad. Sci. U.S.A. 73,* 775 (1976).
30. Thomas, M., White, R. C., and Davis, R. W., *Proc. Nat. Acad. Sci. U.S.A. 73,* 2294 (1976).

Chapter 9

THE ORGANIZATION OF DNA SEQUENCES IN

POLYTENE CHROMOSOMES OF *DROSOPHILA*

A. GAYLER HARFORD

Division of Cell and Molecular Biology
State University of New York at Buffalo
Buffalo, New York 14214

INTRODUCTION

Polytene chromosomes are multi-stranded structures which arise as the result of chromosome duplication in the absence of cell division. The most famous are the giant chromosomes found in the larval salivary glands of Diptera.* Because of their large size, these chromosomes have provided a rare view of the interphase chromosome (for reviews see 2,3,25,40). It is largely from these exquisite structures (Fig. 1) that the notion of the chromomere as a basic unit of function derives. These chromosomes have allowed a precise localization of genes to cytologically visible structures as well as the direct visualization of chromosomal rearrangements. The many elegant studies on chromosomal puffs and Balbiani rings suggest that these chromosomes will have much to tell us about gene regulation and development (25,40).

Intensive work has been done on the sequence components of eukaryotic DNAs and their arrangement with respect to one another. The polytene chromosomes provide us with a valuable opportunity to correlate these findings with the visible structures of these chromosomes. Rather than attempting to review the vast amount of research on polytene chromosomes, this chapter will focus on the organization of their DNA sequences and the relationships of these sequences to cytologically visible structures and, in the process, perhaps provide some insights into the functioning of the eukaryotic chromosome. The discussion will deal primarily with the

---

*One of the orders of insects, including flies, mosquitoes and midges.

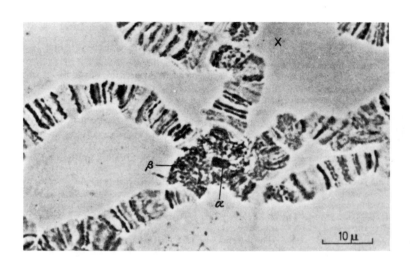

Figure 1. Polytene chromosomes from larval salivary gland of D. virilis. *The banded chromosome arms meet at the chromocenter which consists of* α *and* β*–heterochromatin. From Gall, Cohen and Polan (31) and reprinted by permission of Springer-Verlag.*

polytene salivary gland chromosomes in *Drosophila* larvae
since it is here that we have the greatest chance of integra-
ting the three fields of genetics, cytology and biochemistry.

SIMPLE SEQUENCE DNA AND HETEROCHROMATIN

Simple sequence DNAs are very highly repetitive DNAs
whose sequences may be repeated millions of times in the
genome (for review see 96). These DNAs are often identified
as density satellites on CsCl gradients (5,6,8,10,18,27,28-
31,41,69,70,73,75,85-87,99,55). Others occur as "cryptic"
satellites, so called because their identification on gradients requires the presence of metal ions or drugs which
preferentially bind to certain types of sequences (96,70,65,
30). In general, simple sequence DNAs vary in kind and
amount between different organisms, and large differences
appear even between closely related species (96,30,41).

In general, it is found that simple sequence DNAs are
greatly reduced in the DNA of the third instar salivary
gland (31,18). This is illustrated in Fig. 2 which shows
CsCl gradient profiles of DNA from diploid tissue and salivary glands from *Drosophila melanogaster* and *Drosophila
virilis*. In both species, satellite sequences present in
the diploid tissue are missing from the salivery gland (31).
The significance of this finding was soon grasped (31).

It had long been known that during the development of
the salivary gland the heterochromatic regions of the chromosomes do not replicate to the full extent that the euchromatic regions do. Heterochromatin is the densely staining,
compacted chromatin that occurs in nearly all eukaryotes.
Its highly condensed state is thought to be related to a
general absence of transcriptional activity. Fig. 1 shows
the chromosomes of the salivary gland of a late third instar
larva of *Drosophila virilis*. In the metaphase squash from a
dividing cell, shown for comparison in Fig. 3, large densely
staining heterochromatic blocks around the centromeres are
clearly visible. In the salivary gland, these heterochromatic regions associate to form the densely staining region
called the chromocenter (2,3,26,40). Each of the arms connected to the chromocenter actually consists of two homologous chromosome arms tightly paired (2,3). Each chromosome
has replicated about 9 times to produce these giant arms (3).

Underrepresentation of heterochromatin in polytene
chromosomes was first proposed by Heitz on the basis of
cytological observations (38,39). Quantitative evidence has
been obtained by measuring the DNA content of single cells
spectrophotometrically using the Feulgen stain (80,81).

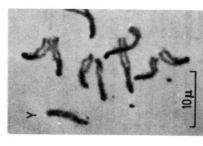

Figure 3. Mitotic chromosomes of D. virilis from larval brain ganglion cell. Note the large heterochromatic regions. This cell is actually 4C, but resembles 2C cells both in chromosome morphology and relative amounts of euchromatin and heterochromatin (4). From Gall, Cohen and Polan (31) and reprinted by permission of Springer-Verlag.

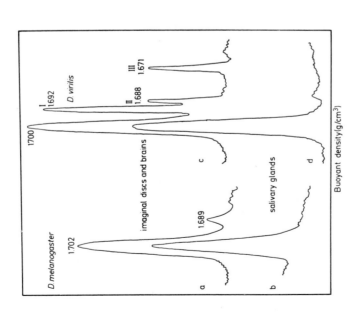

Figure 2. CsCl profiles of DNA from diploid brains and imaginal discs of larvae and from poly-tene salivary glands from two species of Drosophila. The small satellite band in D. melano-gaster actually contains several different simple sequence DNAs. Others occur as cryptic satellites in the main peak. Note the very different profile in D. virilis, where nearly half of the diploid DNA is composed of three satellites. The salivary glands, as discussed in the text, show no satellite bands. From Gall, Cohen and Polan (31) and reprinted by permission of Springer-Verlag.

318

During the development of the salivary gland in *D. melanogaster,* the DNA content of a single nucleus increases in a geometric series $2^N X$ where X is an amount of DNA about 75% of the diploid amount (81). The simplest interpretation of the results is that only about 75% of the DNA in fact replicates. In *D. hydei,* separate measurements of euchromatin and heterochromatin show that the latter remains at a $4C^*$ level while the former achieves a 10C level (68).

As the reader may have guessed, simple sequence DNAs are absent from the salivary gland because they are concentrated in heterochromatin and do not replicate during polytenization. Direct evidence for the location of these DNAs has been obtained using the technique of *in situ* hybridization (31,73, 10,70,41,45,86,70,18). In this technique, ^{3}H-labeled RNA transcribed *in vitro* from purified satellites is hybridized to denatured DNA in cell squashes. Autoradiography reveals grains over the centromeric heterochromatin in metaphase squashes of dividing cells and over the chromocenter in the salivary gland. Suitable controls establish the specificity of the reaction. Although this technique cannot be considered quantitative, it is probably significant that the number of grains over the chromocenter is about the same as over the centromeric heterochromatin of metaphase figures on the same slide (31). Thus the absolute amount of this DNA in the chromocenter is probably roughly the same as in these dividing cells, but because of the vast increase in the rest of the DNA, the simple sequence DNAs cannot be detected in CsCl analytical gradients of salivary DNA (31). Simple sequence DNAs have also been found at various sites along chromosome arms (31,70,41,86). It has been suggested that at least some of these sites represent heterochromatin (41). It is not clear whether the DNA at these sites is fully replicated in the salivary gland. A similar picture is emerging for other Diptera (24,51).

Close examination of autoradiographs shows the grains in the chromocenter confined mainly to a small region called the α-heterochromatin, with many fewer grains over the surrounding more diffuse material called the β-heterochromatin (31, 70,73,41) (see Fig. 1). The α-heterochromatin probably represents the bulk of the heterochromatin which fails to replicate (38,39,31). The β-heterochromatin clearly does replicate as evidenced by the fact that it comprises an estimated 2% of the chromosomal material in the salivary gland (80,81). and incorporates ^{3}H-thymidine (56,57). Gall *et al.* (31) have suggested that the β-heterochromatin actually represents

---

*The C value is the amount of DNA in the haploid genome.

small regions at the junctions between the euchromatin and
heterochromatin and consequently that most of the material in
the chromocenter may not correspond to the bulk of the heterochromatin seen in diploid cells. Indeed, the term heterochromatin may be misleading here since the β-heterochromatin is
apparently active in transcription (56,57). In *D. melanogaster*, sections 19 and 20 of the X chromosome, which were
formerly thought to represent the centromeric heterochromatin,
have now been shown to carry quite a few genes--perhaps as
many as would be expected for a euchromatic region of comparable size (83). In addition, in a variety of chromosomal
rearrangements having breakpoints in the heterochromatin as
viewed in metaphase squashes, sections 19 and 20 are found to
be intact in the salivary gland (83).

The function, if any, of these DNAs remains a mystery.
In *D. melanogaster* and *D. virilis*, they seem to occur in
large blocks (70,30). The sequences which have been determined are exceedingly simple (30,70,28,5,6,86) and argue
against a coding function, particularly in cases where the
repeat unit is not even an integral multiple of the coding
unit. This surmise is consistent with the fact that very few
genes have been found in heterochromatic regions (63). Non-
coding functions have been proposed but none are supported by
more than suggestive evidence.

The failure of simple sequence DNAs to replicate in the
salivary gland suggests that they may be of minimal importance
to the functioning of this tissue. Since underreplication of
satellite sequences is a common feature of many *Drosophila*
tissues (8,85,75,27) and has been found in many, but not all,
polytene tissues in other organisms (for review see 27), the
situation in the salivary gland may be quite general. The
fact that similar organisms have widely different amounts
(53) and the fact that differences are found even among
species which can form viable hybrids (30,41) suggest that
these sequences might be simply an evolutionary byproduct
with no particular function. On the other hand, if these
sequences are simply "junk", then why are they often found
near centromeres in a variety of organisms (53,96) and why
should mechanisms have evolved to regulate their synthesis
independently of the rest of the genome? In this context, it
is interesting that the three satellites of *D. virilis* can
replicate independently of one another in certain cases (8,
85,27).

It is beyond the scope of this chapter to consider these
sequences in detail, but we should not leave the subject
without mentioning how they might have evolved. As a simple
example, the three prominant satellites of *D. virilis* are

each repeating heptamers of closely related sequence: ACAAACT, ATAAACT, and ACAAATT (30). It has been suggested that one satellite could give rise to the other two by a single base pair change in the repeat unit (30). This postulated type of evolution requires some mechanism whereby mutations are either eliminated or rapidly propagated throughout the tandem array. Rolling circle replication (43) or unequal crossing over (88,95,16) have been suggested as possible mechanisms. Digestion of satellites with restriction endonucleases* often yields large fragments of discrete sizes (36,65,87). This finding suggests a long range order which may be related to the evolutionary history of these sequences (36,65,87).

THE CHROMOMERE

The euchromatic arms are not uniform in structure. They consist of a series of compacted regions or bands called chromomeres separated by relatively uncondensed regions called interbands (Fig. 1). The chromomeres, which vary in size and morphology, are arranged in a distinct and recognizable pattern along the chromosome. A skilled cytologist can, from the banding pattern alone, recognize different regions of specific chromosomes. About 95% of the DNA in the arms is in the chromomeres, the remaining 5% being in the interbands (2). When bands can be distinguished in other tissues, the banding pattern is, for the most part, the same as in the salivary gland (for review see 2). For this reason, it is generally assumed that the chromomeric structure seen in the salivary gland is the basic mode of organization of chromosomes in *Drosophila* and probably in other eukaryotes as well.

It was early suggested that one chromomere might correspond to one gene (for reviews see 2,61). Although attractive cytologically, this hypothesis is puzzling from a molecular point of view since the average chromomere contains far more DNA than needed to code for an average polypeptide chain. An average chromomere contains an estimated 3×10^4 base pairs if one takes the haploid DNA content, measured on sperm, to be 1.8×10^8 nucleotide pairs (74) and assumes that about 80% (78,79) of this is in the roughly 5000 chromomeres identified cytologically by Bridges (63). This paradox has its counterpart in other organisms as well (7). The excess DNA problem has stimulated a careful re-examination of the one gene-one chromomere concept.

---

*Endonucleases which cleave specific base sequences.

One of the more thorough studies of this point is that of Judd and co-workers (46,47). Using ingenious genetic techniques, these workers screened for mutants induced by X-rays and various chemical mutagens in a small region of the X chromosome containing 15 chromomeres. Their large collection of mutants fell into only 16 different complementation groups, about the number of bands in the region under consideration. By careful cytological mapping using a variety of chromosomal aberrations, many of the complementation groups could be localized to specific bands. The argument remains unsatisfactory, however, since there is no way to be sure that a large number of genes have not been missed. Most of the mutants were identified by their lethality, but clearly not every gene is lethal when mutated, and other phenotypic changes may not be readily apparent to the observer.

Lefevre has approached this problem differently (60). Rather than looking for mutant progeny, he examined all progeny of X-irradiated males, whether mutant or not, for X chromosome rearrangements and determined the breakpoints. He reasoned that if a band contained one essential gene and a number of inessential genes, then a break within that band should sometimes be lethal and sometimes not. Furthermore, complementation between some rearrangements involving the same band should occur. These predictions are not realized. All breaks in a particular band have the same effect and none complement one another. In cases where visible mutant phenotypes are produced, all breaks in the same band produce the same phenotype.

These results suggest that each chromomere functions as a single indivisible unit. This notion does not fit well with cytogenetic evidence that the white gene occupies only a small portion of a band or perhaps an interband region (90). On the other hand, data on the *Notch* locus have been tentatively interpreted to indicate that this gene is split into two parts which reside at the two ends of the chromomere (101,102). Lefevre's data also suggest that only about half of all bands contain essential genes, a conclusion seemingly at odds with studies showing a nearly one-to-one relationship between bands and essential genes (2,46,47). He has discussed a model to explain his results (61).

The arguments of Lefevre suffer from the difficulty that one cannot be sure that breakage is random or that the breaks are clean and do not involve considerable deletions of DNA within the chromomere. In any case, Lefevre's work represents a refreshing new approach, which may yet yield important insights.

Theories of the chromomere abound (for some examples see

15,19,26,33,62,89,89a,97). I would like to consider in this chapter three general categories of models. The first type of model proposes that a single gene, coding region plus any associated regulatory regions, is tandomly repeated within the chromomere. First proposed as a general model for eukaryotes by Callan (15), this model gained support from the subsequent discovery of repeated sequences in eukaryotes DNAs (13). In order to explain the finding that alleles generally occur in pairs in diploid organisms, it was necessary to postulate mechanisms to insure that the repeats would remain identical in the face of mutational pressure. In the second type of model, the extra DNA plays some role in gene regulation, perhaps as a binding site for regulator molecules (e.g.,11). In the third type of model, the extra DNA has neither a coding role nor a regulatory role. Instead, it serves either some, as yet unspecified, structural role or it is simply junk and has no particular role.

Studies on the arrangement of repetitive and unique DNA sequences in the DNA of the chromomere have narrowed the range of possibilities. As the reader is no doubt aware, DNA which has been denatured will, under appropriate conditions, renature. The kinetics of this reaction is a measure of the sequence complexity of the DNA, that is, a measure of how many different sequences are present (for reviews see 12,13). Prokaryotic DNAs typically renature with simple second order kinetics, a result indicating that all sequences are present in the same relative abundance. In these cases, the $Cot_{1/2}$ (the product of the initial DNA concentration and the time at which one half the DNA has renatured) is found to be directly proportional to genome size as expected. In marked contrast, eukaryotic DNAs typically show complex kinetics indicating that different sequences are present in different relative abundances. A knowledge of the DNA content of the genome allows one to determine the absolute number of times that each of the sequences in the least abundant class occurs in the haploid genome. In general, the answer is about one (52). These sequences are, then, commonly referred to as unique sequences.

In *D. melanogaster* essentially all of the highly repetitive sequences are accounted for by simple sequence DNAs, which are located mainly in the underreplicated heterochromatin in the chromocenter. Thus, we are left with moderately repetitive and unique sequences for the euchromatic regions. Renaturation curves for salivary gland DNA from *D. hydei* (Fig. 4) confirm this idea. Here, 90-95% of the DNA is found in unique sequences with the remaining in moderately repetitive sequences (23).

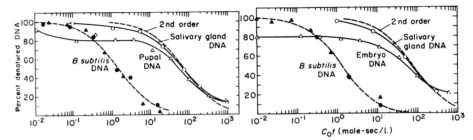

Figure 4. *Renaturation kinetics of* D. hydei *DNA from salivary glands, embryos and pupae. The abscissa represents the
initial DNA concentration multiplied by the incubation time.
The ordinate represents the percent of the input DNA which
remains denatured as measured by hydroxyapatite binding of
labeled DNA samples. The dotted line represents ideal second
order kinetics. DNA has been sheared to a fragment size of
about 500 bases. From Dickson, Boyd and Laird (23) and
reprinted by permission from* Academic Press.

Very few studies on *Drosophila* DNA have been done
directly on salivary gland DNA since it is much easier to
isolate DNA from the whole organism. One might well ask
whether in the formation of polytene chromosomes all the
sequences in the euchromatic regions replicate to the same
extent. Rudkin's measurements showing that during the development of the salivary gland the DNA content of a single
nucleus increases in a geometric series $2^N X$ (where X is about
0.75) suggest that essentially all of the euchromatic DNA
replicates uniformly to form the euchromatic arms in the
mature gland. Direct evidence on this point in *D. hydei* has
been obtained by Laird *et al.* (54). By renaturing trace
amounts of labeled pupae DNA with a large excess of unlabeled
salivary DNA, these workers showed that at least 70% of the
unique sequences present in pupal DNA are uniformly replicated during polytenization. It should be noted that depar-
tures from uniform replication of middle repetitive sequences,
such as those suggested by Woodcock and Sibatani (107), would
not be detected. Disproportionate replication in particular
regions has been noted in other Diptera (see 3 for review).

The finding that the chromomeres contain for the most
part unique sequence DNA would seem to eliminate immediately
any model of the chromomere which requires multiple copies
of genes. It has been argued, and perhaps justifiably so,
that renaturation kinetics alone are not sufficient to eliminate the possibility that what Laird and others have identi-

fied as unique sequences are in fact repeated, say, five times per haploid genome (98). Laird *et al.* have defended their contention (54), but, as persuasive as that defense might be, independent evidence is clearly desirable.

Tandem repeats would be expected to cyclize and form circles after digestion of the ends of randomly fragmented DNA with exonuclease. These circles could then be visualized in the electron microscope. Thomas and his co-workers have exploited this technique in an attempt to demonstrate tandem repeats in a variety of DNAs (98). While there is no doubt that repeats do occur, it is now clear that, in *Drosophila* at least, a large proportion of these circles involve simple sequence DNA, which is essentially absent from the chromomeres (70,82,44). Direct examination of the ability of salivary gland DNA to cyclize has produced conflicting results, one paper claiming that the frequency of circle formation was only slightly reduced from that of DNA from whole flies (59) and a second paper claiming that virtually no circles can be formed with this DNA (44). The reason for the discrepancy is not clear. It should be noted, however, that none of these ring studies rules out the possibility of tandem repeats of greater than 4 kb[*] (44).

Using an independent approach, Hamer and Thomas (36) have digested *Drosophila* DNA with restriction endonucleases, each one of which cleaves the DNA at a particular sequence, typically a symmetric sequence about 4-8 base pairs in length. A random distribution of restriction sites in the DNA should produce a predictable distribution of fragment lengths, which can be measured by gel electrophoresis. However, if the DNA is arranged in tandem repeats, then some very long DNA fragments would be spared from endonucleolytic cleavage since some repeat units will by chance lack the restriction sequence. Hamer and Thomas showed that 85% of the DNA was cleaved in a manner consistent with a random distribution. Many of the spared fragments were shown to be derived from simple sequence DNA. Elegant as this study is, it does not rule out the possibility that there might still be multiple gene copies irregularly spaced throughout the chromomere.

Studies of "cloned" DNA fragments argue against extensive repetition of genes as a general model for the chromomere. Using recently developed techniques, Wensink *et al.* (103) and Glover *et al.* (34) have inserted randomly selected

---

[*]kb stands for kilobases. The term signifies 1000 bases in the case of single stranded nucleic acids and 1000 base pairs in the case of double stranded nucleic acids.

fragments of *Drosophila melanogaster* DNA into bacterial plasmids. The plasmids were then cloned. This procedure allows the production of large quantities of the sequences contained in a single fragment of *Drosophila* DNA. The chromosomal location of the fragment can be readily determined by *in situ* hybridization to salivary gland chromosomes. By examining the reassociation kinetics of the isolated hybrid plasmid DNAs one can determine to what extent the *Drosophila* fragment is internally repetitious. Six clones have been obtained which contain euchromatic DNA. Each of these DNAs hybridizes to a single euchromatic location. These fragments range in length from about 15 to 18 kb. By comparison, the average chromomere contains about 26 kb of DNA. In every case, the renaturation analysis was inconsistent with more than a two-fold repeat within the fragment.

The weight of the evidence indicates, then, that most chromomeres do not contain substantial internal repetition. At this point, I must remind the reader that some of the best studied genes in eukaryotes are tandemly repeated, e.g., the ribosomal genes (14,22,29), the 5S genes (96) and the histone genes (96). The ribosomal genes are located at the nucleolus organizer which is within a heterochromatic region (78,77, 69). The 5S and histone genes, however, are located in euchromatic bands (96). In addition, it has been shown that the DNA of the giant Balbiani rings in *Chironomus tentans* also contains repeated DNA (58). But clearly, the bulk of the evidence argues against a tandem arrangement of sequences in a typical chromomere.

An alternative mode of organization of eukaryotic DNAs has been proposed by Britten and Davidson and their co-workers (for review see 20). In a wide variety of organisms middle repetitive sequences, usually 0.2-0.4 kb, are interspersed among unique sequences, more than 70% of which are less than 3 kb.

Conflicting results have been obtained with *Drosophila melanogaster* (108,64). From DNA sheared to 0.8 kb, Wu, Hurn and Bonner (108) isolated middle repetitive DNA substantially free from highly repetitive and unique DNA using hydroxyapatite columns, which selectively bind only those nucleic acids which have double stranded regions. The resultant 0.8 kb single strands of middle repetitive DNA were reannealed with either themselves (first experiment) or longer pieces from total (unfractionated) DNA (second experiment) under conditions of concentration and time where middle repetitive but not unique sequences can hybridize. The resultant structures were viewed in the electron microscope. In the first experiment, samples were prepared using the conventional protein

326

monolayer technique under aqueous conditions, where single stranded nucleic acids collapse into bushes and only duplex regions can be visualized (21). In the second experiment, the samples were prepared in the presence of formamide, which allows both single and double stranded regions to be visualized and measured (21). From measurements of the length of duplex regions in the first case and the distances between unannealed tails in the second case, they concluded that a prominant mode of organization involved short middle repetitive sequences about 0.1-0.2 kb in length interspersed among longer unique sequences, typically about 0.75 kb.

In a contrasting study, Manning, Schmid and Davidson (64) have concluded that the middle repetitive sequences range in length from 0.5 to 13 kb with a number average of 5.6 kb, considerably longer than the segments described by Wu *et al.* The average interval of intespersion was found to be at least 13 kb. These authors examined in the electron microscope products of renaturation of long strands containing middle repetitive DNA and measured the duplex regions flanked by four single stranded tails (Fig. 5) as well as the single stranded regions separating the duplexes.

Some possible reasons for the conflict can be suggested. In the first experiment of Wu *et al.*, one can imagine that the clumping of single stranded regions that occurs in aqueous spreads may have made it difficult to distinguish duplexes terminated by only two or three strands and thus not including the full length of the middle repetitive region. With regard to the second experiment of Wu *et al.*, Manning *et al.* have suggested that when several 0.8 kb fragments anneal to say a 5.6 kb site, numerous tails are produced because of sequence overlaps. These structures might easily be misinterpreted particularly since no distinction was made in this experiment between single and double stranded regions.[*]

Other data from hydroxyapatite experiments (64) also argues for interspersion, and the results are consistent with the pattern proposed by Manning *et al.* In addition, the structures of the cloned fragments isolated by Wensink *et al.* (103) and Glover *et al.* (34) conform better to the model of Manning *et al.*

Manning, Schmid and Davidson (64) estimate that at least 30% and possibly all of the middle repetitive sequences are interspersed among unique sequences. They make the interesting calculation that if all of the middle repetitive DNA were interspersed, there would be about 2800 such regions in

---

[*]This is because strandedness cannot be reliably determined over short distances.

Figure 5. Electron micrograph showing double forked molecule illustrating interspersion of middle repetitive and unique sequences in D. melanogaster *DNA. Single strands of average length 17.4 kb have been reassociated at a cot appropriate for middle repetitive sequences. DNA samples were prepared for electron microscopy using formamide to spread out single stranded regions. The arrows delimit the duplex region, which is flanked by four single stranded tails. From Manning, Schmid, and Davidson (64) and reprinted by permission from the MIT Press.*

328

the genome, i.e., about one per chromomere. The reader may
wish to speculate on the significance of this calculation.

One final feature of *Drosophila* DNA merits attention:
inverted repeats. An inverted repeat is simply two copies of
a sequence arranged in opposite orientation along the DNA.
If the two members of the pair are sufficiently close together
this DNA can be readily isolated on hydroxyapatite. This is
because a single stranded piece of DNA from an inverted
repeat can renature with itself, a reaction which, in dilute
solutions, will be faster than interstrand renaturation of
even highly repetitive sequences (104,105,35,84). By viewing
the structures in the electron microscope (Fig. 6), the
length of the repeated regions and the spacer region separating them can be readily measured (105,84). The lengths of
the repeated regions vary from very short to greater than
15 kb and the spacer region varies from less than 0.1 kb
(about 20%) to more than 30 kb (84). By digesting the structures isolated on hydroxyapatite with a nuclease specific for
single strands, the DNA present in the repeated sections has
been purified. Highly repetitive, middle repetitive and
unique sequences are all found in this DNA in about the same
proportions as total DNA but with some enrichment for middle
repetitive DNA (84). Inverted repeats have been reported in
a number of eukaryotic DNAs (for review see 105). In *Drosophila melanogaster*, an estimated 2000-4000 such pairs occur
in the genome (84). More might be found if longer strands
were used in the initial hydroxyapatite fractionation (84).

The significance of inverted repeats is not known.
Some speculation focusses on the possibility that these
regions could loop out to form cruciform or cloverleaf structures (105). An interesting recent observation is that the
small linear pieces of DNA containing the ribosomal genes in
the macronucleus of the protozoan *Tetrahymena* each contain
one large inverted repeat (48,113).

A recent paper by Perlman, Phillips, and Bishop concludes
that inverted repeats are a transient feature of the genome
(71). These workers isolated inverted repeats and their
adjacent sequences in *Xenopus* and found that although these
adjacent sequences comprised only 10% of the DNA, they
included every sequence in the *Xenopus* genome. The conclusion seems inescapable that inverted repeats occupy different
positions in different cells.

Thus, we have a picture of a typical chromomere as a
27 kb region of DNA, mostly unique in sequence with perhaps
one middle repetitive region and one inverted repeat. It
contains one coding region perhaps near one end or in the
interband region (2,90). Although the weight of the evidence

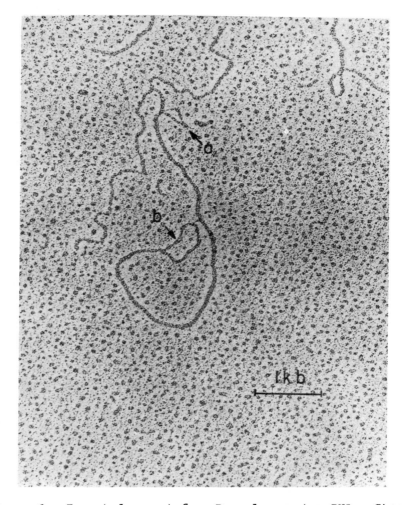

Figure 6. Inverted repeat from D. melanogaster *DNA. Structure has been formed from a single strand which has folded
back on itself. Duplex region is ∿ 7.5 kb. a: internal
nonhomology loop. b: end loop, comprising the DNA that separates the two complementary sequences. From Schmid, Manning
and Davidson (84) and reprinted by permission of the MIT Press.*

is against the first type of model proposed, that of tandemly repeated sequences in the chromomere, the data do not allow us to distinguish between the latter two alternatives posed, namely that the extra DNA has a regulatory role or that it has essentially no role. It is just possible that the number of genes has been underestimated after all, in which case the dilemma may be overemphasized.

Before coming to any conclusions on this matter, it would be well to consider the situation in the ciliated protozoan *Oxytricha* (72). This single-celled organism typically has two types of nuclei: a small micronucleus, which serves as the germline for the organism, and the macronucleus, which makes most of the RNA. It is the micronucleus which is exchanged during conjugation and in which meiosis takes place after fusion of the micronuclei from the two parents. One of the four meiotic products is preserved, and from it a new macronucleus develops, the old one having disintegrated prior to conjugation. In the first phase of the development of the macronucleus, the chromosomal material replicates many times to produce giant polytene chromosomes. These chromosomes then break up into small pieces. Thousands of vesicles form, each one of which apparently contains only one band plus parts of adjacent bands. Most of the DNA in each vesicle is then destroyed leaving only about 7% in the form of 10S linear pieces which then replicate to form the mature macronucleus. Considerable sequence complexity is lost in the process, and the surviving DNA has a complexity only several times larger than that of *E. coli* DNA. Occasionally cells are found which have lost their micronuclei and have only a macronucleus. The ability of these cells to grow and divide for many generations[*] suggests that the eliminated sequences are unnecessary for vegetative growth and perhaps function only in meiosis, if at all.

The relevance of this example to the situation in *Drosophila* is, of course, speculative. In this connection, it should be noted that at least 30% of the unique sequences in *D. melanogaster* are transcribed at some point in the life cycle (100).

THE RIBOSOMAL GENES

The ribosomal genes of eukaryotes have attracted much attention because of their propensity to replicate out of synchrony with the rest of the genome. Probably the most celebrated example is the synthesis of these genes during

---

*but not indefinitely.

oogenesis in the amphibian *Xenopus* (29,14,22). In this case, the ribosomal genes replicate many times to produce thousands of nucleoli. This selective replication, or amplification as it is called, is carried to such an extreme that at its peak, the vast majority of the DNA in the germinal vesicle is actually ribosomal DNA (rDNA). The occurrence of the *Xenopus* genes in a density satellite (29,14,22) has made them particularly easy to isolate and study, and indeed they are among the best studied of eukaryotic genes. The general picture that is emerging is that these genes are tandemly repeated, with an untranscribed spacer region separating the regions which code for the ribosomal precursor (96). Direct study of the *Drosophila* ribosomal genes has been more difficult because they have the same density as the bulk of the DNA. A tandem arrangement similar to that found in *Xenopus* has been inferred from electron micrographs of transcription complexes (66) and from the appearance of fragments of rDNA of discrete sizes after digestion of *Drosophila melanogaster* DNA with restriction endonucleases (103,34,65).

Ribosomal genes also show a certain degree of replicative autonomy in the salivary gland of *Drosophila*. In *D. melanogaster*, these genes represent only about 0.1% of salivary gland DNA although they comprise up to 0.5% of DNA from diploid tissue or adult flies (92,91). A similar situation exists in *D. hydei*, the species in which this effect was first demonstrated (42). Thus, these genes do not replicate to the full extent that the euchromatin does, but they clearly replicate more than the simple sequence DNAs do (69,79). The replication of these genes in the salivary gland is unusual in another way. By appropriate matings, flies can be produced which have one, three or four nucleolus organizers rather than the usual two. In the predominantly diploid larval brains and imaginal discs, the reiteration of these genes, with minor variations, is found to be proportional to the number of nucleolus organizers present (91). In the larval salivary gland, by contrast, no such proportionality is found (91). The same reiteration is found regardless of the number of nucleolus organizers present in the genotype. It would appear, then, that in the salivary gland, the ribosomal genes can replicate independently of the rest of the genome.

In *D. melanogaster*, the weight of evidence favors the idea that all of the ribosomal genes are located at the nucleolus organizers on the sex chromosomes (78,77,69), one in the proximal heterochromatin of the X and another on the short arm of the Y (17). Spear and Gall (92) have suggested that replicative independence of these genes might be related to their location in heterochromatin, probably surrounded on both sides by sequences which do not replicate in the salivary gland. They imagined that rDNA replication might, as a consequence, be uncoupled from that of the euchromatin. Since polyteny with concomitant underreplication of heterochromatin seems to be quite common in *Drosophila* tissues, they theorized that the hitherto puzzling phenomenon of compensation might be the result of widespread independent replication of these genes (93,94). For example, in an XO male or an XX female in which one X lacks the nucleolus organizer region, the reiteration of these genes in DNA from adults is found to be greater than half the value obtained when two nucleolus organizers are present (93,94).

In my laboratory, we have tried to test directly the possibility that in the salivary gland these genes might exist as separate pieces of DNA unconnected to the rest of the DNA as might be expected from the Spear and Gall model. From viscoelastic relaxation measurements of tissue culture or pupal cell lysates, Kavenoff and Zimm (49) have persuasively argued for the presence of chromosome sized pieces of DNA in these cells. Using their lysis procedure, which minimizes shear and nuclease action, we have isolated on sucrose gradients high molecular weight DNA from salivary glands of *D. melanogaster* (110,112). The procedure involves lysing the homogenized tissue at 65°C. in a solution containing detergent, pronase, and a high concentration of EDTA at pH 9. The preparation is then incubated at 50°C with additional pronase. The bulk of the DNA sediments as if it had a molecular weight of at least 5×10^9 daltons. This value, if correct, means that this DNA is the largest to ever be isolated from salivary glands and that the DNA must be continuous over stretches of the chromosome containing many chromomeres. Linear DNA molecules ranging up to 3×10^8 daltons have been isolated from the salivary gland of *Chironomus* by Wolstenhome *et al.* (106) and from *Drosophila* by Derksen and Berendes (22a).

When the fractions from our sucrose gradients are hybridized to labeled ribosomal RNA, it is found that 58% of these genes are located with the bulk of the DNA. The other 42% sediment more slowly and have an estimated molecular weight of roughly 3×10^8. The same results are obtained for a variety of genotypes. These data suggest that at least 42% of these genes occur in separate, unconnected pieces of DNA. By contrast, we do not find low molecular weight genes in diploid brains and imaginal discs except in the case of one unusual female genotype containing one nucleolus organizer. Despite their clear presence in the salivary gland, we

do not find unintegrated genes in adults, except in the case
of female genotypes, X/sc^4 sc^8, containing one nucleolus
organizer (109,111). This is somewhat surprising since adults
are thought to contain considerable amounts of polytene tissue in which underreplication of satellite sequences occurs.
We hesitate to attribute the occurrence of unintegrated genes
to underreplication of surrounding sequences since, as noted
above, we also find them in diploid tissue. Additional work
will be necessary before meaningful models can be constructed.

ACKNOWLEDGEMENTS: The author would like to thank Christine
Zuchowski and Robert Slotnick for reading the manuscript.
Work in the author's laboratory was supported by grants from
NIH (GM21487) and the SUNY Research Foundation (050-7330).

REFERENCES

1. Ashburner, M., *Advances Insect Physiol.* 7, 1 (1970).
2. Beerman, W., in *Developmental Studies on Giant Chromosomes*, Ed. W. Beerman, Springer-Verlag, New York-
   Heidelberg-Berlin, pp. 1-33 (1972).
3. Berendes, H. D., Beerman, W., in *Handbook of Molecular
   Cytology*, Ed. A. Lima-de-Faria, North Holland Publishing
   Co., Amsterdam and New York, pp. 501-519 (1969).
4. Berendes, H. D., and Keyl, H. G., *Genetics 57*, 1
   (1967).
5. Birnboim, H. C., and Sederoff, R., *Cell 5*, 173
   (1975).
6. Birnboim, H. C., Straus, N. A., and Sederoff, R. R.,
   Biochemistry 14, 1643 (1975).
7. Bishop, J., *Cell 2*, 81 (1974).
8, Blumenfeld, M., Forrest, H., *Nature New Biology 239*,
   170 (1972).
9. Boyd, J. B., Berendes, H. D., and Boyd, H., *J. Cell
   Biol. 38*, 369 (1968).
10. Botchan, M., Kram, R., Schmid, C. W., and Hearst, J. E.,
    Proc. Nat. Acad. Sci. U.S.A. 68, 1125 (1971).
11. Britten, R. J., and Davidson, E. H., *Science 165*, 349
    (1969).
12. Britten, R. J., Graham, D. E., and Neufeld, B. R.,
    Methods in Enzymology, Vol. 29, 363 (1974).
13. Britten, R. J., and Kohne, D. E., *Science 161*, 529
    (1968).
14. Brown, D. D., and Dawid, I., *Science 160*, 272 (1968).
15. Callan, H. G., *J. Cell Sci. 2*, 1 (1967).
16. Carroll, D., and Brown, D. D., *Cell 7*, 477 (1976).
17. Cooper, K. W., *Chromosoma 10*, 535 (1959).

18. Cordeiro, M., Wheeler, L., Lee, C. S., Kastritsis, C. D., and Richardson, R. H., *Chromosoma 51*, 65 (1975).
19. Crick, F. H., *Nature 234*, 25 (1971).
20. Davidson, E. H., Galau, G. A., Angerer, R. C., and Britten, R. J., *Chromosoma 51*, 253 (1975).
21. Davis, R. W., Simon, M., and Davidson, N., in *Methods in Enzymology, Vol. 21D*, 413 (1971).
22. Dawid, I., Brown, D. D., and Reeder, R., *J. Mol. Biol. 51*, 341 (1970).
22a. Derksen, J., and Berendes, H. D., *Chromosoma 31*, 468 (1970).
23. Dickson, E., Boyd, J. B., and Laird, C. D., *J. Mol. Biol. 61*, 615 (1971).
24. Eckhardt, R. A., and Gall, J. G., *Chromosoma 32*, 407 (1971).
25. Edström, J.-E., in *The Cell Nucleus, Vol. II*, Academic Press, New York, p. 293 (1974).
26. Edström, J.-E., *J. Theor. Biol. 52*, 163 (1975).
27. Endow, S. A., and Gall, J. G., *Chromosoma 50*, 175 (1975).
28. Endow, S. A., Polan, M. L., and Gall, J. G., *J. Mol. Biol. 96*, 665 (1975).
29. Gall, J. G., *Proc. Nat. Acad. Sci. U.S.A. 60*, 553 (1968).
30. Gall, J. G., and Atherton, D. D., *J. Mol. Biol. 85*, 633 (1974).
31. Gall, J. G., Cohen, E. H., and Polan, M. L., *Chromosoma 33*, 319 (1971).
32. Gall, J. G., and Pardue, M. L., *Genetics (Supp. 1) 61*, 1 (1969).
33. Gersh, E. S., *J. Theor. Biol. 84*, 413 (1975).
34. Glover, D. M., White, R. L., Finnegan, D. J., and Hognes, D. S., *Cell 5*, 157 (1975).
35. Hamer, D. H, and Thomas, C. A., Jr., *J. Mol. Biol. 84*, 139 (1974).
36. Hamer, D. H., and Thomas, C. A., Jr., *Chromosoma 49*, 243 (1975).
37. Hamkalo, B. A., Miller, O. L., and Bakken, A. H., *Cold Spring Harbor Symp. 38*, 915 (1973).
38. Heitz, E., *Z. Zellforsch. 20*, 237 (1934).
39. Heitz, E., *Biol. Zbl. 54*, 588 (1934).
40. Hennig, W., in *The Cell Nucleus, Vol. II*, Academic Press, New York, p. 333 (1974).
41. Hennig, W., Hennig, I., and Stein, H., *Chromosoma 32*, 31 (1971).
42. Hennig, W., and Meer, B., *Nature New Biol. 233*, 70 (1971).

43. Hourcade, D., Dressler, D., and Wolfson, J., *Proc. Nat. Acad. Sci. U.S.A. 70*, 2926 (1973).

44. Hutton, J. R., and Thomas, C. A., Jr., *J. Mol. Biol. 98*, 425 (1975).

45. Jones, K. W., and Robertson, F., *Chromosoma, 31*, 331 (1970).

46. Judd, B. H., Shen, M. W., and Kaufman, T. C., *Genetics 71*, 139 (1972).

47. Judd, B. H., and Young, M. W., *Cold Spring Harbor Symp. 38*, 571 (1973).

48. Karrer, K., and Gall, J. G., *J. Mol. Biol. 104*, 421 (1976).

49. Kavenoff, R., and Zimm, B., *Chromosoma 41*, 1 (1973).

51. Kunz, W., and Eckhardt, R. A., *Chromosoma 47*, 1 (1974).

52. Laird, C. D., *Chromosoma 32*, 378 (1971).

53. Laird, C. D., *Ann. Rev. Genetics 7*, 177 (1973).

54. Laird, C. D., Chooi, W. Y., Cohen, E. H., Dickson, E., Hutchinson, N., and Turner, S. H., *Cold Spring Harbor Symp. 38*, 311 (1973).

55. Laird, C. D., and McCarthy, B., *Genetics 63*, 865 (1969).

56. Lakhotia, S. C., *Chromosoma 46*, 145 (1974).

57. Lakhotia, S. C., and Jacob, J., *Exp. Cell Res. 86*, 253 (1974).

58. Lambert, B., *J. Mol. Biol. 72*, 65 (1975).

59. Lee, C. S., and Thomas, C. A., Jr., *J. Mol. Biol. 77*, 25 (1973).

60. Lefevre, G., Jr., *Cold Spring Harbor Symp. 38*, 591 (1973).

61. Lefevre, G., Jr., *Ann. Rev. Genetics 8*, 51 (1974).

62. Lima-de-Faria, A., *Hereditas 81*, 249 (1975).

63. Lindsley, D. L., and Grell, E. H., *Genetic Variations of Drosophila Melanogaster*, Carnegie Institution of Washington Publication No. 627 (1968).

64. Manning, J. E., Schmid, C. W., and Davidson, N., *Cell 4*, 141 (1975).

65. Manteuil, S., Hamer, D. H., and Thomas, C. A., Jr., *Cell 5*, 413 (1975).

66. Meyer, G. F., and Hennig, W., *Chromosoma 46*, 121 (1974).

67. Miller, O. L., Jr., and Beatty, B. R., *Genetics 61 (Suppl.)*, 133 (1969).

68. Mulder, M. P., van Duijn, P., and Gloor, H. J., *Genetica 39*, 385 (1968).

69. Pardue, M. L., Gerbi, S. A., Eckhardt, R. A., and Gall, J. G., *Chromosoma 29*, 268 (1970).

70. Peacock, W. J., Brutlag, D., Goldring, E., Appels, R., Hinton, C. W., and Lindsley, D. L., *Cold Spring Harbor*

Symp. 38, 405 (1973).

71. Perlman, S., Phillips, C., and Bishop, J. O., *Cell 8,* 33 (1976).

72. Prescott, D. M., and Murti, K. G., *Cold Spring Harbor Symp. 38,* 609 (1973).

73. Rae, P. M. M., *Proc. Nat. Acad. Sci. U.S.A. 67,* 1018 (1970).

74. Rasch, E. M., Barr, H. J., and Rasch, R. W., *Chromosoma 33,* 1 (1971).

75. Renkawitz-Pohl, R., and Kunz, W., *Chromosoma 49,* 375 (1975).

76. Ritossa, F. M., *Proc. Nat. Acad. Sci. U.S.A. 60,* 509 (1968).

77. Ritossa, F. M., Atwood, K. C., and Spiegelman, S., *Genetics 54,* 819 (1966).

78. Ritossa, F. M., and Spiegelman, S., *Proc. Nat. Acad. Sci. U.S.A. 53,* 737 (1965).

79. Rodman, T. C., *J. Cell Biol. 42,* 575 (1969).

80. Rudkin, G. T., *Genetics 52,* 470 (1965).

81. Rudkin, G. T., *Genetics, Suppl. 61,* 227 (1969).

82. Schachat, F. H., and Hogness, D. S., *Cold Spring Harbor Symp. 38,* 371 (1973).

83. Schalet, A., and Lefevre, G., Jr., *Chromosoma 44,* 183 (1973).

84. Schmid, C. W., Manning, J. E., and Davidson, N., *Cell 5,* 159 (1975).

85. Schweber, M. S., *Chromosoma 44,* 371 (1974).

86. Sederoff, R., Lowenstein, L., and Birnboim, H. C., *Cell 5,* 183 (1975).

87. Shen, C.-K. J., Wiesehahn, G., Hearst, J. E., *Nucl. Acids Res. 3,* 931 (1976).

88. Smith, G. P., *Cold Spring Harbor Symp. 38,* 507 (1973).

89. Sorsa, V., *Hereditas 81,* 77 (1975).

89a. Sorsa, A., *Hereditas 82,* 63 (1976).

90. Sorsa, V., Green, M. M., and Beerman, W., *Nature New Biol. 245,* 34 (1973).

91. Spear, B. B., *Chromosoma 48,* 159 (1974).

92. Spear, B. B., and Gall, J. G., *Proc. Nat. Acad. Sci. U.S.A. 70,* 1359 (1973).

93. Tartof, K. D., *Science 171,* 294 (1971).

94. Tartof, K. D., *Genetics 73,* 57 (1973).

95. Tartof, K. D., *Proc. Nat. Acad. Sci. U.S.A. 71,* 1272 (1974).

96. Tartof, K. D., *Ann. Rev. Genetics 9,* 355 (1975).

97. Thomas, C. A., Jr., *Cold Spring Harbor Symp. 38,* 347 (1973).

98. Thomas, C. A., Jr., Pyeritz, R. E., Wilson, D. A., Dancis, B. M., Lee, C. S., Bick, M. D., Huang, H. L., and Zimm, B. H., *Cold Spring Harbor Symp. 38,* 353 (1973).

99. Travaglini, E., Petrovic, J., and Schultz, J., *Genetics 72,* 419 (1972).

100. Turner, S. H., and Laird, C. D., *Biochem. Genetics 10,* 263 (1973).

101. Welshons, W. J., *Genetics 76,* 775 (1974).

102. Welshons, W. J., and Keppy, D. O., *Genetics 80,* 143 (1975).

103. Wensink, P. C., Finnegan, D. J., Donelson, J. E., and Hogness, D. S., *Cell 3,* 315 (1974).

104. Wilson, D. A., and Thomas, C. A., Jr., *Biochim. Biophys. Acta 331,* 333 (1973).

105. Wilson, D. A., and Thomas, C. A., Jr., *J. Mol. Biol. 84,* 115 (1974).

106. Wolstenholme, D. R., Dawid, I. B., and Ristow, H., *Genetics 60,* 759 (1968).

107. Woodcock, D. M., and Sibatani, A., *Chromosoma 50,* 147 (1975).

108. Wu, J. R., Hurn, J., and Bonner, J., *J. Mol. Biol. 64,* 211 (1972).

109. Zuchowski, C., and Harford, A. G., *Biophys. J. 16,* 226a (1976).

110. Zuchowski, C., and Harford, A. G., *J. Cell Biol. 70,* 101a (1976).

111. Zuchowski, C. I., and Harford, A. G., *Chromosoma* (1976) in press.

112. Zuchowski, C. I., and Harford, A. G., *Chromosoma* (1976) in press.

113. Engberg, J., Andersson, P., Leick, V., and Collins, J., *J. Mol. Biol. 104,* 455 (1976).

Chapter 10

LAMPBRUSH CHROMOSOMES

HERBERT C. MACGREGOR

Department of Zoology
School of Biological Sciences
University of Leicester
Leicester LE1 7RH, England

In a recent publication on lampbrush chromosomes, Mott
and Callan (1) start out with a number of astute and fundamental observations regarding these structures. They say
that lampbrushes are choice material for the study of chromosome structure, that because they can be isolated and are
large, we have been able to learn much about their basic
structure and chemistry without resorting to complex technology, and that they present certain interesting challenges
insofar as certain parts of them are active in the transcriptional sense. In Callan's opinion, heartily endorsed by
myself and all others who have worked with lampbrush chromosomes ... "The cytologist must take advantage of what Nature
provides, and there is no doubt that lampbrush chromosomes
are one of Nature's kindest provisions..."
 In this chapter, I shall try to use a strategy that will
provide a basic knowledge of the main features of lampbrush
chromosomes, and then I shall highlight what I consider to
be the 2 major questions that we must ask about these structures today.
 From the historical standpoint, lampbrushes were first
seen in sections of salamander oocytes by Fleming in 1882
(2), and then again 10 years later, they were described in
oocytes of a dogfish by Ruckert (3). The name "lampbrush"
comes from Ruckert, who likened the objects to the 19th
century lampbrush, equivalent to the late 20th century test-
tube brush.
 The lampbrush type of chromosome is now known to be
characteristic of growing oocytes in the ovaries of most
kinds of animal, including man (4). The chromosomes are in a
highly extended form, sometimes reaching lengths of a millimetre or more, and they are essentially bivalents that
persist through the extended diplotene phase of the first

meiotic division, a phase that can last as long as 6 months
or more. In the biological sense, these chromosomes are odd
only insofar as they are confined to meiotic prophase in
germ cells, but they are a general phenomenon insofar as they
can be found in nearly all animals.

Lampbrushes are exceedingly delicate structures, and no
further progress beyond the pioneer studies of Flemming and
Ruckert was possible until a technique could be devised for
dissecting them out of their nuclei and examining them in a
life-like condition, separated from the remainder of the
nuclear contents. Such a technique was developed by Gall in
1954 (5). Essentially, the oocyte is punctured with a
needle, the nucleus is gently squeezed out of the hole,
picked up in a drawn out Pasteur pipette, and transferred to
saline in a chamber constructed by boring a hole through a
microscope slide and sealing a coverslip across the hole
with wax. The nuclear envelope is then removed and the
nuclear contents, including the lampbrush chromosomes lie
flat, and hopefully unbroken and well displayed. Then, by
using a phase contrast microscope with an inverted optical
system, the chromosomes can be examined in a fresh and
unfixed condition with the highest resolution obtainable by
light microscopy.

A lampbrush chromosome may be described as a row of
granules of deoxyribonucleoprotein (DNP), the chromomeres,
connected by an exceedingly thin thread of the same material
(Fig. 1). Chromomeres are $\frac{1}{4}$ - 2µm in diameter and spaced
approximately 2µm from centre to centre along the chromosome.
They are not necessarily round; indeed, their shapes can be
highly irregular. Each chromomere has 2 or some multiple of
2 loops associated with it. The loops have a thin axis of
DNP surrounded by a relatively bulky matrix of ribonucleoprotein (RNP). Loops are variable in length, and during the
period of oogenesis when they are maximally developed, they
extend from 5 to 50µm laterally from the chromosome axis.
They are also variable in appearance, and some loops with
exceptionally distinctive appearances can be used in chromosome identification (6). I should stress, however, that the
vast majority of loops have the same general appearance,
varying only in the coarseness of the granularity of their
matrix. The basis of such variation in loop morphology has
most recently been discussed by Malcolm and Sommerville (7).
Sister loops arising from the same chromosome have the same
lengths and structural characteristics.

In the newt *Triturus cristatus,* the chromosomes are
relatively short and contracted at the end of pachytene in
the female. They then become greatly extended and assume

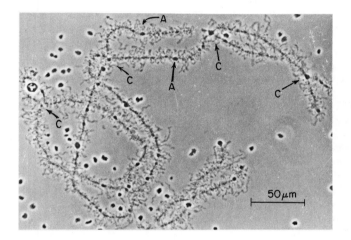

*Figure 1. Phase contrast photomicrograph of a complete
lampbrush bivalent from the Californian newt* Taricha granu-
losa. *This bivalent shows the chromomeric organization of
the chromosomes' axes, the variable lengths of the lateral
loops, a number of prominent "axial granules" that are aggregates or ribonucleoprotein (A), and 5 chiasmata (C). There
are terminal fusions at both ends of the bivalent. Photo-
graph kindly provided by Dr. J. Kezer.*

341

the lampbrush form. The actual process of development of a
lampbrush chromosome has never been described, but it seems
likely that in small, yolky oocytes of newts and other
amphibia, the lampbrushes are at the peak of their development in that the chromosomes and their loops are at their
longest. Subsequently, the loops and chromosomes become
shorter, the chromomeres become larger, and eventually, as
the oocyte nears maturity, the chromosomes revert to a loop-
less contracted state, and assume the form of condensed
diplotene bivalents. The pattern, then, is one of extension
followed by retraction of the loops and a clear inverse relationship between loop length and chromosome size. The longer
the loop, the smaller the chromomere, and *vice versa*. Two
more simple observations are important. First, most lateral
loops have an asymmetrical form. They are thin at one end of
insertion into the chromomere and thick at the other (Fig. 2).
Hopefully, the significance of this observation will be
apparent later. Secondly, when one stretches a lampbrush
chromosome, either deliberately or accidentally, breaks first
occur transversely across chromomeres in such a way that the
resulting gaps are spanned by the loops that are associated
with the chromomeres (Fig. 3). Breaks of this kind produce
what Callan terms "double bridges" (6, 8). Clearly, the
phenomenon of double bridge formation indicates that there
must be a line of weakness separating the two halves of each
chromomere, and more importantly, it indicates a structural
continuity between the main axis of the chromosome - the
interchromomeric fibre - and the axes of the loops.

The last 2 basic points that I wish to make about lampbrush chromosomes and their loops are very important ones
indeed, since they serve as foci for the 2 major questions
that I shall wish to examine later. First, the loops are
sites of active RNA synthesis, and in the vast majority of
cases, RNA is synthesized simultaneously all along the length
of the loop (9). In the crested newt, *Triturus cristatus*,
there are about 20,000 loops per oocyte!

Secondly, it has become clear that within a sub-species,
particular loops may be present or absent in homozygous or
heterozygous combinations, and if one examines, as Callan
and Lloyd (6) have done, the frequencies of combinations
within and between bivalents with respect to presence or
absence of particular loops, then we find that these characteristics assort and combine like pairs of Mendelian alleles.
In other words, there is evidence to suggest that a chromo-
mere/loop pair complex may behave as a Mendelian unit.
Succintly, a chromomere and its loops may represent a gene.

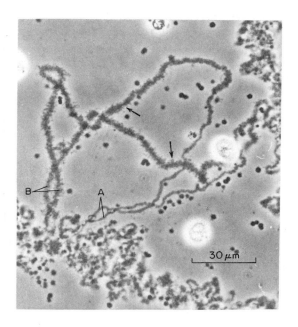

Figure 2. Phase contrast photomicrograph of a part of a
lampbrush bivalent from the salamander Plethodon cinereus
showing a pair of very long (240 μm) loops, both of which
arise from the same chromomere. The loops have a fairly
granular matrix of ribonucleoprotein, and in some places
(arrowed), the loop axis is faintly discernible. Each loop
has a thin end or "insertion" (A) and grows gradually thicker
towards the thick end (B), a phenomenon that is characteristic of most loops and referred to as "loop asymmetry".

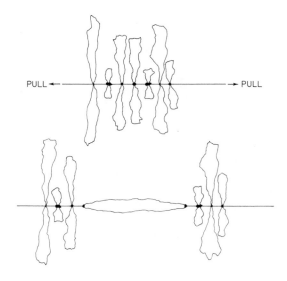

Figure 3. A representation of the formation of a "double bridge" break which is produced when a lampbrush chromosome is stretched to the point at which a break occurs transversely across a chromomere, so that the resulting gap becomes spanned by the 2 loops that originally sprang from that chromomere. In such cases, it can be shown conclusively that the polarity of asymmetry (see Fig. 2) of sister loops arising from one chromomere is always the same.

All the facts and figures that I have given so far were known by 1960, just 6 years after Gall inverted his phase contrast microscope and made the detailed study of lamp-brushes possible. What then emerged was a model and 2 extremely interesting and related hypotheses. The model presents a lampbrush chromosome as consisting of 2 DNA duplexes running alongside one another in the interchromo-meric fibre, compacted into chromomeres at intervals, and extending laterally from a point within each chromomere to form loops where RNA transcription takes place (Fig. 4). Each duplex is considered to represent one chromatid.

The 2 hypotheses that followed this model have not stood the test of time or experiment, but both have stimulated a vast amount of thought and research from which there has been a remarkable spin-off of truth and understanding. In this sense, we are reminded of the encouraging fact that a hypothesis does not have to be correct to be useful.

In the sequence in which they were developed, the first hypothesis took account of

(1) The asymmetry of the loops,
(2) the inverse relationship between chromomere size and loop length, and
(3) the synthesis of RNA along the lengths of the loops.

It was referred to as the spinning out and retraction hypothesis (6, 9), and it said that during the lampbrush stage of oogenesis all DNA is progressively spun out from one side of the chromomere and the loop extends to become longer and longer. Subsequently, and at times simultaneously, loop axis DNA is retracted back into the other side of the chromo-mere, at which point it ceases to support RNA transcription. According to this hypothesis, all the chromosomal DNA, or more specifically, all chromomeric DNA, forms part of a loop and is involved in transcription at some time in oogenesis. Loop asymmetry is accounted for by supposing that the portion of the loop that has been involved in transcription for the longest time will have the most material associated with it, and that will be the thick end of the loop; whereas the por-tion of the loop that is newly released from the chromomere in the spinning out process will have the least material associated with it, and that will be the thin end.

The second hypothesis, which in modern terms was much more fundamental, made the point that the spinning out and retraction explanation for loop asymmetry included the assumption that there was no genetic diversity within indi-vidual lampbrush loop/chromomere complexes, and that the

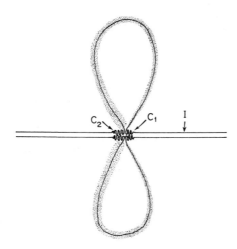

*Figure 4. Diagrammatic representation of the supposed gen-
eral organization of the DNA components - the 2 chromatids -
that make up a single lampbrush chromosome. Each horizontal
line represents a single chromatid that consists of a single
DNA duplex. Each chromatid is thought to run relatively
straight through the interchromomeric region (I), become
locally compacted in one side of the chromomere (C_1), pass
out into a loop, where it is again in a relatively extended
state, return to the compacted state in the other side of
the chromomere (C_2), and then proceed along the interchromo-
meric fibril to the next chromomere/loop complex.*

information carried in any one of these complexes was serially repeated along the DNA that was located in the loop and its chrommere. As we shall see later, this assumption was unnecessary. Callan believed that the notion of serially repeated sequences was incompatible with the fact that many phenotypically expressed mutations resulted from changes in only a few nucleotides, and he found it hard to see how a mutation could possibly be expressed if it were not simultaneously imprinted in all copies of a repetitive gene sequence. To overcome this dilemma, he produced the famous Master/Slave hypothesis (10), in which a Master sequence imprinted itself upon all its associated repeats or Slaves once per generation. The only mutational events that would then be detectable would be those that affected the form of the Master. Ingeniously he explained how the formation of a lampbrush loop might reflect the actual event of sequential rectification of Slaves by the Master, and in so doing he detracted attention from RNA synthesis as the prime role of a lampbrush chromosome, and suggested that lampbrush development might have a genetic rather than a synthetic significance (10).

There, then, is the historical background. Now we are left with 2 major questions, and it seems likely that they are broadly the same questions that were in the minds of Flemming and Ruckert almost a century ago. First, what's going on? Second, what is a chromomere and its associated loops in molecular and genetic terms? Let's see if we can find the answers.

To find out what's going on, we have to look at a variety of recent findings. First, more than 90% of the chromosomal DNA is in the chromomeres and their loops, and all chromomeres have loops. So if we think in terms of chromomeres as cytological units, then we must accept that at least part of every such unit is expressed through transcription. It is worth mentioning at this point that 5 genera of Mexican and Central American plethodontid salamanders are known to have variable numbers of supernumerary chromosomes (11), and in oocytes of these animals, the supernumeraries, like all the other chromosomes, assume the lampbrush form; yet they seem to be dispensable from the standpoint of normal development.

Secondly, the mRNA that is synthesized on lampbrush chromosomes and conserved beyond the lampbrush stage of oogenesis is transcribed from less than 5% of the genome (12-15). If we examine this 5% more carefully, then the outcome is surprising. According to Hough and Davidson (13), about 40% of the genome of *Xenopus laevis* is repeated

347

sequence DNA, but only 3.5% of this class of DNA, amounting
to just 1.4% of the genome, is transcribed into RNA that is
conserved beyond the lampbrush stage. According to Rosbash,
Ford and Bishop (15), in *Triturus cristatus*, in which the
majority of the DNA consists of repeated sequences, virtually
all the mRNA that is transcribed by lampbrush chromosomes,
has poly-A associated with it, and is conserved beyond the
lampbrush stage, is transcribed preferentially from sequences
that are represented just once per genome. Indeed, Rosbash
and his colleagues even go so far as to say that single copy
DNA may be the <u>exclusive</u> template for synthesis of stored
mRNA in the oocyte. Therefore, if I may interject a comment
on Callan's Master/Slave hypothesis, there is no need for a
rectification mechanism, and if the loops are sites of transcription from serially repeated sequences, then such transcripts are never available for translation.

This is a perplexing situation, because we know that,
for example, in *X. laevis* about 40 - 50% of the genome is
repeated sequence DNA. In *Triturus*, the animal at the centre
of the problem, at least 60% of the genome consists of
repeated sequences (15), and probably much more than that.
Indeed, Rosbash and my own laboratory are in agreement that
there is really very little detectable single copy DNA in the
Triturus genome. Nonetheless, it is this small diverse component that makes survivable and supposedly useful mRNA for
the oocyte and for early embryonic development.

It helps a little to resolve this problem if we take a
brief look at a few specific examples of known relationships
between loops or chromomeres and uniformly repeated sequences.
Eukaryotic DNA may be considered to fall into 4 different
classes, based on sequence repetition and function. The
properties of these classes that are significant in the
present context are shown in Table 1.

First the satellites: these are concentrated for the
most part at or around the centromeres of all chromosomes,
and, at least in salamanders of the genus *Plethodon*, they
occupy what are essentially large, irregularly shaped, loop-
less aggregates of DNP (16-18). Satellite DNA almost certainly has no function whatever in the transcriptive/trans-
lative sense.

Then we have the functional repetitive sequences. The
nucleolus organizer in a newt or salamander is represented
in its lampbrush chromosome by a more or less normal loop
pair/chromomere complex, as in *Triturus cristatus* (19, 20),
by a conspicuous but otherwise normal loop pair and chromomere as in the axolotl (21), or by a highly complex collection of loops and chromomeres, as in certain plethodontid

Table I.

	Sequence size range: nucleotide pairs	Copies per genome	Function	Chromosomal distribution	Selection for conservation
Satellites	2 (crab) (39) to 1,400 (calf) (40)	Up to 40,000,000	?	Heterochromatin (18, 38)	Very low
"Functional Repetitive"	Several thousand (41,42,44) (rDNA - 14,000)	100 to 24,000 (41,42,44)	Ribosomal RNA tRNA Histone RNA	Localised 20,43,44 (e.g.,nucleolus organizer)	Very high
"Middle Repetitive"	100 low (27,29,30,45) 300 average 4,000 high	Up to 10,000 (27,30)	?	General (?)	Low (27)
Single copy	600 to several thousand	1	mRNA transcription	General (?)	High (27)

salamanders (16). So at least a part of the nucleolus organizer, which is moderately repetitive and clearly functional, is transcribed and expressed in the form of a more or less normal lampbrush loop. It is not, however, transcribed all at once; of that we can be quite sure. The rDNA of a newt or axolotl is sufficient to make a large chromomere or a loop at least a millimetre long. What we see at the nucleolus locus in the axolotl is as expected: a largish chromomere and a relatively large loop, perhaps 50 - 100µm long at the most (21).

Of the other obviously functional sequences we can say little, mainly because the amphibia in which they can be studied most easily from the molecular standpoint have very poor and unworkable lampbrush chromosomes, and those animals with good lampbrush chromosomes are, for a variety of reasons, unsuitable for molecular studies. However, we can say with reasonable certainty from the observations of Barsacchi and her associates (22) that the 5S sequences in *Triturus* can be detected in a single chromomere and its loop pair, and these sequences are, of course, highly repetitive (23).

Before we move on to consider the middle repetitive and single copy material, let us focus in for a moment on a highly remarkable discovery made recently by Gould, Thomas and Callan (24). The discovery involves the American newt *Triturus viridescens* and the giant loops that are found near the centromere on the second longest chromosome, and it also involves the use of restriction endonucleases that cleave DNA at highly specific sites. Two related questions were posed.

(1) Do restriction endonucleases cleave the DNA axes of all loops - to be expected if the loops contain sequence diversity?

(2) Do some loops survive specific endonucleases - to be expected if they contain uniformly repeated sequences in which the enzyme sensitive site is lacking from the repeated unit?

The questions are easy to answer. All one has to do is to dissect lampbrush chromosomes into solutions of different endonucleases and see what breaks and what doesn't. The investigators used the following 4 enzymes: Endo R *hae*, Endo R *Eco* B, Endo I, and Pancreatic DNase. The last 3 produced total, fast and unequivocal destruction of the chromosomes and all their loops. Endo R *hae* did the same, but left the giant loops on the second longest chromosome intact: a dramatic result to the third in a series of historic experiments

(24-26). The conclusion is that the sequence 5' GGCC, which is specifically attacked by Endo R *hae*, is absent from the giant loops. Chance dictates that this sequence should occur every 256 base pairs or every 0.1 μm of DNA. The loops in question are over 200 μm long. Gould, Thomas and Callan argue that their DNA axes must therefore consist of a large number of repeats of a unit sequence 200 - 300 nucleotides long at the most.

Against this experimental background, let us see what more can be said about the middle repetitive class of sequence (Table 1), and the single copy sequences in lampbrush chromosomes. There are 2 main points that can be made. First, in an animal like the common red-backed salamander that has respectable lampbrush chromosomes, the middle repetitive DNA has an average repetitive frequency of around 7,500x (27), and an average sequence size of 300 - 500 nucleotide pairs, and it represents about half the genome. The genome has 20×10^{-12}g of DNA (27), and a haploid set of lampbrush chromosomes has about 3,500 chromomeres (17, 28). It is easy to calculate on this basis that there is room in an average chromomere for at least 1 family of middle repetitive sequences and more besides. In other words, you only need some reasonable experimental data and some simple arithmetic to prove that there has to be substantial sequence diversity in most chromomeres and their loops.

Secondly, we can learn something from looking at the rate of change of different classes of sequence during evolution, on the principle that there will be greater pressure of natural selection for conservation of those sequences whose transcripts or translates are biochemically important, than for conservation of sequences that are either not transcribed or not translated into useful macromolecules. The last column in Table I shows the situation. The implication is that some repetitive and some single copy sequences are highly important and strongly conserved. Others are of little biochemical importance and are left free by the forces of natural selection to change and diversify as they please; and in many organisms, these "liberated sequences" represent a majority of the chromosomal DNA.

Finally, we add to all this the insistence of Davidson and his co-workers (14, 29, 30) that there is commonly widespread intermingling of single copy and middle repetitive DNA in the eukaryotic genome.

So what's the conclusion? At best it is a messy one and a compromise. Some lampbrush loops and their chromomeres are clearly uniformly repetitive, like the loops of the nucleolus organizer and the 5S genes, and the Endo R *hae* resistant

loops in *T. viridescens*, but the vast majority of loop
chromomere complexes have sequence diversity that probably
reflects a liberal mixture of middle repetitive and single
copy sequences. Loop RNA is transcribed as long mixed molecules, and it is immediately complexed with protein to form
RNP particles of widely varying size that can be seen with
the light microscope in the loop matrix (7). The loop RNA
is subsequently degraded into components that are kept to
perform a biochemical function, and components that are lost.

The answer to our question - what's going on" - may be
summarized as follows. There is widespread transcription of
many sequences on the loops of lampbrush chromosomes. The
primary transcript is large. Only a small part of it is
retained, and apart from clearly repetitive situations such
as the ribosomal, 5S and histone genes, the retained material
is transcribed from single copy sequences. This fits with
most of the available evidence. It does not tell us any-
thing about regulation. It does not explain the fact that at
least a part of every chromomere is transcribed, and it has,
in my opinion, one further shortcoming. It focuses attention
back onto RNA transcription as the primary purpose of a lampbrush chromosome. Surely there is more to lampbrushes than
this!

With regard to Callan's spinning out and retraction
hypothesis, we are left in doubt. Either the loop axis is
stationary and represents part of the chromomeric DNA that is
selected for transcription, or it moves and all the chromomeric DNA is sequentially transcribed. In the first case,
you transcribe only bits of the genome that have the necessary sequences: in the other case, you transcribe the whole
genome and select from the transcripts. At the moment, we
cannot tell what happens, but insofar as the nucleolus
organizer forms a chromomere and a loop and possesses a high
degree of repetitive uniformity, it seems hard to understand
how only a part of that complex might be selected for loop
formation and transcriptive activity. The same might be said
of the giant loops on chromosome 2 of *T. viridescens*. And in
this connexion we cannot ignore the observations and experiments of Callan and Lloyd (6), Gall and Callan (9), and Snow
and Callan (31), all of which indicate that some component of
the loop moves, and some of which point strongly towards a
continuous spinning out and retraction of the loop axis such
as would be needed to allow transcription of all the chromomeric DNA.

Now let us turn to the second major question. What is a
lampbrush chromomere? In the simplest possible terms, it is
an aggregate of deoxyribonucleoprotein (DNP). In fine struc-

ture it is not very informative, and curiously it shows no
sign whatever of the transverse plane of weakness that must
exist to account for the formation of double bridges (1).
In genetic terms, the matter is not so straightforward. It
was in 1935 that Muller and Prokofyeva (32) postulated a one
gene band relationship with regard to the giant polytene
chromosomes of dipteran larvae. Since then, cytogenetic and
recombination studies have provided arguments in favour of
this notion. In *Drosophila*, 3 groups of investigators have
shown that there is one essential function expressed by the
genetic information in each chromomere or band (33-35), and
in *Triturus*, we have seen evidence to suggest that individual
peculiarities of loop morphology are transmitted in a regular
Mendelian fashion to offspring (6, 36).

Now the principle of one gene to one chromomere holds in
Drosophila irrespective of the amount of DNA that the chromomere contains, and the average chromomere in this organism
has enough DNA for about 40 genes each 1,000 nucleotide pairs
long. The problem becomes much more conspicuous in amphib-
ians where each chromomere has as much DNA as is contained
in the entire genome of *E. coli*. So we have a discrepancy
here between single function and the length of DNA housed by
one chromomere. It can be reconciled in 2 ways. First we
can say that a chromomere includes several genetic functions,
both structural and regulatory, and this is the idea sub-
scribed to by Judd and his colleagues (34). They propose
that a chromomere contains a short length of "structural DNA"
together with a variable length of "regulatory DNA".
Secondly, we can resort to the Master/Slave hypothesis, and
think in terms of chromomeres and repetitive gene sequences
that are periodically rectified.

Both Judd's and Callan's hypotheses emphasise that a
chromomere is a genetic unit whose size is dependent upon the
number of copies of a nucleotide sequence in that unit
(Callan), or the amount of regulatory DNA that is present
(Judd). Accordingly, a difference in C value (the amount of
DNA per haploid chromosome set) between related species would
most likely signify a difference in the sizes of genes or
their internal repetitiveness, rather than a difference in
the number of genes.

The question can be put concisely and objectively with
respect to 3 animals that have been the subjects of a number of cytological and molecular studies in this laboratory
over the past 5 years. The animals concerned are *Plethodon
cinereus*, the red-backed salamander, *P. vehiculum*, the western red-backed salamander, and *P. dunni* or Dunn's salamander. These 3 animals belong to the same genus, they have

the same chromosome number (N = 14), and the relative dimensions of their chromosomes are nearly identical, yet the C values of *P. vehiculum* and *P. dunni* (38.8 pg and 36.8 pg respectively) are almost twice that of *P. cinereus* (20.0 pg) (27, 28). All these animals have excellent lampbrush chromosomes in their oocytes. Do the lampbrush chromosomes of *P. dunni* and *P. vehiculum* have bigger chromomeres than those of *P. cinereus*, or do they have more chromomeres? The answer is clearly more chromomeres. Estimates of the number of chromomeres per haploid lampbrush chromosome set (17, 28) have shown that there are 55% more chromomeres per haploid lampbrush set in *P. vehiculum*, in which C = 36.8 pg (29), than there are in *P. cinereus*, in which C = 20 pg. Similarly, *P. dunni*, with a C value of 38.8 pg has 59% more lampbrush chromomeres than *P. cinereus*. What does this mean?

Our studies of DNA sequence organization in *Plethodon* (27) have shown that the difference in genome size between eastern and western species is largely attributable to moderately repetitive DNA. About half of the genome of *P. cinereus* reassociates by Cot 10^3, whereas more than 80% of the genomes of *P. dunni* and *P. vehiculum* have reassociated by Cot 10^3 under comparable reaction conditions. The same studies have shown that the level of conservation of moderately repetitive sequences in the genus *Plethodon* is low. Species such as *P. cinereus* and *P. vehiculum* despite their obvious likenesses have less than 10% of their repetitive DNA in common. This indicates a low selection pressure favouring conservation of general repetitive sequences from species to species. The similarity in karyotypes, and the quantitative and qualitative differences in repetitive DNA in *Plethodon* species have led to the proposal that the genomes of western plethodons grew through saltatory replications of certain sequences, followed by diversification of the replication products (27). Since moderately repetitive sequences are scattered throughout the genome, growth of the genome has been balanced over the whole karyotype, leaving the relative dimensions of the chromosomes unchanged.

In *Plethodon*, the observed increase in the total number of repetitive sequences is accompanied by an increase in chromomere number. Certain general considerations follow from this observation.

Most of the *Plethodon* genome consists of repetitive DNA. Observations on interspersion of repetitive and single copy DNA suggest that moderately repetitive DNA is likely to be more or less evenly distributed along the chromosomes. Therefore most chromomeric DNA must be moderately repetitive. About 90% of the DNA in a lampbrush chromosome is in its

354

chromomeres. In *P. cinereus*, 40% of the genome consists of single copy sequences. Therefore chromomeres <u>must</u> contain single copy sequences. The amount of unique sequence DNA per haploid genome in western species is not conspicuously greater than in eastern species. Now, if the uniformity of interspersion is the same in both species groups, and there is nothing to suggest otherwise, then it follows that in western species the spacing between unique sequences must be substantially greater than in eastern species. Western species have many more chromomeres than eastern species. Therefore there cannot be a direct numerical relationship between chromomeres and unique sequences. In terms of Davidson's and Britten's model (37) and in the light of present day knowledge of the role of unique or single copy DNA sequences, this is to say that there cannot be a relationship between chromomeres and structural genes in lampbrush chromosomes.

Beyond these remarks I can only speculate about the nature of chromomeres. I would suggest that chromomeric material is essentially heterochromatic in the same general sense that a chromocentre is heterochromatic, and that it owes its compactness to a commonness in the sequences of its DNA. I consider the classical image of a chromomere as a granule possessing genetic unity to be misleading. Instead I propose that an average chromomere is little more than a gathering together of a stretch of DNA that is rich in a particular repetitive sequence and its near descendants, together with such unique sequences as may have found their way into the repetitive cluster. On account of its repetitive commonness, this DNA clumps, and complexed with protein, it forms a granule that is visible as a chromomere. Where there are clusters of conserved repetitive sequences, such as are found at ribosomal and 5S loci, then chromomeres will form whose DNA is highly uniform. Where selection pressure favouring conservation of a particular sequence is low, and the sequence is relatively non-functional in the transcriptional sense, then chromomeric DNA will be less uniform.

Implicit in these suggestions is a relationship between chromomeres and families of repetitive sequences, and the ideas would gain much support from a demonstration that the diversity or variety of moderately repetitive sequences in a large genome such as that of *P. vehiculum* is substantially greater than that in the smaller genome of *P. cinereus*. I hope that evidence along these lines may soon be forthcoming.

REFERENCES

1. Mott, M. R., and Callan, H. G. *J. Cell Sci.* 7, 241 (1975).
2. Flemming, W. "Zellsubstanz, kern, und zell theilung" Leipzig: F. C. W. Vogel (1882).
3. Ruckert, J. *Anat. Anz.* 7, 107 (1892).
4. Stahl, A., Luciani, J. M., Gagne, R., Devictor, M., and Capodano, A. M. "Heterochromatin, micronucleoli, and RNA containing body in the diplotene stage of the human oocyte" In *Proc. 5th Int. Chromosome Conf.* Leiden (1974).
5. Gall, J. G. *J. Morph.* 94, 283 (1954).
6. Callan, H. G., and Lloyd, L. *Phil. Trans. R. Soc.* 243B, 135 (1960).
7. Malcolm, D. G., and Sommerville, J. *Chromosoma (Berl.)* 48, 137 (1974).
8. Callan, H. G. "Recent work on the structure of cell nuclei". In *Symposium on the Fine Structure of Cells, I. U. B. S. publ. series B. 21,* 89 (1955).
9. Gall, J. G., and Callan, H. G. *Proc. Nat. Acad. Sci. U.S.A. 40,* 562 (1962).
10. Callan, H. G. *J. Cell Sci. 2,* 1 (1967).
11. Kezer, J. Department of Biology, University of Oregon, Unpublished observations.
12. Davidson, E. H., and Hough, B. R. *J. Mol. Biol. 56,* 491 (1971).
13. Hough, B. R., and Davidson, E. H. *J. Mol. Biol. 70,* (1972).
14. Davidson, E. H., and Britten, R. J. *Cancer Res. 34,* 2034 (1975).
15. Rosbash, M., Ford, P. J., and Bishop, J. O. *Proc. Nat. Acad. Sci. U.S.A.* 71, 3746 (1974).
16. Kezer, J., and Macgregor, H. C. *Chromosoma (Berl.) 42,* 427 (1973).
17. Vlad, M., and Macgregor, H. C. *Chromosoma (Berl.) 50,* 327 (1975).
18. Macgregor, H. C., and Kezer, J. *Chromosoma (Berl.) 33,* 167 (1971).
19. Mancino, G., Nardi, I., and Ragghianti, M. *Experientia 28,* 856 (1972).
20. Hennen, S., Mizuno, S., and Macgregor, H. C. *Chromosoma (Berl.) 50,* 349 (1975).
21. Callan, H. G. *J. Cell Sci. 1,* 85 (1966).
22. Barsacchi, G. P., Nardi, I., Batistoni, R., Andronico, F., and Beccari, E. *Chromosoma (Berl.) 49,* 135 (1975).

23. Brown, D. D., Wensink, P. C., and Jordan, E., *Proc. Nat. Acad. Sci. (Wash.) 68*, 3175 (1971).

24. Gould, D. C., Callan, H. G., and Thomas, C. A., *J. Cell Sci. In Press.*

25. Callan, H. G., and Macgregor, H. C., *Nature 181*, 1479 (1958).

26. Gall, J. G., *Nature 198*, 36 (1963).

27. Mizuno, S., and Macgregor, H. C., *Chromosoma (Berl.) 48*, 239 (1974).

28. Macgregor, H. C., Mizuno, S., and Vlad, M., "Chromosomes and DNA sequences in salamanders". In *Proc. 5th Int. Chromosome Conf. Leiden, 1974. In press* (1975).

29. Davidson, E. H., Hough, B. R., Amenson, C. S., and Britten, R. J., *J. Mol. Biol. 7*, 1 (1973).

30. Davidson, E. H., Galau, G. A., Angerer, R. C., and Britten, R. J., *Chromosoma (Berl.) 51*, 253 (1975).

31. Snow, M. H. L., and Callan, H. G., *J. Cell Sci. 5*, 1 (1969).

32. Muller, H. J., and Prokofyeva, A. A., *Proc. Nat. Acad. Sci. (Wash.) 21*, 16 (1953).

33. Rayle, R. E., and Green, M. M., *Genetics 39*, 497 (1968).

34. Judd, B. H., Shen, M. W., and Kaufman, T. C., *Genetics 71*, 139 (1972).

35. Shannon, M. P., Kaufman, T. C., Shen, M. W., and Judd, B. H., *Genetics 72*, 615 (1972).

36. Callan, H. G., and Lloyd, L., *Nature 178*, 355 (1956).

37. Britten, R. J., and Davidson, E. H., *Science 165*, 349 (1969).

38. Barsacchi, G., and Gall, J. G., *J. Cell Biol. 54*, 580 (1972).

39. Beattie, W. G., and Skinner, D. M., *Biochim. Biophys. Acta 281*, 169 (1972).

40. Botchan, M. R., *Nature 251*, 288 (1974).

41. Brown, D. D., and Sugimoto, K., *Cold Spring Harbor Symp. Quant. Biol. XXXVIII*, 501 (1973).

42. Nardelli, M. B., Amaldi, F., and Lava-Sanchez, P. A., *Nature New Biol. 238*, 134 (1972).

43. Pardue, M. L., Brown, D. D., and Birnstiel, M. L., *Chromosoma (Berl.) 42*, 191 (1973).

44. Ritossa, F. M., and Spiegelman, S., *Proc. Nat. Acad. Sci. (Wash.) 53*, 737 (1965).

45. Strauss, N. A., *Proc. Nat. Acad. Sci. U.S.A. 68*, 799 (1971).

7
8
9
0
1
2
3
4
5